FLAT ROOFING
A GUIDE TO GOOD PRACTICE

Produced by Briggs Amasco Limited
with the assistance of
Coolag Limited
Permanite Limited

Sponsored by Tarmac Building Products Limited

First edition

ISBN 0 9507919 0 3

Published by Tarmac Building Products Limited
Designed by Technical Literature Service Limited, London
Production by Gilroy Print Limited
Printed in the Netherlands by Roto Smeets

Distributed by RIBA Publications Limited
Finsbury Mission, Moreland Street, London EC1V 8VB

FOREWORD

The greatest difficulty facing us all in the building industry today is the pervasive influence of science. It has shifted what was a craft industry on to a knowledge basis and made it into a moving technology for which all of us need the relevant inputs from research and feed-back from experience.

Probably the single most influential change has been the upgrading of the criteria for thermal insulation in response to the rise in energy costs. The incorporation of insulants into walls and roofs has had the effect of increasing greatly the temperature range to which everything outside the insulation becomes subject, and this in turn creates much enlarged changes of dimension, stress and pressure, a quickening of degradation due to frost and solar heat, and frequently misbehaviour of moisture vapour.

These have produced - individually or together - many of the building failures of recent years, but by the same token they have stimulated manufacturers to try to develop products that stand up to the new demands or which enable us to side-step some of them.

This has been the case particularly with flat roofs which, because they are the most severely exposed parts of the building envelope and must be so perfectly waterproof, present the most exacting tests of design and workmanship.

But it is characteristic of building that whatever the merits of any individual product, it can never be considered alone. It is always part of a sub-system, and whether it has the right properties or not depends on its role in relation to the other products and how they all function together.

For this reason a discussion of weatherproofing membranes - or insulants, or vapour barriers or decks - on their own must be of limited value; but most of what has been published about flat roofs in Britain has been of this character - fragmentary.

From my own work I know how much effort is required to extract what really matters from research findings and to distil from experience the practicalities needed to put it all into perspective; and for that reason I welcome the initiative that has led to the production of this comprehensive guide to good practice, putting together as it does the principles and data required for the design of all the elements in flat roof systems.

The incidence of flat roof failures has been unacceptably high but there is no need for this. Reliability is not unattainable and what this book provides is the information needed to achieve it. I am sure that architects, quantity surveyors, builders and all those involved in the construction of flat roofs are going to find it an excellent reference.

William Allen, CBE LLD RIBA
London, January 1982

Author Francis March BSc

Managing editor Alan Rogers

Technical advisor David Roy

Acknowledgements

The publishers gratefully acknowledge the assistance given by the following individuals and organisations:

W W L Chan BSc, PhD, DIC, CEng, FICE, FIStructE, FIWSc, MBIM

The British Standards Institution

The Garston and Princes Risborough Laboratories of The Building Research Establishment

The University of Bristol Department of Mechanical Engineering

The Hydraulics Research Station, Wallingford

CONTENTS

ERRATA

Page 15. 2nd column, 5th paragraph should read:
"...............of -5°C. Table 1.4 gives data for a high emissivity internal surface and Table 1.5 gives data for a low emissivity internal surface. **Low** emissivity surfaces include white and bright aluminium. Other surfaces should be taken as being of **high** surface emissivity and Table 1.4 should be used.

ERRATA

Page 18: 2nd column, 4th paragraph should read: "Recent work on the amount of ventilation required suggests that openings should be equivalent to **0.4%** of the plan roof area."

NOTE: The change from 4% to 0.4% should also be made under the heading Thermal Design on pages 94, 100, 132 and 136.

SECTION 1 DESIGN FACTORS

1.1 FALLS AND DRAINAGE

INTRODUCTION

It is generally accepted as good practice for flat roofs to be designed to clear surface water as rapidly as possible and it would be exceptional for a roof to be designed without falls.

The ponding of rainwater is frequently observed on flat roofs. As well as being unsightly and increasing the dead load on the roof, the consequences of waterproofing failure are obviously more serious if the area involved is not properly drained and allows a reservoir of water to collect, ready to feed into the building.

Falls may be formed in the structure or can be created within the specification above the deck. Falls in the structure can be achieved by adjusting the height of supporting beams or purlins, by using tapered supports, or by the addition of firring pieces before the deck is laid. The latter method is normally used with decks such as woodwool, timber, precast concrete and metal decking. In the case of an in-situ cast concrete slab, falls are normally provided by the use of a screed.

Preformed tapered insulation boards also provide a useful method of forming falls on a level roof deck though they may not be suitable if a complex pattern of falls and cross falls is required.

Ponding is frequently observed on flat roofs

DESIGN OF FALLS

It is generally accepted that flat roofs should be constructed to a minimum fall of 1 in 80, and that to achieve this the designer needs to adopt a design fall which will allow for the deflections and inaccuracies in construction.

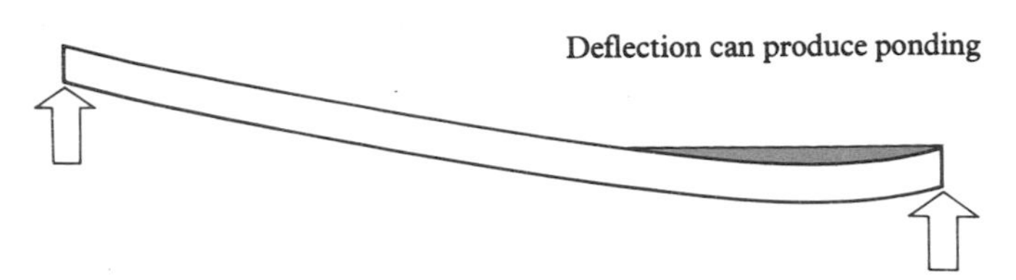

Design fall 1 in 80

Some designers arbitrarily double the finished fall and adopt 1 in 40 as the design fall, assuming that this will always produce a finished fall of at least 1 in 80. An alternative approach is to choose an intermediate figure of 1 in 60.

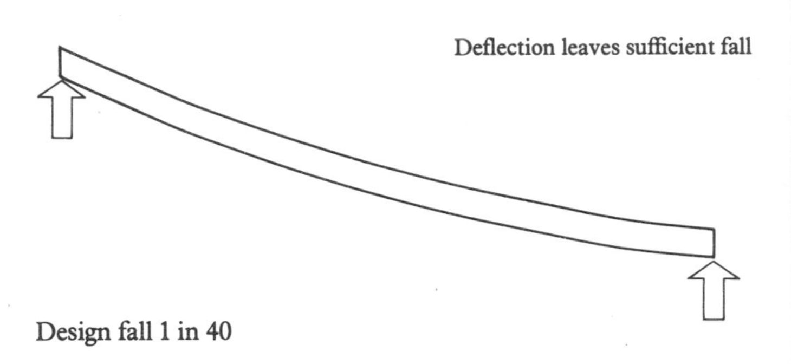

Design fall 1 in 40

On many occasions it is both practical and economic to design falls to 1 in 40, but on some buildings it will prove an unnecessarily severe design criterion. With screeded roofs in particular, doubling the screed depth at the highest points merely to allow for inaccuracies in the construction could cause an unnecessary increase in the thickness and cost of the roof system.

As an alternative, the designer should consider the accuracy and deflection of the roof in question and may find a reasonable compromise would be to take 1 in 80 as the finished fall, and add an arbitrary adjustment for construction inaccuracies, such as 25mm for concrete roofs or 15mm for metal decks.

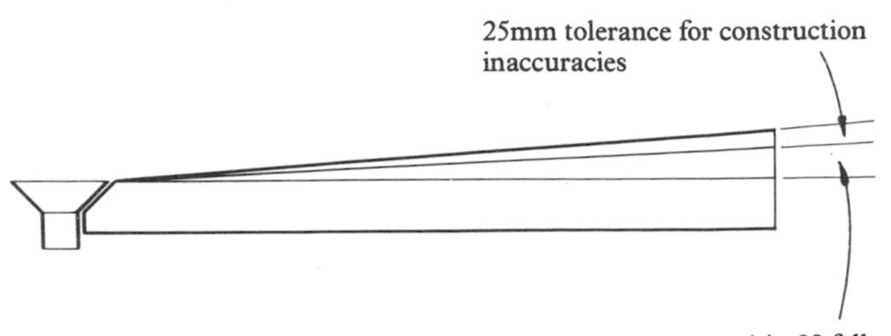

Having chosen a minimum finished fall and an allowance for inaccuracies, consideration should then be given to the effects of deck deflection which may have a favourable or adverse effect on drainage flow.

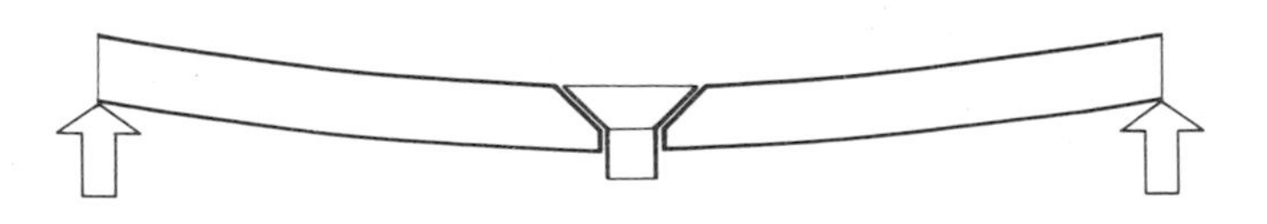

Mid-span deflection of deck aids drainage.

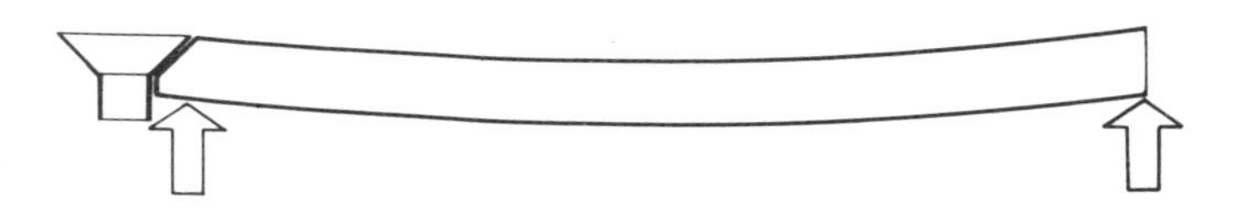

Mid-span deflection of deck restricts drainage.

Outlets in the central area of the roof may be positioned at or near the point of maximum deflection of the deck, and any deflection would therefore assist the drainage flow. In practice, however, there is usually a need to position internal downpipes against columns or walls for support and protection, and this will mean that the outlets will be positioned far away from the natural low point of roof deflection. Under these circumstances, the effect of mid-span deflection will be to reduce the fall to the outlet, and this should be taken into account when calculating the design fall.

When allowing for these deflections it should not normally prove necessary to consider the full imposed load on the roof. The design will ensure the removal of standing water and it is only necessary to take account of the dead load deflection.

Assuming that the deck takes a circular shape when deflecting, a reverse fall will be avoided entirely by raising one end of the deck by four times the deflection. For example a typical deflection for metal deck under dead load is span/650 in which case an additional fall of 4/650 or approximately 1 in 160 will compensate for deflection adverse to drainage. Some decks however are so stiff that their deflection due to dead load can be ignored.

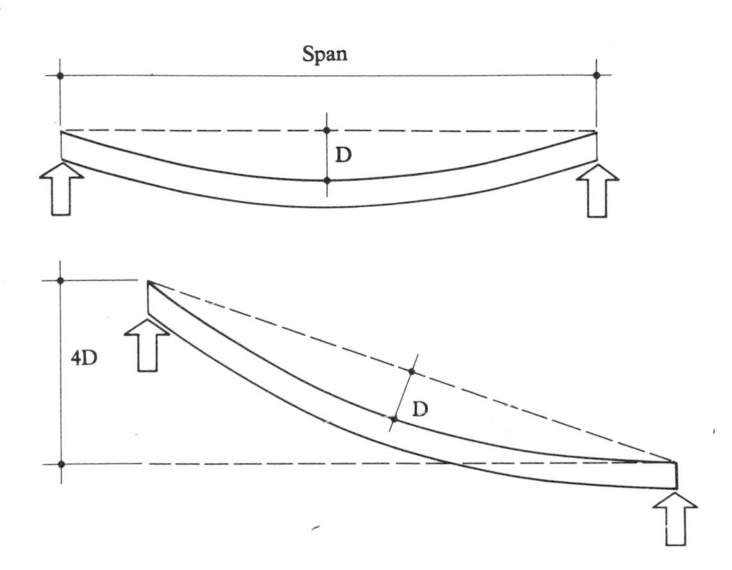

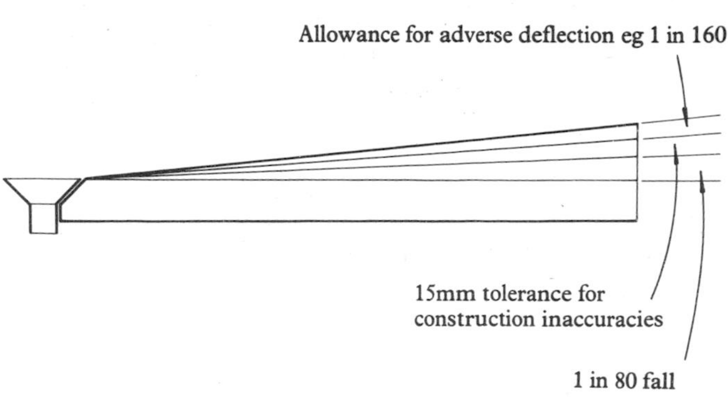

Adjustment of fall for adverse deflection

When the falls are provided by screeding, the deck deflection can be taken out by the application of the screed and no allowance need be made for deflection.

Where deflection is favourable to drainage, it should only be necessary to include an allowance for construction inaccuracies. The design fall could be reduced in line with the anticipated deflection but this would not be wise unless the designer is completely confident that the dead load deflection can be accurately predicted and the work carried out within design tolerances.

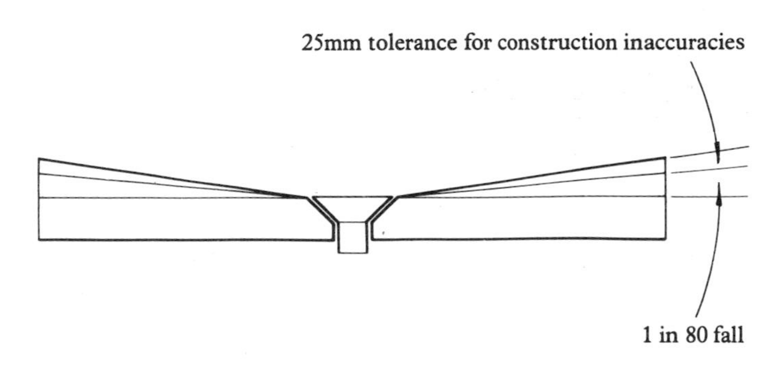

Adjustment of fall for favourable deflection.

Cross falls

At the junction of two sloping roof surfaces, a valley will be formed, known as a cross fall, and the effective slope of this will be less than that for the main falls.

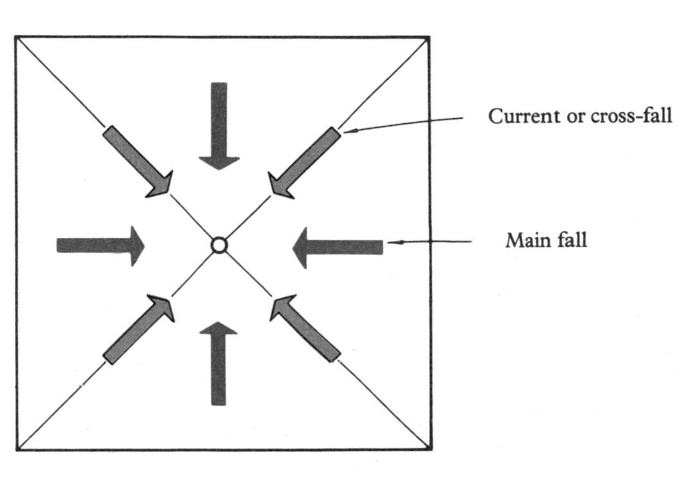

Many designers favour the adoption of 1 in 80 for the cross fall, which on a square roof produces a main fall of 1 in 56. Similarly, if 1 in 40 is adopted for the cross fall, the main fall will be 1 in 28. The implications of this approach are a substantially increased volume and cost of screed and an increased parapet height to accommodate the extra depth of screed.

The alternative approach is to accept that a small reduction in the cross fall will not impair the efficiency of drainage. Assuming a finished fall of 1 in 80 to the main area, the cross fall will be 1 in 113 which is unlikely to cause any great volume of residual water after rainfall.

Conversion table for falls
The fall is most commonly expressed as a ratio, such as 1 in 80, or as an angle, although it is sometimes convenient to describe it in terms of a percentage slope where by definition 1 in 100 is 1%. This is convenient for calculation as it expresses the fall in centimetres per metre run.

The relationship between falls, angles and percentage slope is indicated in the table below.

Fall ratio	Slope angle	% slope
1:120	0° 29'	0.8
1:110	0° 31'	0.9
1:100	0° 34'	1.0
1:90	0° 38'	1.1
1:80	0° 43'	1.25
1:70	0° 49'	1.4
1:60	0° 57'	1.7
1:50	1° 09'	2.0
1:40	1° 26'	2.5
1:30	1° 55'	3.3
1:20	2° 52'	5.0
1:11	5°	8.75
1:9	6°	10.5
1:8	7°	12.3
1:7	8°	14.1
1:6	9°	15.9
1:5	10°	17.6

ROOF DRAINAGE

The design of falls and drainage patterns will have a considerable influence on the depth of the total roof construction or roof zone, which should be a fundamental consideration at the very earliest stages of conception of a building. It is only after assessing the depth of roof zone that the designer can decide the levels of all other aspects of construction above the level of the flat roof.

It is a common mistake to under-estimate the depth of the roof zone, and only too often it is found on site that skirtings under windows and thresholds are too low and falls are inadequate. Unfortunately, designers tend to compromise on these aspects rather than increase the height of the higher level construction or decrease the size of windows or doors to ensure that good design principles can be adopted for an adjoining flat roof.

Flat roofs may be drained by two basic methods: towards the outer edges and into external gutters, or towards gutters or outlets within the main roof area. Straight falls to external gutters are simple to form by sloping the roof deck, by screeding or by using tapered insulation boards. Internal drainage will be achieved by straight falls to gutters or a pattern of falls and cross falls to outlets.

When the falls are created by a screed, it should always be possible to drain the whole roof efficiently, with falls and cross falls to outlets and without the use of gutters. If the falls are formed in the structure, a pattern of falls and cross falls will be difficult to achieve and straight falls to a gutter or to outlets will normally be incorporated, but falls between outlets can be provided by the addition of tapered firrings to the purlins between outlets or by introducing a fall in the purlins themselves.

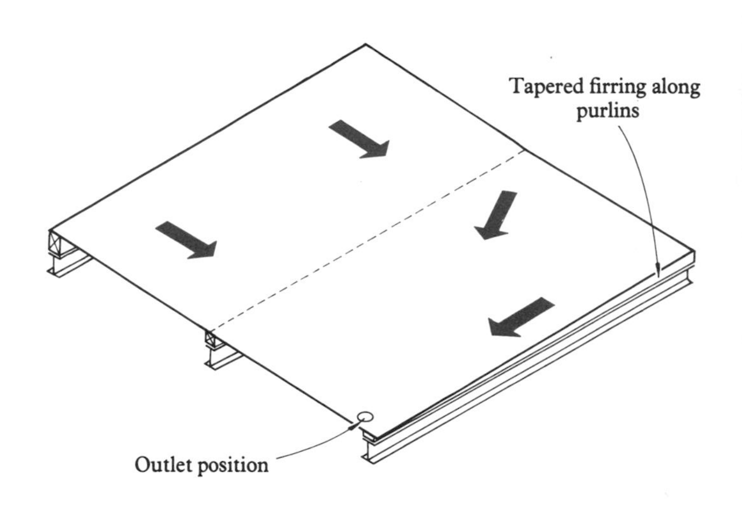

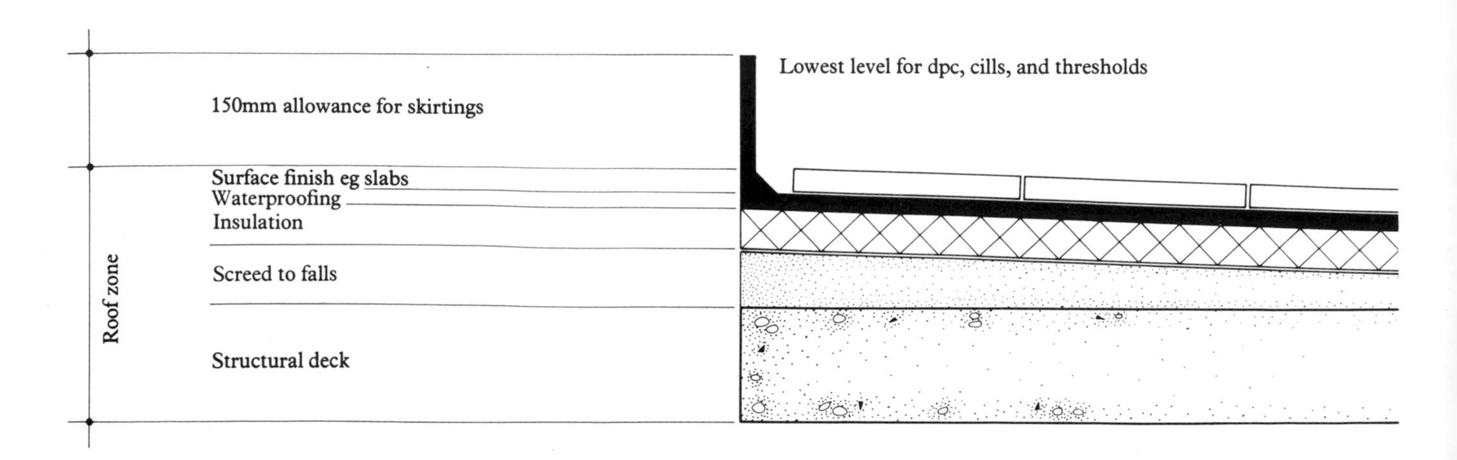

Where internal gutters are to be used, they should also be laid to falls and this may lead to a considerable depth of gutter at the low point. Dead level gutters are not normally recommended as they can hold a considerable quantity of standing water. It is normally better to omit the gutter and accept a construction which has flat sections of roof between outlets. Indeed, one of the advantages of flat roofs is the opportunity to avoid gutters and include a continuous wall-to-wall waterproof covering. As a generalisation, a well designed flat roof will contain a good number of outlets and no internal gutters.

Internal rainwater pipes are usually positioned against the main columns and the options for positioning outlets will be limited. The outlets should be positioned to divide the roof into convenient drainage areas so far as this is possible.

If the level at the outlets is taken as zero, then the pattern of drainage can be drawn and the levels at the high points of the roof calculated.

There are many different approaches to the design of drainage patterns. The three examples below show typical solutions for the design of drainage for a rectangular roof with two outlets.

For illustration purposes dimensions are based on a finished fall of 1 in 80 and any allowance for construction tolerances and deflections will depend on the type of specification used.

Comparing the second and third examples, it can be seen that the effect of the gutter is to increase the height of the roof zone by 62.5mm.

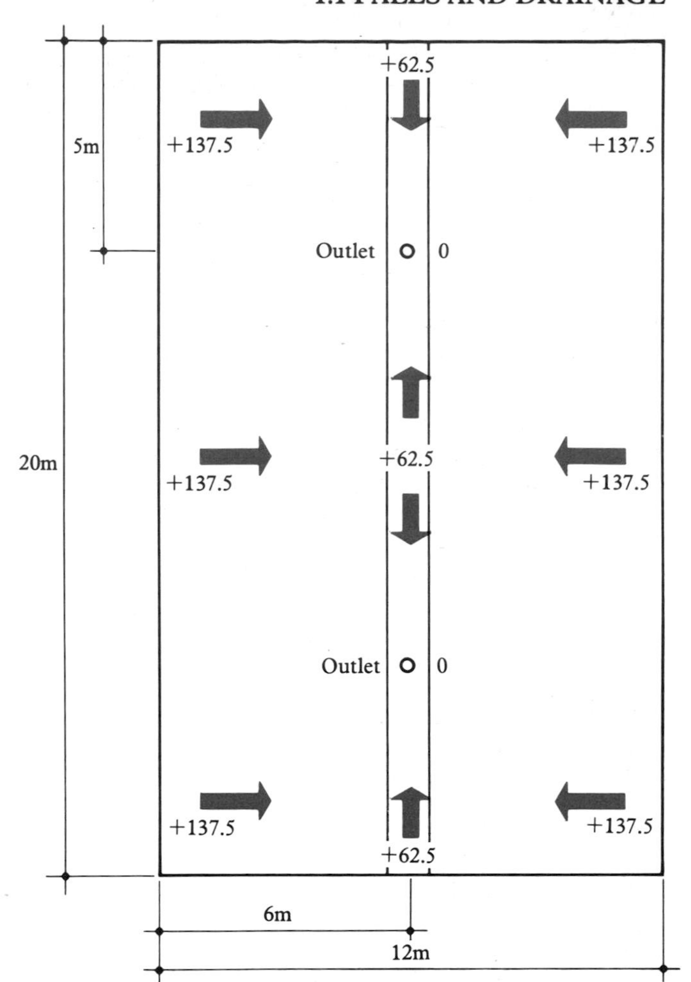

2. Straight fall to internal gutter. Gutter should also be to falls.

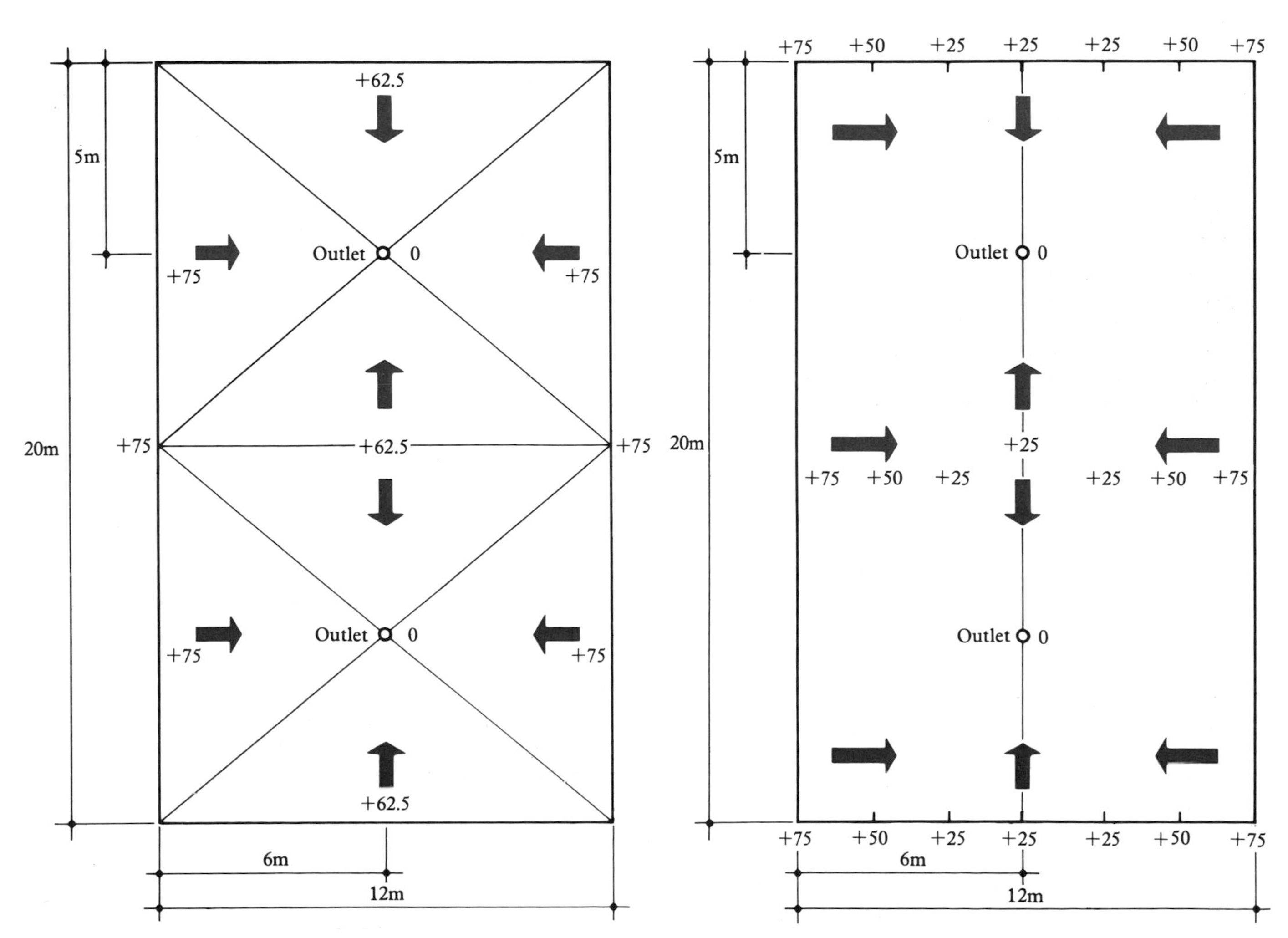

1. Screeded roof with main falls and cross falls

3. Metal deck roof with cross fall formed by packing on purlins.

There can be little doubt that dead level gutters add nothing to the efficiency of drainage and only increase the risk of problems arising on the roof.

RAINWATER OUTLETS

CP 308:1974, BRE Digest 189 and the Institute of Plumbing Handbook give advice on calculation methods employed to design the roof drainage. The Code of Practice is in the process of revision and is likely to take into account the latest recommendations of the Hydraulic Research Station. These recommendations have also been used in the preparation of data given in the following pages.

It is normal to adopt a rainfall rate of 75mm per hour as the basis of design in the UK provided that overflow will not cause damage to the building or its contents. The likelihood of this rate of rainfall occurring for two minutes is shown in rainfall map 1. It can be seen that there is less likelihood of this rate of rainfall being exceeded in Northern Ireland, Wales, Scotland and the north of England than in the rest of England. From the rainfall maps, it can be seen that, surprisingly, it is in the drier areas of the UK that the intensity of short bursts of rainfall is greatest.

The rainwater will flow over the roof area as a relatively thin surface film, perhaps only a few millimetres thick, depending on the length of run to the outlet, the texture of the roof surface and the fall. The recommended fall of 1 in 80 will ensure that the water remains a thin layer on the roof if suitable outlets are provided.

Rainwater discharges into the outlets at a rate depending on the head of water at the outlet. It will collect in the gutter or on the roof until the head of water at the outlet has built up sufficiently to discharge the rainwater as fast as it falls on the roof. A small increase in the head of water will produce a substantial increase in the rate of flow and it does not matter whether the head is produced by a dead level roof, a local collection near the outlet of a roof to falls, or collection in gutters or sumps.

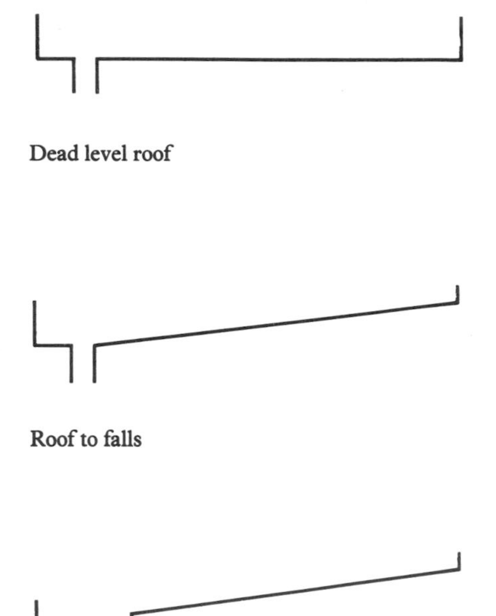

Gutter or sump

The form of flow of the water into the outlet can be of two types; weir flow or orifice flow. Weir flow is the free flow of water over an edge with an unrestricted drop. The flow of water into outlets will be by weir flow when the water is relatively shallow, and can be assumed to act when the depth of water does not exceed half the top diameter of the outlet. For greater depths of water, weir flow is prevented and orifice flow takes over.

Tapered outlets are more efficient than those with a uniform diameter. If the vertical dimension of the taper is at least equal to the top opening and if the diameter of the top opening is not more than one and a half times the downpipe size, the calculation of flow can be based on the top diameter.

If the taper of the outlet is greater than that given above, calculations for flow should be based on a maximum design outlet diameter of one and a half times the diameter of the downpipe. The only reasonable alternative is to subject the outlet to hydraulic tests to establish the relationship between the rate of flow of the water and the depth of water above the outlet.

Rainfall map 1: period in years between events of 75mm per hour for 2 minutes

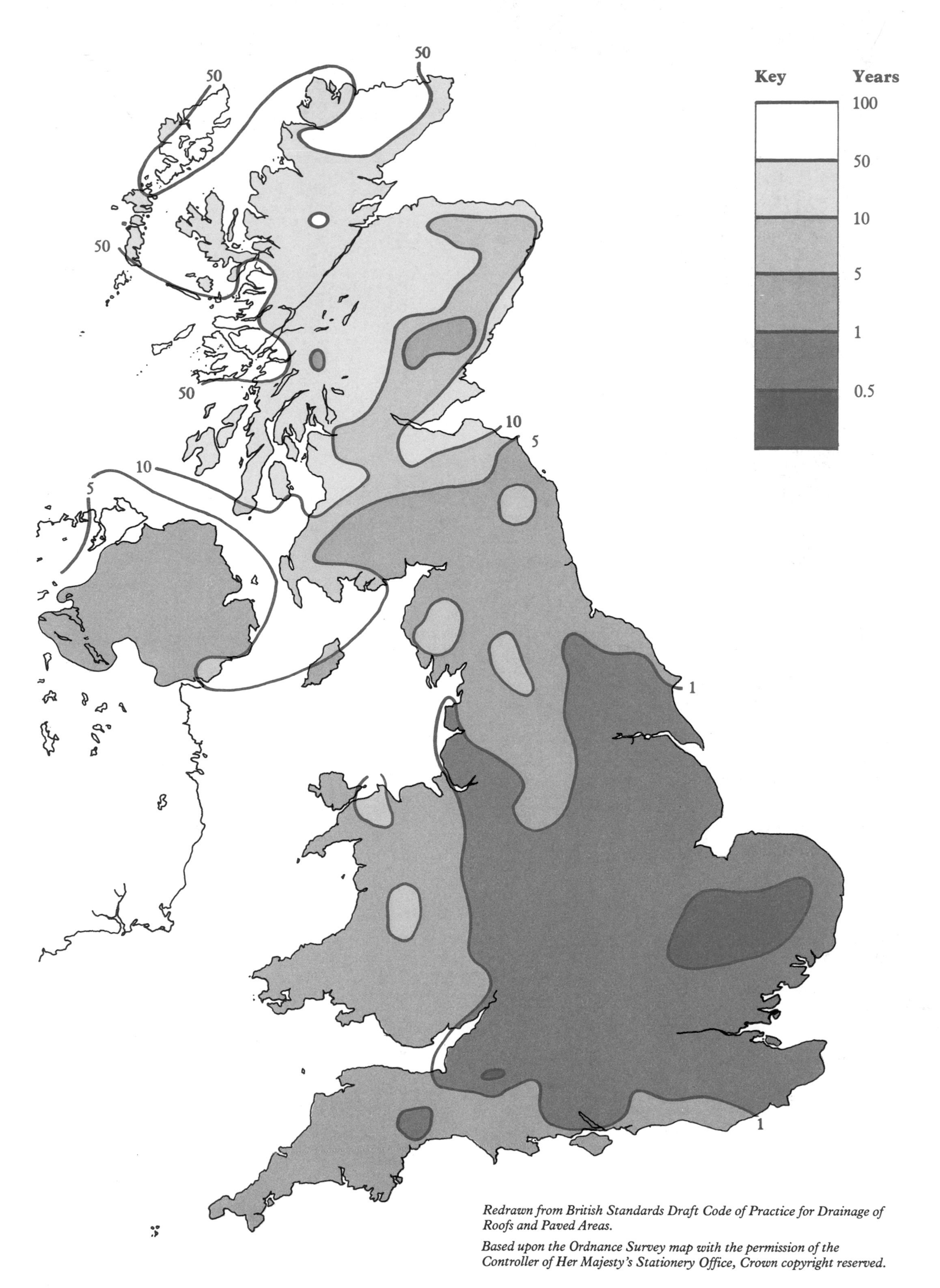

Redrawn from British Standards Draft Code of Practice for Drainage of Roofs and Paved Areas.

Based upon the Ordnance Survey map with the permission of the Controller of Her Majesty's Stationery Office, Crown copyright reserved.

Rainfall Map 2: period in years between events of 150mm per hour for 2 minutes

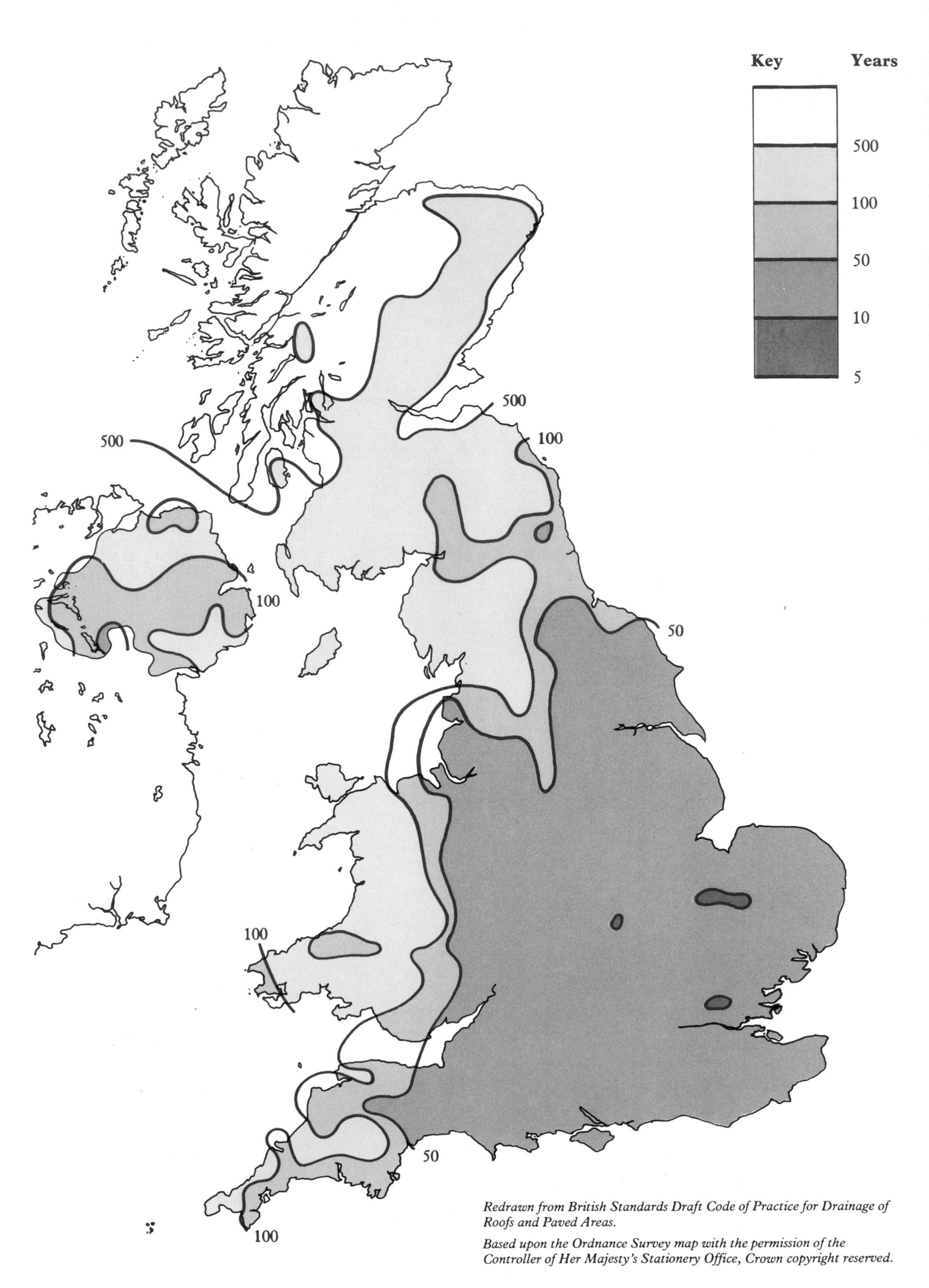

Redrawn from British Standards Draft Code of Practice for Drainage of Roofs and Paved Areas.

Gravel guards
Gravel guards are normally necessary on all outlets where the downpipe size is less than 150mm diameter. If the downpipe is 150mm diameter or more and it discharges as a straight drop from the outlet with a single bend to the main surface water drainage system, it may well be regarded as a self-cleansing system. Although gravel guards will normally be used they are not necessarily required and may be omitted if the downpipe can be regarded as self-cleansing.

The provision of gravel guards introduces the need for routine inspection and cleaning.

All internal downpipes must have rodding eyes at floor level positioned so that a blockage between the downpipe and the surface water drainage system can be cleared by rodding.

The junction between the outlet and the internal downpipe should be sealed or caulked as a precaution against backing up but experience suggests that these seals are not always effective in the long term. The surface water drains must be of sufficient size to carry away all the water from the roofs immediately or there will be a danger of water backing up the downpipes.

Area drained by outlets
The majority of flat roofs are drained to rooftop outlets only, and the crucial aspect of design is the depth of water at the outlet. On a roof to falls during a storm, the water will collect over the roof area local to the outlet to form a natural sump, and a head will be formed.

If the fall is 1 in 80, a head of 50mm will be provided by a natural sump which extends 4 metres from the outlet. It should be appreciated that this will only occur for a few minutes during the part of the storm which gives rain at 75mm per hour.

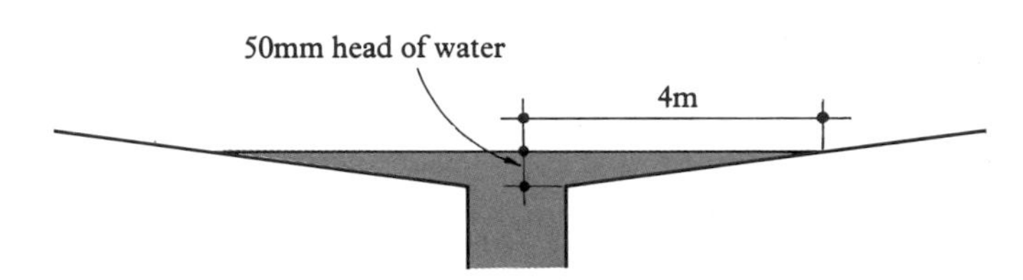

As an aid to design, Tables 1.1 and 1.2 give the roof areas which will be drained by straight drop outlets or tapered outlets. They are based on a rainfall rate of 75mm per hour taking into account weir and orifice flow as appropriate and the effect of swirl. For a rainfall rate of 150mm per hour, the values in the tables should be halved.

Where there is a substantial area of wall projecting above the level of the roof and draining onto the roof, this must be allowed for when calculating the total area to be drained. In the case of only one wall, the effective additional area can be taken as half the exposed vertical area of the wall. Further guidance for other configurations of walls is given in CP 308.

The tables also show the effect of the use of gravel guards and give two cases representative of the various types available.

The calculations are based on weir flow only through the gravel guard and for this the height of the slots in the gravel guards must be greater than the depth of water assumed for the design head. If the water were to rise above the slots, a complex flow pattern would result and a full calculation or test would be required.

The flow of water through outlets will be reduced if water cannot approach them from all directions. Tables 1.1 and 1.2 can be taken to represent the case of an outlet which is placed sufficiently far from a wall or sides of a gutter to allow a flow of water between the wall and the outlet.

It will be seen from these tables that the roof area to be drained is influenced more by the head of water at the outlet than by other factors. It is not always practical to limit the design head of water at the outlet to 13mm as recommended in CP 308 as large numbers of outlets would need to be installed and practice shows this to be unnecessary. It is recommended that 35mm be used for the design head of water at outlets with downpipes up to 100mm and 50mm design head for downpipes of 125mm to 150mm.

A dead level roof would normally be designed to have the same number and size of outlets as would an equivalent roof laid to falls. In this case the whole roof will therefore form a shallow sump but there is no question of a full design head of say 50mm developing, as this would require 50mm of rainfall with all outlets blocked. In practice however there is little build up of water above the level of the outlets on a dead level roof when the outlets are designed to a notional 35mm or 50mm head.

SUMPS AND GUTTERS

Sumps and gutters allow a shallow flow of water over a long periphery into a deep collection space. The depth of water can then build up to form a substantial head at the outlet to drive the maximum of water into the rainwater drainage system. In order to achieve weir flow, the depth of the sump or gutter must be equal to the design head of water at the outlet plus at least 25mm.

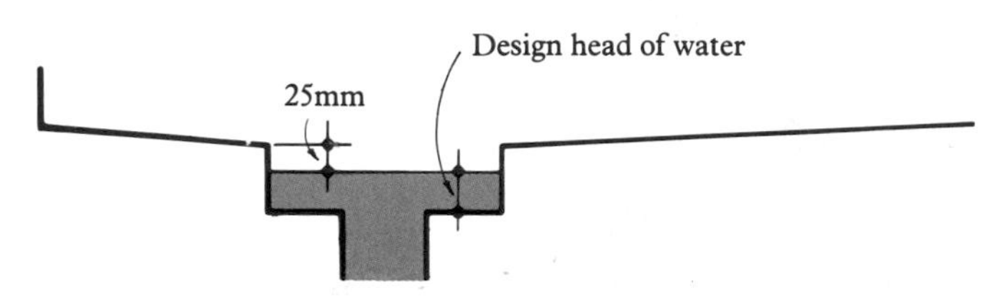

During a 75mm per hour storm lasting for a few minutes, sumps and gutters will fill extremely fast until the depth of water is equal to the design head. At this stage an equilibrium will form and the water will discharge down the outlets as quickly as it arrives.

Area drained by sumps
The area drained by an outlet/sump assembly needs to be considered in two stages. The area drained by the outlet itself must first be calculated or taken from Tables 1.1 and 1.2. Secondly, the sump size must be chosen so that the rate of drainage into the sump is matched to the rate of drainage through the outlet.

Table 1.3 gives the area drained by a given head and periphery of sump.From this it is possible to judge the periphery which is necessary to discharge water into the sump at the same rate as the outlet discharges water into the downpipe. There is no reason why sumps should be made larger than this and no increased flow will result from the outlet/sump assembly, unless the size of the outlet is also increased.

TABLE 1.1 STRAIGHT DROP OUTLETS

Area of roof (m^2) drained by one straight drop outlet without gravel guard

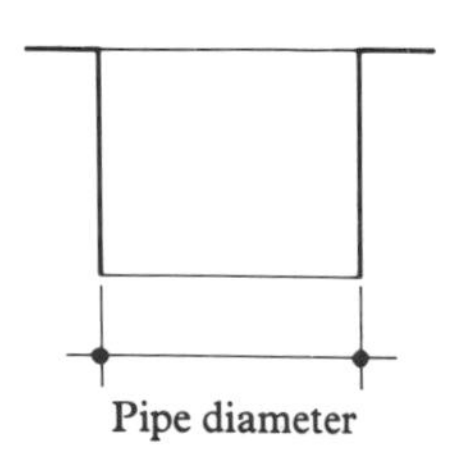

Pipe diameter mm	Head of water mm 5	10	15	20	25	30	35	40	50	60	70	80	90	100
65	5	13	24	37	52	68	80	86	96	105	113	121	128	135
75	5	15	28	43	60	79	99	114	127	139	151	161	171	180
100	7	20	37	57	80	105	133	162	226	248	268	286	304	320
150	11	30	56	86	120	158	199	243	339	446	562	644	683	720

Area of roof (m^2) drained by one outlet with gravel guard having 50% of circumference open

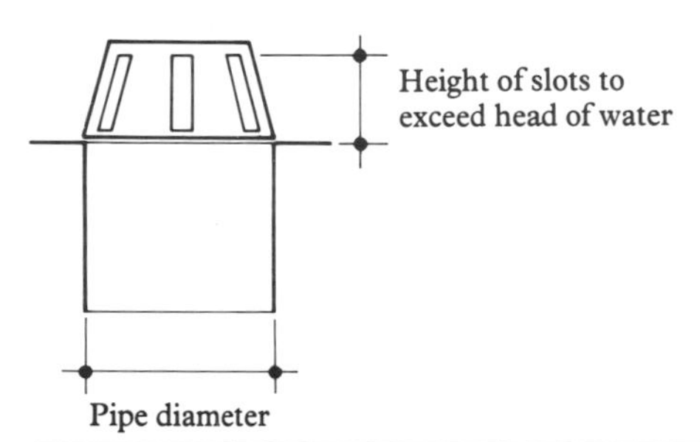

Pipe diameter mm	Head of water mm 5	10	15	20	25	30	35	40	50	60	70	80	90	100
65	2	6	12	18	26	34	42	52	72	95	113	121	128	135
75	3	7	14	21	29	39	49	60	83	110	138	161	171	180
100	4	10	18	28	39	52	65	79	111	146	184	225	268	314
150	5	15	27	42	59	77	98	119	167	219	276	337	402	471

Area of roof (m^2) drained by one outlet with gravel guard having 75% of circumference open

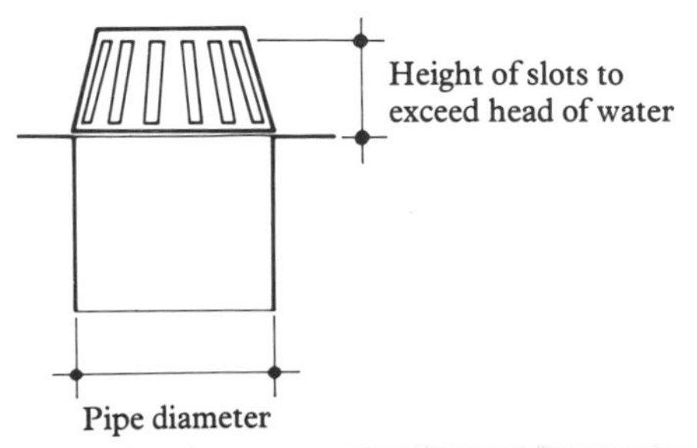

Pipe diameter mm	Head of water mm 5	10	15	20	25	30	35	40	50	60	70	80	90	100
65	3	10	18	27	38	50	63	77	96	105	113	121	128	135
75	4	11	21	32	44	58	73	89	125	139	151	161	171	180
100	5	15	27	42	59	77	98	119	167	219	268	286	304	320
150	8	22	41	63	88	116	146	179	250	329	414	506	604	707

TABLE 1.2 TAPERED OUTLETS

Area of roof (m^2) drained by one tapered outlet without gravel guard

Top diameter not to be less than 150% of pipe diameter

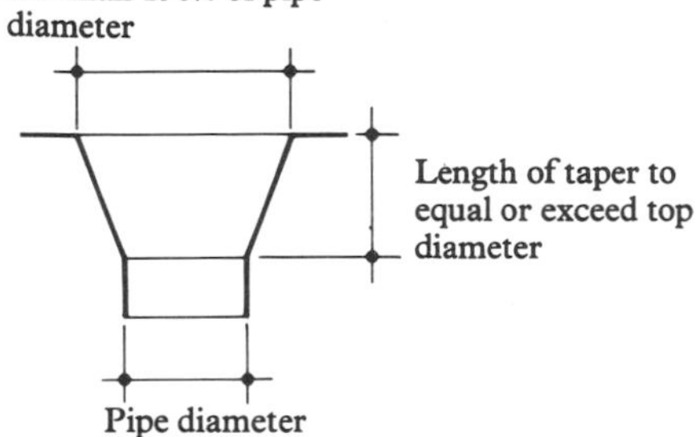

Pipe diameter mm	Head of water mm 5	10	15	20	25	30	35	40	50	60	70	80	90	100
65	7	20	36	56	78	103	129	158	215	236	255	272	288	304
75	8	23	42	64	90	118	149	182	255	314	339	362	384	405
100	11	30	56	86	120	158	199	243	339	446	562	644	683	720
150	16	46	84	129	180	237	298	364	509	669	843	1030	1229	1440

Area of roof (m^2) drained by one tapered outlet with gravel guard, having 50% of circumference open

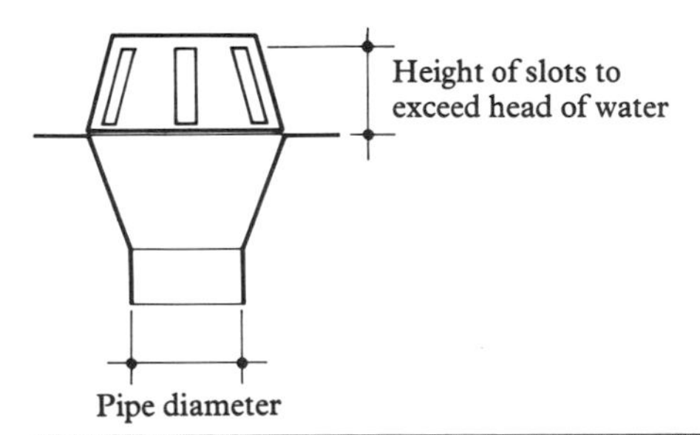

Pipe diameter mm	Head of water mm 5	10	15	20	25	30	35	40	50	60	70	80	90	100
65	3	10	18	27	38	50	63	77	108	142	179	219	262	304
75	4	11	21	32	44	58	73	89	125	164	207	253	302	353
100	5	15	27	42	59	77	98	119	167	219	276	337	402	471
150	8	22	41	63	88	116	146	179	250	329	414	506	604	707

Area of roof (m^2) drained by one tapered outlet with gravel guard having 75% of circumference open

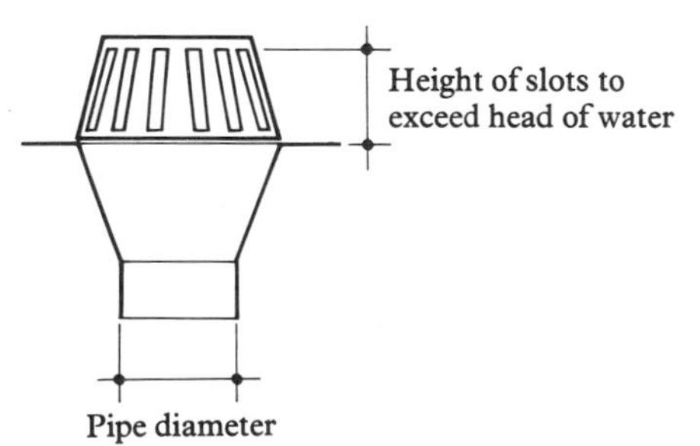

Pipe diameter mm	Head of water mm 5	10	15	20	25	30	35	40	50	60	70	80	90	100
65	5	15	27	41	57	75	95	116	162	214	255	272	288	304
75	6	17	31	47	66	87	110	134	187	246	310	379	453	530
100	8	22	41	63	88	116	146	179	250	329	414	506	604	707
150	12	34	62	95	133	174	220	268	375	493	621	759	905	1060

The periphery of the sump should be taken as that part of the edge which is exposed to the flow of water. For example, if a sump is close up against a wall there will only be three sides which discharge water into the sump.

As with rooftop outlets, the designer must decide what design head he will allow for flow of water into sumps and again this is a rather arbitrary decision. The 13mm head of water recommended in CP 308 is likely to be revised to 15mm. This may generally be satisfactory but Table 1.3 gives the area of roof which may be drained into a sump for design depths of water up to 25mm.

The areas of roof which will drain to sumps of the given sizes assume that the flow of water is from all directions.

When the sump is positioned in such a way that the flow of water to one or more sides is obstructed, the effective perimeter of the sump will be reduced pro-rata and reference should be made to the effective periphery column.

Table 1.3 Area of roof (m^2) drained into sumps

Sump size mm	Effective periphery mm	Head of water 5mm	10mm	15mm	20mm	25mm
300 x 300	1200	27	76	139	215	300
400 x 400	1600	36	101	186	286	400
500 x 500	2000	45	126	232	358	500
600 x 600	2400	54	152	279	429	600
700 x 700	2800	63	177	325	501	700
800 x 800	3200	72	202	372	572	800
900 x 900	3600	81	228	418	644	900
1000 x 1000	4000	89	253	465	716	1000

Gutters

Gutters have such a long edge that shallow weir flow over the side will always take place regardless of the roof area to be drained and there is no need to check this by calculation. It is only necessary to calculate the size of outlet to discharge water at the required rate.

Lined gutters are no more than a waterway and the size of the gutter is immaterial other than the provision of a suitable depth to provide the design head of water plus 25mm to ensure free weir flow into the gutter. If lined gutters are thought to be necessary, it is important to make sure that they are suitably shaped for the installation of a satisfactory waterproofing. It is recommended that the sole of the gutter is at least twice the maximum depth of the gutter and not less than 300mm wide after insulation and lining.

Parapet gutters should have the skirting height against the parapet at least 75mm higher than the main roof area to accept overfilling of the gutter.

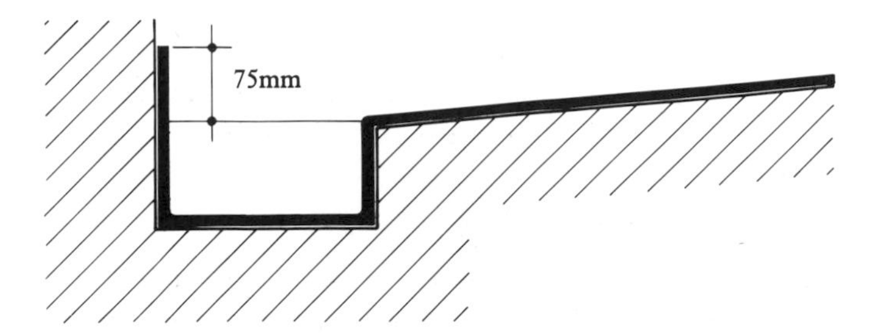

It must be remembered that if designing for 75mm per hour, the design depth is likely to be exceeded occasionally in the southern part of England where 150mm per hour will occur from time-to-time. In these areas a reserve of head will occasionally prove useful and a flat roof will always provide this by forming a reservoir if the gutter overflows. The level of water building up on the roof is only likely to rise a few millimetres, not enough to cause any concern and not enough to call the height of upstands into question.

If gutters and sumps are independent of the flat roof and the roof membrane is not formed continuously through the gutter, water may enter the building in the event of overflow. Under these circumstances a design rainfall of 150mm per hour should be used as the flat roof cannot be used as a reservoir for excessive rainfall. Such gutters and sumps should be designed in accordance with CP 308 and include an emergency overflow to ensure that in the event of overfilling, the overflow of water is to the outside of the building in a position which will cause no harm.

Sumps and gutters may tend to block from silting up or from blown or washed leaves, twigs or industrial residues and should only be used when a clean roof system and regular maintenance inspection are assured.

1.2 THERMAL DESIGN

INTRODUCTION

Thermal design is concerned with the flow of both heat and water vapour through the roof construction, and the effect of these on the performance of the roof and on the various components in the roofing system. The designer has two separate areas for consideration: the amount of thermal insulation required, and the risk of condensation occurring.

All materials used in roof construction possess thermal insulating properties to varying degrees and, in certain cases, such as woodwool slabs or aerated concrete units, the deck component alone can provide significant thermal insulation.

Generally, however, a lighter and more cost effective roof is obtained by adding a separate insulation layer, usually in the form of a rigid board insulation above the deck, or a fibrous quilt immediately above the ceiling.

Roof constructions can normally be categorised as either warm roofs or cold roofs, depending on the position of the principal thermal insulation layer in relation to the deck.

Warm roof
Warm roof or warm deck construction has the principal thermal insulation placed above the structural deck.

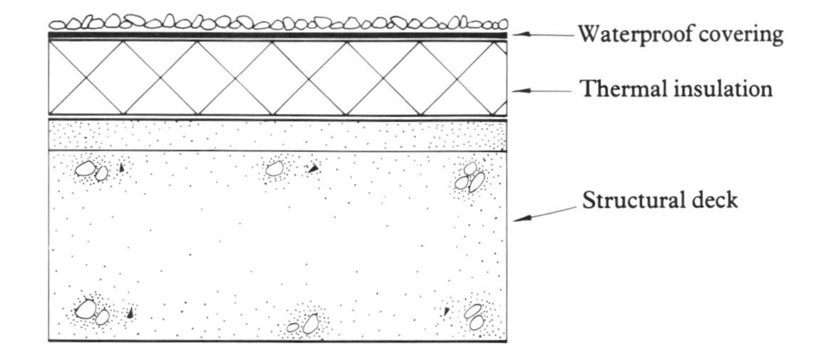

Protected membrane roof
A form of warm roof construction where the principal thermal insulation is placed above the waterproof covering. This system is also referred to as an inverted roof, or upside-down roof.

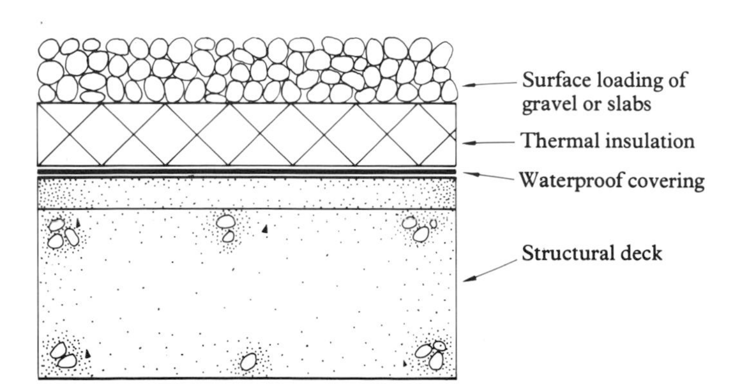

Cold roof
Cold roof or cold deck construction has the principal thermal insulation layer below the structural deck and the concept is usually concerned with roof structures which include an independent ceiling enclosing an air space between the deck and ceiling.

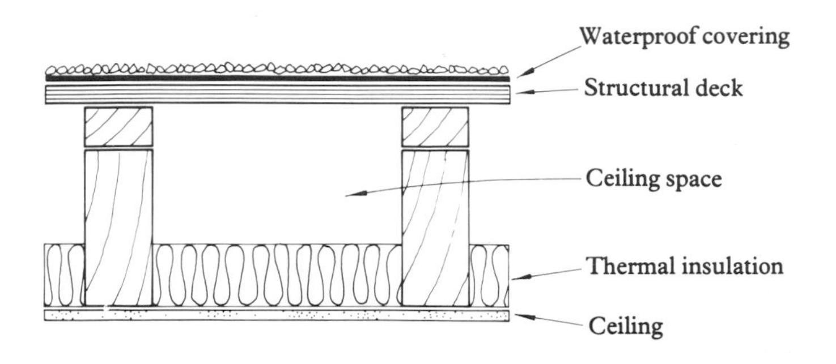

Hybrid roof
Not all roof constructions will fall within the warm or cold categories and the concept can be misleading if the exceptions are not recognised. Some structural decks are themselves composed of insulating materials, for example woodwool, and it is difficult to place the final roof construction into a warm roof or cold roof category.

Cases also arise where insulation is added above the deck in addition to insulation at ceiling level, and again it is difficult to categorise the construction. These exceptions are sometimes called hybrid roofs.

THERMAL PROPERTIES

The rate of flow of heat through the roof is determined by the thermal conductivity of the elements making up the roof system. Regulations set out mandatory standards for the thermal transmittance or U-value of the roof. The associated terms and methods of calculation are as follows:

Thermal conductivity (k)
Thermal conductivity is a measure of the rate at which heat will flow through a material when a difference exists between the temperatures of its surfaces. It is expressed in W/m°C.

For example, cork insulation with a k-value of 0.042W/m°C will allow more heat to pass through than the same thickness of polyurethane foam insulation with a k-value of 0.022W/m°C. A good insulant therefore has a low k-value.

Thermal resistance (R)
As the thickness of a material increases, its resistance to heat flow increases in direct proportion and can be calculated as:

$$\text{Thermal resistance } R = \frac{t}{k}$$

Where t is the thickness in metres and k the thermal conductivity. The thermal resistance (R) is expressed in m^2°C/W.

For example, the thermal resistance of 40mm of polyurethane insulation board ($k = 0.022W/m°C$) is:

$$\frac{t}{k} = \frac{0.04}{0.022} = 1.81m^{2}°C/W$$

Similarly, for 40mm mineral wool slab ($k = 0.034W/m°C$) the thermal resistance is:

$$\frac{t}{k} = \frac{0.04}{0.034} = 1.17m^{2}°C/W$$

At an exposed surface, the resistance to heat transfer by radiation and convection can also be regarded as thermal resistance, generally termed surface resistance. The surface resistance value depends on the emissivity of the surface, the direction of the flow of heat and additionally for the external surface, on the degree of exposure. Part FF of the Building Regulations takes a standard value of $0.15m^{2}°C/W$ as the total combined internal and external surface resistance for roofs. The individual values are normally taken to be $0.045m^{2}°C/W$ for the external surface resistance, and $0.105m^{2}°C/W$ for the internal surface resistance.

The thermal resistance of airspaces depends on the size and ventilation of the cavity, the direction of the flow of heat, and on the emissivity of the surfaces of the cavity. For unventilated cavities, with an upward heat flow, the following values should be used:

Cavities with low surface emissivity $R = 0.18m^{2}°C/W$

Cavities with high surface emissivity $R = 0.32m^{2}°C/W$

The total thermal resistance (R) of a roofing system is the summation of all the individual thermal resistances, taking into account the resistance of all the components of the roof including surface resistance and the resistance of cavities.

Thermal transmittance (U)
The thermal transmittance of the roof is defined as the quantity of heat that flows through unit area in unit time, per unit difference in temperature. It is expressed in $W/m^{2}°C$ and is the reciprocal of the total thermal resistance of the roof:

$$U = \frac{1}{R} \ W/m^{2}°C$$

The U-value provides an easy method of assessment of the heat loss through the building structure and is not only required by the heating engineer in his calculations for heating systems, but also allows the designer to compare thermal performances of alternative roof constructions. The smaller the U-value, the better the insulation.

The various UK Building Regulations define acceptable levels of heat loss from buildings by requiring maximum U-values for walls and roofs. The Building (First Amendment) Regulation 1978, Part FF (Conservation of Fuel and Power in Buildings other than Dwellings) came into force on 1st June 1979, and specifies in general terms a maximum U-value of $0.7W/m^{2}°C$ for heated factories and storage buildings, and $0.6W/m^{2}°C$ for most other heated buildings. The current standard for dwellings, Part F, specifies a maximum U-value of $0.6W/m^{2}°C$ for roofs. It is likely that the trend towards higher insulation standards will continue.

U-Value calculation method
To summarise the above, the U-value is obtained from the total thermal resistance (R) of the roof structure which is calculated from the individual thermal resistance of each component of the roof.

$$U = \frac{1}{Rsi + Rso + Rcav + R_1 + R_2 + R_3 \text{ etc}} \ W/m^{2}°C$$

where Rsi = internal surface resistance
Rso = external surface resistance
Rcav = resistance of any cavity from standard thermal resistance values
R_1, R_2, R_3 etc = thermal resistance of material, calculated from t/k where t is the thickness of material and k the thermal conductivity of the material.

A number of calculated examples are given in Appendix A, together with standard thermal properties of materials used in roof construction. The thermal properties are those adopted by the Felt Roofing Contractors Advisory Board, and are generally accepted within the roofing industry.

For roofs incorporating rooflights or areas with different thermal properties, it is permissible in heat loss calculations to take a mean thermal transmittance value.

$$\frac{\text{Area A x U-value of area A}}{\text{Total area}} + \frac{\text{Area B x U-value of Area B}}{\text{Total area}}$$

= Mean U-value for total roof

CONDENSATION

Moisture producing activities take place in most buildings. Some manufacturing processes clearly release a large amount of water vapour into the internal air. Bathrooms, kitchens, laundries and swimming pools are also sources of high humidity and the combustion products of gas, oil and paraffin are rich in water vapour.

Air has a limited capacity for carrying water vapour and when it can take up no more water it is said to be fully saturated. The moisture vapour in air exerts a pressure, as does any gas, and this is known as the vapour pressure. The ratio between the vapour pressure of moisture in the air and the saturated vapour pressure at the same temperature is termed the relative humidity (RH) and is expressed as a percentage.

The temperature at which the air becomes fully saturated with moisture, ie 100% RH, is called the dew point. When warm moist air meets a cold surface it is cooled, and if its temperature drops below the dew point it will give up moisture in the form of surface condensation.

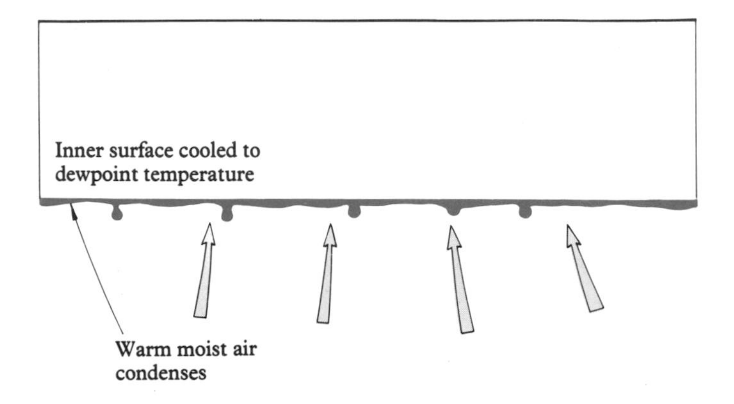

Surface condensation

The air in a building normally contains more water vapour than the external air and so has a higher vapour pressure. This creates a vapour drive from the areas of high pressure to areas of low pressure and therefore the water vapour will try to escape by all available routes to the low pressure conditions outside the building.

Moisture vapour is also present in all the permeable materials of a building, including the roof construction. and, as the vapour pressure inside a building is continually changing, there is a constant flow of water vapour in and out of the roofing materials. The movement continues until the vapour pressure in the materials is in equilibrium with the vapour pressure inside the building.

In cold weather the temperatures under the waterproofing will fall and can create a zone where the temperatures are below the dew point. Moisture will condense in this zone to form interstitial condensation. When interstitial condensation is occurring, the vapour pressure in the condensation zone will be less than the vapour pressure inside the building and the resulting pressure difference causes a vapour drive into the zone of condensation.

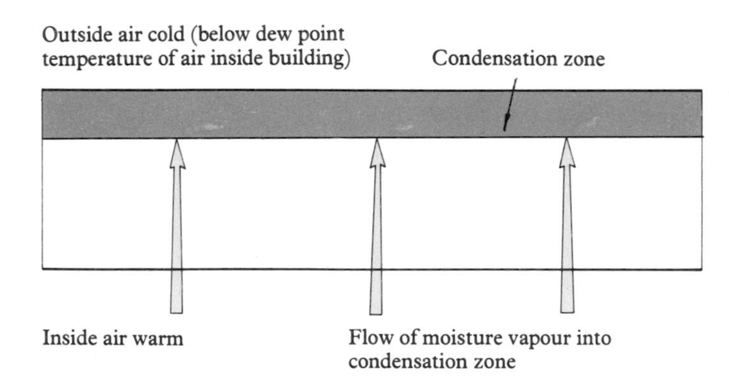

In warmer weather the temperature of the roof materials will not fall below the dew point. No interstitial condensation will occur and any remaining condensate will dry out by evaporation. During evaporation, the vapour pressure in the zone will rise above the vapour pressure inside the building and moisture vapour will be driven out of the roofing materials.

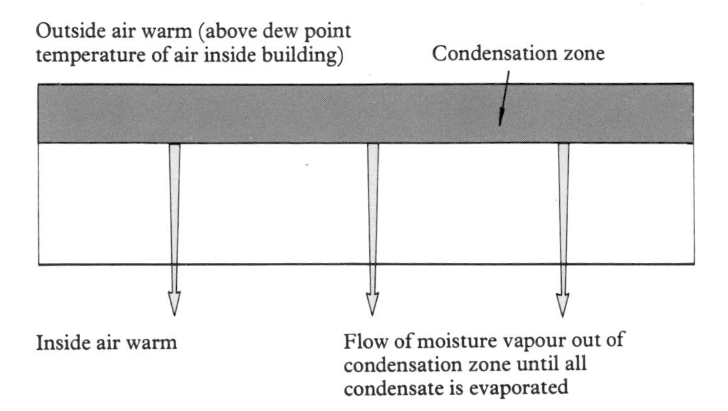

The flow of water vapour through the roof takes place slowly and continuous periods of condensation are necessary before interstitial condensation becomes a problem. Normally the periods of condensation will be relatively short and the condensate will dry out again without causing any harm.

Continual wetting of materials due to condensation can, however, lead to fungal growth, decay in structural timbers and the accelerated corrosion of metal components and fixings. There is also the risk that insulation boards containing organic fibres can decay from the growth of fungus, lose their strength and suffer a reduction in their insulating efficiency. These effects may be taking place within the structure but with no visible indication of problems appearing at the ceiling level.

SURFACE CONDENSATION

The overall level of thermal insulation has a direct influence on the likelihood of surface condensation. If sufficient thermal insulation is provided to keep the ceiling or soffit temperature above dewpoint, surface condensation will not occur.

In order to check for surface condensation, it is necessary to calculate the ceiling or soffit surface temperature and the dewpoint temperature, for the prevailing internal temperature and relative humidity. A comparison of these temperatures will then indicate whether surface condensation will occur. Calculation procedures are set out in Appendix A.

As an alternative to the calculation, Tables 1.4 and 1.5 show the internal relative humidity and temperatures at which surface condensation will occur on the underside of a roof construction for a specified U-value and at an external temperature of -5°C. Table 1.4 gives data for a high emissivity internal surface and Table 1.5 gives data for a low emissivity internal surface. High emissivity surfaces include white and bright aluminium. Other surfaces should be taken as being of low surface emissivity and Table 1.5 should be used.

The tables indicate that for the insulation values normally specified, the relative humidity at which surface condensation occurs is well in excess of normal working conditions.

When the U-value varies across the roof because of the use of tapered insulation or insulating screed to falls, the condensation risk must be calculated at the point of minimum U-value. The average U-value is only used for heat loss calculation.

It should be noted that an increased danger of surface condensation arises on cold water pipes which are positioned between the ceiling and the deck. Such pipes need to be insulated and wrapped with a moisture resistant material to act as a vapour check.

RELATIVE HUMIDITY (%) AT WHICH SURFACE CONDENSATION WILL OCCUR

Table 1.4 High emissivity internal surface assuming internal surface resistance of 0.105m^2 °C/W

U-value	Internal temperature °C															
W/m^2 °C	15	16	17	18	19	20	21	22	23	24	25	26	27	28	29	30
0.4	94	94	94	94	93	93	93	93	93	92	92	92	92	92	92	91
0.6	92	91	91	91	90	90	90	90	89	89	89	89	88	88	88	88
0.7	90	90	90	89	89	89	88	88	88	87	87	87	86	86	86	86
1.0	87	86	86	85	85	84	84	83	83	83	82	82	81	81	81	80
1.5	81	80	80	79	78	78	77	76	76	75	75	74	73	73	72	72
2.0	75	75	74	73	72	71	70	70	69	68	68	67	66	66	65	64

Table 1.5 Low emissivity internal surface assuming internal surface resistance of 0.218m^2 °C/W

U-value	Internal temperature °C															
W/m^2 °C	15	16	17	18	19	20	21	22	23	24	25	26	27	28	29	30
0.4	89	88	88	88	87	87	86	86	86	85	85	85	84	84	84	83
0.6	84	83	83	82	82	81	80	80	79	79	78	78	77	77	76	76
0.7	81	81	80	79	79	78	78	77	76	76	75	75	74	74	73	73
1.0	75	74	73	72	71	70	70	69	68	67	67	66	65	65	64	63
1.5	64	63	62	61	60	59	58	57	56	55	54	53	52	51	51	50
2.0	55	54	53	51	50	49	48	47	46	45	44	43	42	41	40	39

The thermal mass of the structure also has a bearing on the likelihood of condensation. In general, roofs may be of heavy construction, such as a dense concrete slab and screed, or of lightweight materials, such as metal decking and insulation boards. Whilst each may be designed to the same U-value, the heavy structure has a high thermal mass and will therefore heat up or cool down slowly, while the reverse is true of lightweight, low thermal mass systems. The heating system must therefore be suited to the construction. High thermal mass combined with intermittent heating is more likely to lead to surface condensation in cold weather, as the internal air temperature will increase much faster than the internal surface temperature, which may remain below the dewpoint for long periods.

INTERSTITIAL CONDENSATION

Warm roofs

Under certain conditions too much vapour will rise into a warm roof system and it will prove necessary to protect the components, in particular the thermal insulation, from excessive interstitial condensation.

Movement of the moisture in a warm roof is mostly by diffusion and, as diffusion rates are fairly easy to predict, the moisture movements and moisture gain in the roof materials can be calculated.

A useful method of predicting the probability of condensation in a warm roof is known as moisture gain analysis. This involves an assessment of the resistance to diffusion of moisture vapour for all the materials of the roof system including an allowance for gaps between the insulation material and penetrations in the vapour barrier, if present.

The resistance to diffusion is used to calculate the moisture vapour flow into and through the system with a final calculation of the weight of water which collects in the system during a typical severe cold spell in winter. The weight of water which can reasonably be expected to dry out in a typical spell of summer weather is also determined.

The analysis then checks that the water collection in the winter spell does not exceed certain stated limits. It also checks that no more than a trivial amount of moisture remains in the system after the dry spell in the summer. It should be appreciated that these calculations are by their nature arbitrary, and represent a prediction of extreme circumstances in one year. There is no likelihood of a build-up of moisture of the same severity year after year, because the calculation assumes exceptionally adverse conditions which will not be repeated in the succeeding years.

Moisture gain analysis enables the designer to decide whether or not a vapour barrier is necessary, whether sufficient thermal insulation has been allowed for, and whether it is suitably positioned with regard to ceiling spaces, vapour barriers and other components of the roof construction.

An example of moisture gain analysis is given in Appendix A and the results of the analysis for typical warm roof constructions are given in the Vapour Barrier Design Guide, Section 1.3.

A warm roof system will be designed to avoid the need for ventilation of closed air spaces, and this aspect may be checked by calculation to ensure that air spaces do not present a surface condensation risk or conditions sufficiently humid to support mould or fungus growth in timber products. Such a space is usually between the structural deck and a suspended ceiling. Suspended ceilings usually contain joints which allow a measure of convection through them, and it will be prudent to assume convection is taking place from the inside of the building into the ceiling space.

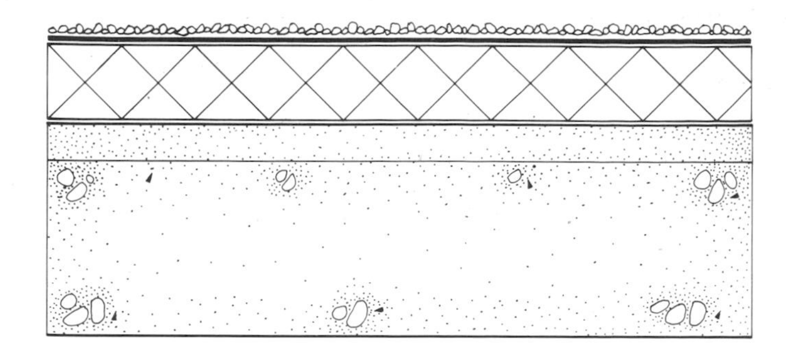

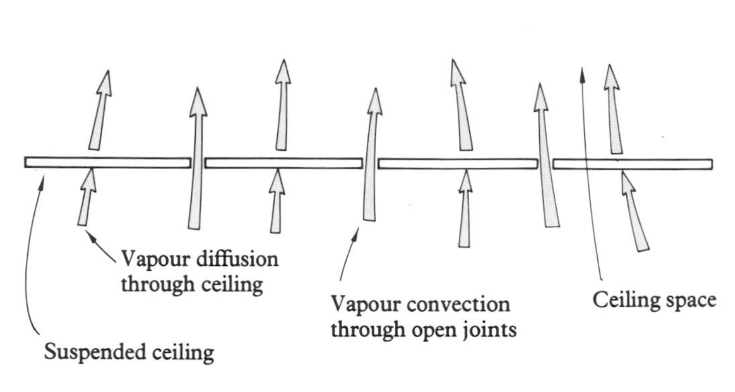

The space forms a large reservoir for moisture vapour which is kept continuously and speedily replenished by convection and, the amount of condensation taking place will be far greater than that indicated by calculation based on diffusion alone. The flow of air into the ceiling space can be assumed to be sufficient to give the same moisture content, and consequently the same vapour pressure above the ceiling as below. The temperature in the ceiling space will however be lower than the internal temperature due to the insulating effect of the ceiling and it is important to demonstrate that no condensation occurs in the ceiling space, even for short periods of time.

The occurrence of surface condensation in the ceiling space is most likely on the underside surface of the deck, and will depend on the amount of insulation above this surface compared with the insulation below. This is commonly referred to as the balance of insulation and can be checked by reference to table 1.12 in the Vapour Barrier Design Guide, Section 1.3.

If the check shows a risk of condensation under the roof deck, the insulation on top of the roof deck should be increased or the insulation of the ceiling should be reduced in order to achieve a satisfactory balance of insulation. The addition of ventilation of the ceiling space to the outside of the building would also be a possible solution, but this would significantly increase the heat loss from the building, and it would be better to adjust the insulation system.

The passage of water vapour through the ceiling into the ceiling space is sometimes prevented by the installation of a plenum system in the ceiling space, which increases the pressure inside the space by powered ventilation to ensure that it exceeds the pressure inside the building. The movement of air in the ceiling space will then be through the ceiling and into the building below and will prevent air from inside the building reaching the ceiling space.

Vented insulation system

The conventional warm roof construction can be modified to form a vented insulation system by the introduction of venting channels on the underside of the insulation coupled with breather vents in the waterproofing.

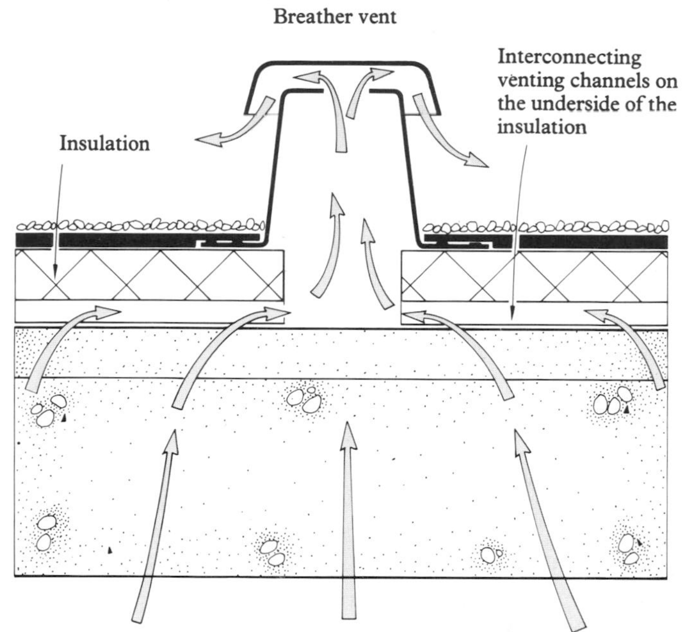

Vented insulation system

Although a vented insulation system would be properly described as an example of a warm roof, it has a strong similarity in principle to the cold roof. The water vapour passes through the venting channels on the underside of the insulation rather than through the insulation, allowing the water vapour to be discharged to the outside of the building.

Vented insulation is effective provided the whole mechanism of operation is understood by the designer, and certain necessary precautions are taken. Water vapour passes through the system more rapidly and in greater quantities than is the case for a normal warm roof, and the efficiently of the venting is crucial. Insufficient venting may lead to an increased risk of condensation because unvented channels will allow water vapour to accumulate ready to form condensation, rather than be dispersed by the venting system.

The breather vents must be carefully positioned and sufficient in number to ensure that efficient venting is achieved. It should be noted that the venting system could break the suction on the underside of the membrane that would otherwise hold it in position against wind forces. The instructions for the attachment of the insulation and membrane and the placing of the breather vents must therefore take into account any additional hazard which might be formed with respect to wind. Only those proprietary systems where the design and installation procedures have been proven and are adequately described should be used.

Protected membrane roofs

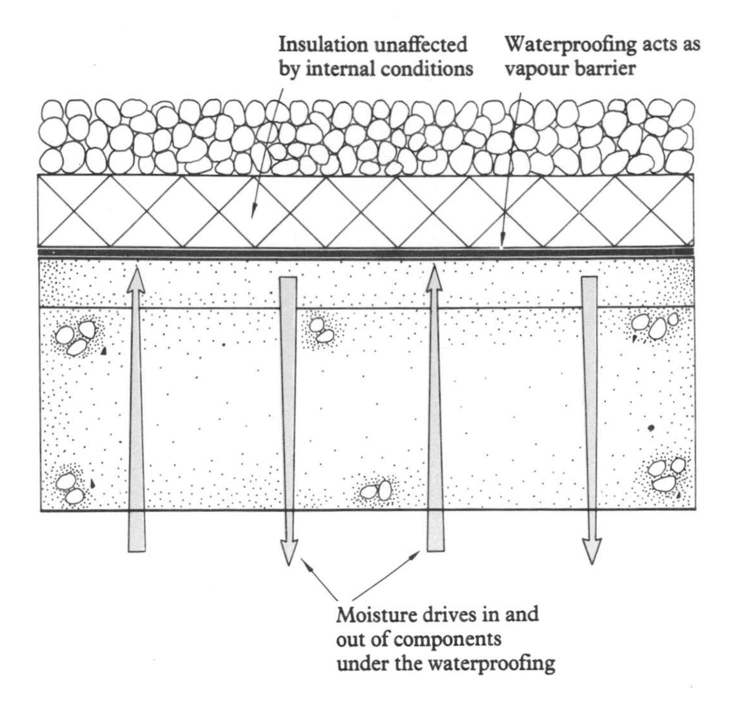

Moisture movements in the insulation layer of a protected membrane roof are virtually unaffected by internal conditions. Below the level of the waterproofing, the moisture vapour from inside the building will flow in and out of all the components and there is a possibility of surface condensation due to the flow of cold rain below the insulation on roof decks of low thermal mass.

A method of checking for this is to assume a temperature of 0°C at the waterproofing level and carry out the surface condensation check calculation as given in Appendix A.

Cold roofs

Cold roofs usually include a ceiling space which is also cold as it has little or no insulation above it. Unless it is ventilated, the ceiling space or cavity acts as a reservoir for water vapour, held ready to condense when the temperature drops.

A cold roof will always suffer the risk of condensation if the ceiling construction allows moisture vapour into the cavity by convection through open joints and if cavity ventilation is not provided.

A typical pattern of moisture movement in a cold roof is for penetration of moisture vapour to the cavity by the combined effects of diffusion, convection and air leakage through joints, followed by dispersal of the moisture vapour by means of ventilation to the external air.

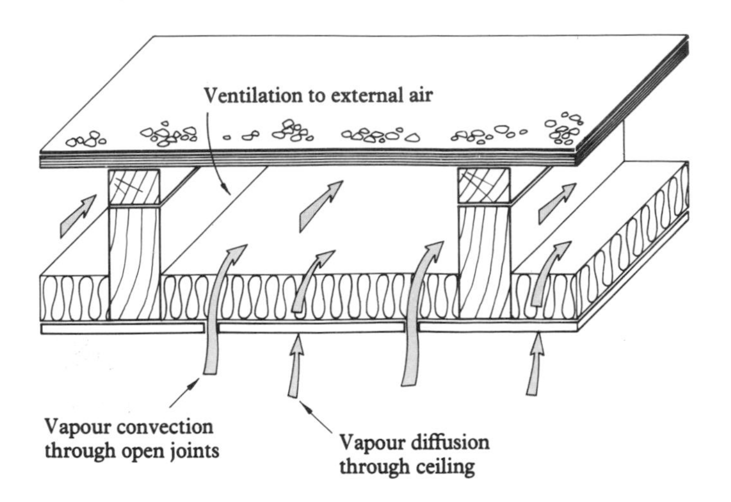

Under these complex circumstances, there is no practicable method for predicting moisture movement, and the guide lines for the design of ventilation are based on experience and trials. Unfortunately, there is only limited research on the subject and recommendations for the size of openings to provide ventilation are somewhat arbitrary. A broad recommendation is made by most authorities to ventilate the cavity of all cold roof constructions to the outside of the building, and in the majority of cases this is wise advice.

For natural ventilation to be useful, there must be an adequate open area on each side of the cavity and there should be a free path for the flow of air in between. Through ventilation may often be obstructed by sound insulation, cavity fire barriers, service installations or bracing, and it may be necessary to provide ventilation through the roof covering to some areas when no other form of ventilation is possible.

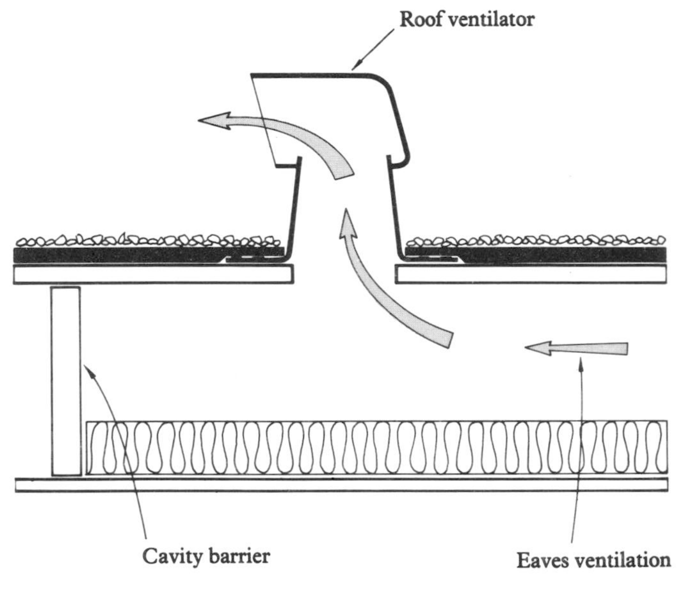

Natural cross-ventilation depends on the windspeed, but the air is often static at the time that ventilation is most needed, during long periods of still, cold weather. Wind-induced ventilation is not the only mechanism of vapour escape, however, as the vapour pressure in the ceiling space will produce a vapour drive through the openings. The stack effect from rising warm air will also drive the air to the outside.

Recent work on the amount of ventilation required suggests that the openings should be equivalent to 4% of the plan roof area. The width of the opening should be approximately 10mm to allow for unrestricted flow of air and to prevent the entry of birds. Larger openings will require screens.

The insulation in a cold roof construction is usually in the form of a fibrous quilt or a loose-fill material above the ceiling. With high ventilation rates, some insulation materials can become displaced, and when high winds are combined with low temperature, there may be a reduction in the effectiveness of the insulation caused by the fast passage of air over and through the insulation. This effect is minimal with closed cell insulation materials, but open cell or open fibre insulation materials may require a top surfacing of breather paper. This will reduce the movement of air through the insulation, but still allow moisture vapour to pass relatively uninterrupted into the ventilated ceiling space to be dispersed to the outside air.

In cases where it is difficult to achieve enough through ventilation to clear away the moisture vapour from the cavity, it may be helpful to provide a layer of vapour resisting material at ceiling level to form a vapour check and slow the passage of moisture vapour into the cavity.

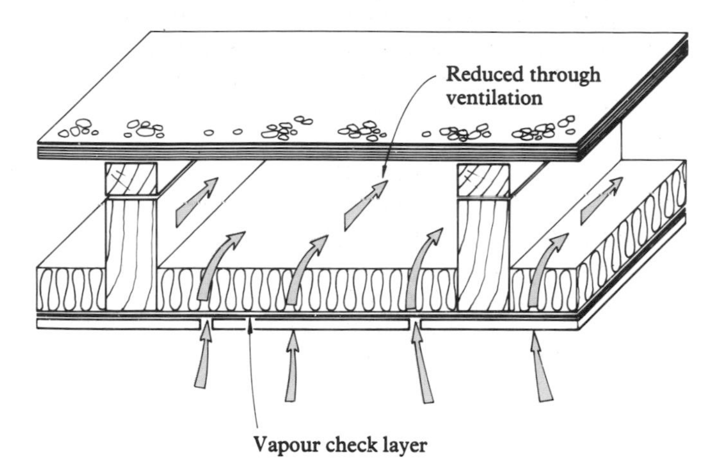

Restricted ventilation compensated by vapour check layer to reduce flow of vapour to cavity

Any such layer would have to accommodate a large number of penetrations for services and with the numerous problems of application it will be difficult to achieve continuity of the layer. It will not be possible to form an efficient vapour barrier and at best it will only be possible to provide a vapour check, normally by polythene sheeting or by using foil or polythene faced ceiling boards. In many cases a better method of condensation control would be to change the design to a warm roof construction.

SUMMARY

Design principles for moisture control
Whatever the type of roof construction - warm, cold, or hybrid -there are three basic design rules for moisture control:

*All constructions which allow moisture movement by diffusion alone can be analysed by calculation, and if necessary, adjustments can then be made to produce a design which is satisfactory.

*All constructions which allow an element of convection in addition to diffusion can be checked by calculation to determine whether condensation is likely in any spaces where air may be circulated by convection.

*In the event that ventilation of a ceiling space is required, guidelines for the determination of the area of openings are available and, although arbitrary, should be followed unless definitive analysis can be tackled by expert designers.

1.3 VAPOUR BARRIER DESIGN GUIDE

INTRODUCTION

The guidance available to designers on the subject of vapour barriers in a warm roof construction is generally unscientific and is all too often inconsistent or unsatisfactory.

One approach is that there should always be a vapour barrier under the insulation regardless of the deck or the conditions within the building, and with little concern for the efficiency of the barrier. The principle falls down when the structural deck also provides the insulation, such as woodwool or lightweight concrete roof units. In these cases, the principle of vapour barrier below insulation is conveniently ignored. Similarly, a vapour barrier is not applied to concrete roofs covered with an insulating cement-based screed because the screed is normally applied direct to the concrete and keyed to it. The principle is, therefore, inconsistent, and in practice vapour barriers are only applied when it is practical to install them, otherwise they are tacitly ignored.

Another approach is represented by the phrase 'if in doubt, leave it out'. This recognises that a vapour barrier can sometimes do more harm than good, or can be an unnecessary expense. This is probably true, but the approach is hardly scientific and does not give guidance on how to resolve the doubts. Yet another approach is to install a vapour barrier only in certain classifications of buildings where the relative humidity exceeds, perhaps, 60% for long periods at a time. An efficient vapour barrier is then installed and a restriction is placed on the choice of insulation materials to those least vulnerable to breakdown in the event of leakage of vapour into the system. The approach is useful but is over-simplified, and it leaves some awkward decisions when conditions are marginal.

MOISTURE GAIN ANALYSIS

The recent introduction of moisture gain analysis described earlier in Section 1.2 provides a more scientific approach to the subject. This form of analysis has been adopted by the British Standards Institution for their forthcoming Code of Practice for flat roofs with continuously supported coverings. The analysis is normally on a trial and error basis. First of all the designer proposes a specification which appears to be satisfactory and the moisture vapour movement is checked by moisture gain analysis. If this proposal is unsatisfactory, the specification is modified and then re-checked.

The first proposal need not include a vapour barrier unless it is clear that one will be required, as analysis may show that a vapour barrier is not required. A disadvantage of including a vapour barrier when not strictly necessary is the unavoidable formation of a waterproof lining beneath the insulation, which acts as a reservoir and will collect water in the event of leakage of the membrane. Under these circumstances, a small fault in the membrane can lead to widespread damage to the insulation. Roof systems which can be constructed without a vapour barrier have an advantage in allowing leakage to show up immediately and near to the point of entry. The omission of an unnecessary vapour barrier will also reduce the cost of the roofing system.

If a vapour barrier is shown to be required, it will be necessary to consider its efficiency. Rather than attempting to try to predict a precise resistance to vapour diffusion for each of the many possible specifications for a vapour barrier, it is more convenient to consider the barrier in two categories of efficiency: a fully efficient vapour barrier, or a penetrated vapour barrier referred to in this guide as a vapour check.

GUIDANCE TABLES

As an aid to design, the results of moisture gain analysis for typical warm roof constructions have been calculated and are given in Tables 1.6 to 1.11. These show the relative humidity at given internal temperatures at which a vapour check or vapour barrier become necessary.

Correct use of the tables requires an assessment of the average internal conditions of the building under consideration over a typical 60-day period. In the absence of specific data, the indoor conditions representing average temperature and relative humidity given in the following table should be assumed.

Type of building	Temperature °C	Relative humidity %
Houses and flats	20	55
Offices	20	40
Schools	20	50
Factories and heated warehouses	15	35
Textile factories	20	70
Swimming pool halls	25	70

For the purposes of calculation, a vapour check is taken to be a single layer of bitumen felt and a vapour barrier as two layers of bitumen felt. If a single layer metal lined vapour barrier is to be used, then the vapour resistance of this can be taken as being equivalent to that of a two layer bitumen felt vapour barrier. The vapour resistance of these layers used to calculate the tables is as follows:

Vapour check resistance
Metal deck 100MNs/g
All other decks 300MNs/g

Vapour barrier resistance
All decks 500MNs/g

The above assumes that a single layer of felt used as a vapour check on metal deck is likely to be damaged or penetrated and the vapour resistance of this layer is taken as one third that of a fully supported layer.

Where the insulation board itself has a high vapour resistance the effect of joints is significant and is taken into account in determining a realistic figure for calculations.

The vapour resistance of the waterproofing membrane is so large compared with the resistance of the other components in the system, that its magnitude has very little effect on the calculation of the amount of condensate and it is assumed to be impermeable.

The moisture vapour flow assumed for the calculation is based on a 60-day winter period, with external conditions of -5°C and a 60-day summer period with external conditions of 18°C. The amount of condensate in winter and the amount of evaporation in summer have been taken into account and the calculation method used is set out in Appendix A. The tables are relevant for a range of U-values from 0.3 to 1.0W/m^2°C.

VAPOUR CHECK/VAPOUR BARRIER REQUIREMENTS

TABLE 1.6 IN-SITU CAST CONCRETE DECK

Insulation	Temperature °C															
	15	16	17	18	19	20	21	22	23	24	25	26	27	28	29	30
Glass fibre roofboard																
Mineral wool slabs																
% RH at which a vapour check is required	76	71	67	63	59	56	53	50	47	45	42	40	38	36	34	32
% RH at which a vapour barrier is required			94	89	83	79	74	70	66	62	59	56	52	50	47	44
Wood fibreboard																
% RH at which a vapour check is required	76	71	67	63	60	56	53	50	47	45	42	40	38	36	34	32
% RH at which a vapour barrier is required			94	89	84	79	74	70	66	62	59	56	53	50	47	44
Expanded polystyrene with 13mm wood fibreboard																
% RH at which a vapour check is required	76	72	68	64	60	57	53	50	48	45	43	40	38	36	34	32
% RH at which a vapour barrier is required			95	89	84	79	75	70	66	63	59	56	53	50	47	45
Cork																
% RH at which a vapour check is required	77	73	68	65	61	57	54	51	48	46	43	41	38	36	34	33
% RH at which a vapour barrier is required					95	90	84	80	75	71	67	63	60	56	53	50
Polyurethane																
Polyisocyanurate																
% RH at which a vapour check is required	78	73	69	65	61	58	54	51	49	46	43	41	39	37	35	33
% RH at which a vapour barrier is required					96	90	85	80	75	71	67	63	60	57	54	51

TABLE 1.7 PRECAST AERATED CONCRETE DECK

Insulation	Temperature °C															
	15	16	17	18	19	20	21	22	23	24	25	26	27	28	29	30
Glass fibre roofboard																
Mineral wool slabs																
% RH at which a vapour check is required	47	44	42	39	37	35	33	31	29	27	26	24	23	22	21	20
% RH at which a vapour barrier is required		98	92	87	81	77	72	68	64	61	57	54	51	48	46	43
Wood fibreboard																
% RH at which a vapour check is required	51	47	45	42	40	37	35	33	31	29	28	26	25	23	22	21
% RH at which a vapour barrier is required		98	92	87	82	77	72	68	64	61	57	54	51	48	46	43
Expanded polystyrene with 13mm wood fibreboard																
% RH at which a vapour check is required	64	60	56	53	50	47	44	42	39	37	35	33	31	29	28	26
% RH at which a vapour barrier is required		98	92	87	82	77	73	69	65	61	58	54	51	49	46	44
Cork																
% RH at which a vapour check is required	66	62	58	54	51	48	45	43	40	38	36	34	32	30	28	27
% RH at which a vapour barrier is required				98	92	87	82	77	73	69	65	61	58	55	52	49
Polyurethane																
Polyisocyanurate																
% RH at which a vapour check is required	74	70	66	62	58	55	52	49	46	44	41	39	37	35	33	31
% RH at which a vapour barrier is required				98	93	87	82	78	73	69	65	62	58	55	52	49

VAPOUR CHECK/VAPOUR BARRIER REQUIREMENTS

TABLE 1.8 PLYWOOD DECK

Insulation	Temperature °C															
	15	16	17	18	19	20	21	22	23	24	25	26	27	28	29	30
Glass fibre roofboard																
Mineral wool slabs																
% RH at which a vapour check is required	66	62	58	55	52	49	46	43	41	38	36	34	32	30	29	27
% RH at which a vapour barrier is required		98	92	87	82	77	73	69	65	61	58	54	51	49	46	44
Wood fibreboard																
% RH at which a vapour check is required	72	67	63	60	56	53	50	47	44	42	39	37	35	33	31	29
% RH at which a vapour barrier is required		98	92	87	82	77	73	69	65	61	58	55	52	49	46	44
Expanded polystyrene with																
13mm wood fibreboard																
% RH at which a vapour check is required	74	70	66	62	58	55	52	49	46	44	41	39	37	35	33	31
% RH at which a vapour barrier is required		99	93	88	82	78	73	69	65	61	58	55	52	49	46	44
Cork																
% RH at which a vapour check is required	74	70	66	62	58	55	52	49	46	44	41	39	37	35	33	31
% RH at which a vapour barrier is required				99	93	87	82	78	73	69	65	62	58	55	52	49
Polyurethane																
Polyisocyanurate																
% RH at which a vapour check is required	75	70	66	62	59	56	52	49	47	44	42	39	37	35	33	32
% RH at which a vapour barrier is required				99	93	88	83	78	74	69	65	62	58	55	52	49

TABLE 1.9 TONGUED AND GROOVED BOARDED DECK

Insulation	Temperature °C															
	15	16	17	18	19	20	21	22	23	24	25	26	27	28	29	30
Glass fibre roofboard																
Mineral wool slabs																
% RH at which a vapour check is required	42	40	38	35	33	31	30	28	26	25	23	22	21	20	19	18
% RH at which a vapour barrier is required		98	92	86	81	77	72	68	64	61	57	54	51	48	46	43
Wood fibreboard																
% RH at which a vapour check is required	48	45	42	40	38	35	33	31	30	28	26	25	24	22	21	20
% RH at which a vapour barrier is required		98	92	87	82	77	72	68	64	61	57	54	51	48	46	43
Expanded polystyrene with																
13mm wood fibreboard																
% RH at which a vapour check is required	72	68	64	60	56	53	50	47	44	42	39	37	35	33	31	30
% RH at which a vapour barrier is required		98	92	87	82	77	73	69	65	61	58	55	52	49	46	44
Cork																
% RH at which a vapour check is required	64	60	57	53	50	47	44	42	39	37	35	33	31	29	28	26
% RH at which a vapour barrier is required				98	92	87	82	77	73	69	65	61	58	55	52	49
Polyurethane																
Polyisocyanurate																
% RH at which a vapour check is required	74	70	60	62	58	55	52	49	46	44	41	39	37	35	33	31
% RH at which a vapour barrier is required				98	93	87	82	77	73	69	65	61	58	55	52	49

VAPOUR CHECK/VAPOUR BARRIER REQUIREMENTS

TABLE 1.10 WOODWOOL DECK

Insulation	Temperature °C															
	15	16	17	18	19	20	21	22	23	24	25	26	27	28	29	30
Glass fibre roofboard Mineral wool slabs																
% RH at which a vapour check is required	42	40	37	35	33	31	29	28	26	25	23	22	21	20	18	17
% RH at which a vapour barrier is required		98	92	86	81	77	72	68	64	61	57	54	51	48	46	43
Wood fibreboard																
% RH at which a vapour check is required	45	43	40	38	36	33	32	30	28	26	25	24	22	21	20	19
% RH at which a vapour barrier is required		98	92	86	81	77	72	68	64	61	57	54	51	48	46	43
Expanded polystyrene with 13mm wood fibreboard																
% RH at which a vapour check is required	59	55	52	49	46	43	41	38	36	34	32	30	29	27	26	24
% RH at which a vapour barrier is required		98	92	87	82	77	73	68	65	61	58	54	51	49	46	43
Cork																
% RH at which a vapour check is required	58	55	51	48	45	43	40	38	36	34	32	30	28	27	25	24
% RH at which a vapour barrier is required				98	92	87	82	77	73	69	65	61	58	55	52	49
Polyurethane Polyisocyanurate																
% RH at which a vapour check is required	74	70	66	62	58	55	52	49	46	44	41	39	37	35	33	31
% RH at which a vapour barrier is required				98	93	87	82	77	73	69	65	61	58	55	52	49

TABLE 1.11 METAL DECK

Insulation	Temperature °C															
	15	16	17	18	19	20	21	22	23	24	25	26	27	28	29	30
Glass fibre roofboard Mineral wool slabs																
% RH at which a vapour check is required	68	64	60	57	53	50	47	45	42	40	37	35	33	31	30	28
% RH at which a vapour barrier is required	84	79	74	70	66	62	59	55	52	49	47	44	42	39	37	35
Wood fibreboard																
% RH at which a vapour check is required	74	69	65	62	58	55	51	48	46	43	40	38	36	34	32	30
% RH at which a vapour barrier is required	84	79	74	70	66	62	59	55	52	49	47	44	42	39	37	35
Expanded polystyrene with 13mm wood fibreboard																
% RH at which a vapour check is required	74	70	66	62	59	55	52	49	46	44	41	39	37	35	33	31
% RH at which a vapour barrier is required	85	80	75	71	67	63	59	56	53	50	47	44	42	40	38	36
Cork																
% RH at which a vapour check is required	74	70	66	62	59	55	52	49	46	44	41	39	37	35	33	31
% RH at which a vapour barrier is required	89	84	79	74	70	66	62	59	55	52	49	47	44	42	40	37
Polyurethane Polyisocyanurate																
% RH at which a vapour check is required	75	70	66	63	59	56	52	50	47	44	42	39	37	35	33	32
% RH at which a vapour barrier is required	90	84	79	75	70	66	63	59	56	53	50	47	44	42	40	38

GENERAL GUIDANCE

Decks

For insulated timber decks, woodwool and aerated concrete slabs, the water vapour resistance of the decks taking into account their joints is relatively low and does not contribute very much to the total resistance of the construction. A vapour check between these decks and the insulation would normally be required.

Concrete decks have a relatively high vapour resistance and so the temperatures and humidities at which a vapour barrier is required will normally exceed the internal conditions to which concrete is likely to be subjected. It should however be remembered that concrete contains water during construction and an underlayer of bitumen felt is required to act both as a damp proof layer and temporary vapour check while the concrete dries out.

For metal decking, the tables indicate that on the majority of buildings, where humidity is not excessive, a vapour check or vapour barrier is not necessary. However, an underlay may sometimes be required to provide a support and bonding layer for the insulation.

Generally, for an internal temperature of 20°C, a full two-layer vapour barrier is only required on buildings with high continuous humidity. On metal deck the relative humidity at which a vapour barrier is required is in the order of 60% and for other decks 75%.

Ceilings

The addition of ceilings does not alter the internal temperature and humidity levels at which a vapour check or vapour barrier is required, but may introduce the possibility of surface condensation under the deck. The temperature in the ceiling space will be reduced if the ceiling has a significant insulation value but the vapour pressure will remain the same as in the building below. It is necessary to check that the amount of insulation above the deck is sufficient to prevent condensation occurring on the underside of the deck. In order to do this, Table 1.12 can be used. The procedure is to calculate the thermal resistance below the deck and multiply this by the appropriate factor in the table for the given internal conditions. This will give the thermal resistance required above the deck to prevent condensation occurring. An example of this procedure is shown in Appendix A.

Types of building

The following general recommendations on the need for a vapour barrier or vapour check are made according to anticipated conditions of building use.

Swimming pools, textile mills, paper mills

In these buildings the conditions will be at or above 20°C and 70% relative humidity and the conditions are generally continuous as opposed to intermittent, which makes them much more severe in terms of searching out weaknesses in vapour barriers.

It will be found that such conditions always call for an efficient vapour barrier on metal decks and the choice of insulation should be restricted to those which are least affected by leaking water vapour.

Even with a two-layer vapour barrier, which is regarded as fully efficient, it would be unwise to assume that there will be no local leakage of water vapour. Experience has shown that cork is a suitable choice under these conditions. It is resistant to moisture and it can be installed full bonded between the vapour barrier and the waterproofing with no intervening air spaces or part bonded layers. It is therefore certain that moisture will only leak into the system by diffusion. Insulations which require the main waterproofing to be part bonded are rather more at risk because it is possible that leakage of water vapour could arise through convection.

TABLE 1.12 BALANCE OF INSULATION

Ratio of thermal resistance above the deck to thermal resistance below the deck, based on external temperature of -5°C.

% RH	Internal temperature °C															
	15	16	17	18	19	20	21	22	23	24	25	26	27	28	29	30
30	0.2	0.2	0.2	0.3	0.3	0.4	0.4	0.5	0.5	0.6	0.6	0.6	0.7	0.7	0.8	0.8
35	0.3	0.4	0.4	0.5	0.5	0.6	0.6	0.7	0.7	0.8	0.8	0.9	0.9	1.0	1.0	1.0
40	0.5	0.5	0.6	0.7	0.7	0.8	0.8	0.9	1.0	1.0	1.1	1.1	1.2	1.2	1.3	1.3
45	0.7	0.8	0.8	0.9	1.0	1.0	1.1	1.2	1.2	1.3	1.4	1.4	1.5	1.5	1.6	1.6
50	0.9	1.0	1.1	1.2	1.3	1.3	1.4	1.5	1.6	1.6	1.7	1.8	1.8	1.9	2.0	2.0
55	1.2	1.3	1.4	1.5	1.6	1.7	1.8	1.9	1.9	2.0	2.1	2.2	2.3	2.3	2.4	2.5
60	1.6	1.7	1.8	1.9	2.0	2.1	2.2	2.3	2.4	2.5	2.6	2.7	2.8	2.9	3.0	3.1
65	2.1	2.2	2.3	2.4	2.6	2.7	2.8	2.9	3.0	3.2	3.3	3.4	3.5	3.6	3.7	3.8
70	2.7	2.8	3.0	3.1	3.3	3.4	3.6	3.7	3.9	4.0	4.1	4.3	4.4	4.5	4.6	4.8
75	3.6	3.7	3.9	4.1	4.3	4.5	4.7	4.8	5.0	5.2	5.3	5.5	5.6	5.8	6.0	6.1
80	4.9	5.1	5.3	5.6	5.8	6.0	6.3	6.5	6.7	6.9	7.1	7.3	7.5	7.7	7.9	8.1
85	7.0	7.3	7.7	8.0	8.3	8.6	8.9	9.2	9.5	9.8	10.1	10.4	10.7	10.9	11.2	11.5
90	11.3	11.8	12.3	12.8	13.3	13.8	14.3	14.7	15.2	15.6	16.1	16.5	17.0	17.4	17.8	18.2

Housing
The majority of families maintain a satisfactory level of ventilation in their homes to clear away steam and water vapour from washing, cooking and crowded conditions. Unfortunately there is a minority who live in high humidity conditions with little ventilation, either through thoughtlessness or in the belief that ventilation will increase the cost of heating. For this reason housing must be considered a fairly high humidity situation. The humid conditions are usually intermittent however and a vapour check would normally suffice.

Some families use oil heaters continuously without ventilation and they will be producing conditions equivalent to a textile mill or swimming pool. They would be well advised to introduce substantial ventilation or make sure that the heaters are used only intermittently to allow plenty of time for drying out.

Sports halls and communal showers, baths, toilets and kitchen accommodation
This group of buildings or parts of buildings produce high humidity but normally for intermittent periods, so that the average conditions for design are substantially less than the maximum that will be observed in service. The designer will normally have the opportunity to introduce substantial ventilation which will control the relative humidity but it should be borne in mind that the management of some buildings can be careless with regard to the use of ventilation systems provided.

Analysis will usually show that a vapour check is necessary but a fully efficient vapour barrier may not be required. It is often extremely difficult to predict a suitable average humidity for calculation purposes and if the designer is in doubt, he will be well advised to allow for a vapour check as a precaution.

Classrooms
Classrooms sometimes move towards high humidity because of the number of people in a relatively confined space. Re-circulating heating systems are common with little ventilation unless the teacher opens a window. Nevertheless, large stocks of schools have been built with no vapour barrier or vapour check over the classrooms, corridors or halls and these have amply demonstrated a successful result.

Factories with normal dry processes, warehouses, shops, shopping centres, commercial buildings and offices
These are representative of a group of buildings which are operating under normal dry conditions and analysis will normally show that a vapour barrier or vapour check is not required. There may, however, be local parts of the building which are at high humidity. Small kitchen facilities and well ventilated toilets may usually be ignored but kitchens for commercial catering or shower rooms which are in frequent use need to be considered separately and will usually require a vapour check.

Change of building use
Moisture gain analysis will also help those who wish to make allowance for occupancy changes during the life of the building by allowing for future high humidity operations. An efficient vapour barrier is an expensive precaution, but a penetrated vapour barrier or vapour check is unlikely to be good enough in the event of a change to high humidity conditions. The building owner should specify the worst conditions to be alllowed for in the design. If, for example, this is to allow for the possibility of installing a textile mill in an advance factory, the design conditions will probably be set at 20°C and 70% relative humidity. This will almost certainly call for a full vapour barrier preferably with an insulation which is not moisture sensitive.

The cost of making such an allowance as a precaution against change of occupancy can be analysed, and the building owner can consider the options of spending extra money for the extra facility, against the alternative cheaper construction which would be appropriate for normal usage.

VAPOUR BARRIER/VAPOUR CHECK INSTALLATION

A fully efficient vapour barrier will require bonded overlaps and penetrations must be effectively sealed. It must be applied with precautions to ensure that it is isolated from building movement where necessary, so that it maintains integrity in service. This will sometimes lead to the use of a partially-bonded two-layer barrier.

A vapour check is constructed from a single layer and partial bonding is not necessary as a small amount of damage from building movement would make little difference to the effectiveness of the vapour check. Unsealed laps and penetrations are accepted, and a small amount of physical damage can be tolerated.

An underlay beneath the insulation may be necessary as a bonding or isolating layer, or as the critical layer of wind attachment, but this would not be for the purpose of providing a vapour check, and it should not be referred to as such. It is generally referred to merely as an underlay.

There is, however, a half-way case between a vapour check and an underlay. This arises when applying an insulation to a wet deck, such as concrete or lightweight screed or screeded surfaces and where calculation has shown that no permanent vapour check is required. These constructions are likely to present a wet surface for the insulation and an underlay would then be required beneath most insulations to prevent dampness rising into the insulation. The underlay in this case acts as a temporary damp proof course and vapour check until the concrete and screeds dry out, and it is only during the drying out that the underlay will be performing a useful function. There is no satisfactory calculation for an assessment of moisture movement during this period, but experience has shown that an efficient vapour check is not necessary. After a year or two, the roof construction will have dried and the underlay will have served its purpose and becomes redundant.

The materials forming the vapour barrier or vapour check and their method of attachment will be determined by the nature of the deck.

In-situ cast concrete decks

Vapour barrier
The vapour barrier may be formed by two layers of BS 747 felt type 2B or 3B fully bonded in bitumen, though preferably one of the layers should be a high performance membrane to reduce the possibility of damage. Alternatively, a single layer of specialised vapour barrier material with high resistance to moisture vapour can be used, bonded in bitumen. Some proprietary materials contain a layer of metal foil to ensure a high resistance to moisture transmission.

Vapour check
A vapour check on concrete decks may be formed by a single layer of BS 747 felt type 2B or 3B, fully bonded in bitumen. The insulation is then bonded to the vapour check, again in a continuous coat of bitumen.

If a vapour barrier or vapour check are not necessary, an underlay is usually required to provide a temporary damp proof course and temporary vapour check until the concrete deck and screeds have dried out with the natural process of drying of the building. A bonded layer of BS 747 type 2B or 3B roofing will suffice.

Precast concrete, woodwool and plywood decks

Vapour barrier
If high humidity conditions warrant a vapour barrier on precast concrete, woodwool or plywood decking it may be necessary to partially bond the vapour barrier to isolate it from movement of the roof deck. The first layer should be BS 747 type 3G perforated gritted glass base roofing. This should be followed by a second layer of roofing which could be BS 747 type 3B or 2B, fully bonded in bitumen, but should preferably be a high performance membrane to provide additional strength. One of the lighter grades of polyester base roofing would suffice. Where there is no likelihood of significant joint movement, a fully bonded vapour barrier will be suitable. The designer must use his own judgement on this, bearing in mind the temperatures within the building and the stability of the roof structure as a whole.

Vapour check
A vapour check on concrete decks and woodwool may be formed by a single layer of BS 747 felt type 2B or 3B, fully bonded in bitumen. The insulation is then bonded to the vapour check in a continuous coat of bitumen. It is possible that damage could be caused to the vapour check by movement, as the specification is fully bonded, but it is not likely to occur to the extent that it will seriously impair the effectiveness of the roofing as a vapour check.

On plywood, the vapour check can be a fully bonded layer of type 2B or 3B felt, or the same materials loose-laid and attached with the insulation to the deck by screw fixings and large washers.

Timber decks

Vapour barrier
Timber is not usually incorporated in a construction above high humidity areas, although certain timbers are resistant to humidity and have been used with success. The known properties of the timber should be compatible with the conditions which are likely to be encountered in service.

The first layer of the vapour barrier must be nailed and a second layer will be required to cover the penetrations in order to form an efficient vapour barrier. It is recommended that a high performance first layer be nailed and then sealed by a BS 747 felt type 3B fully bonded to the first layer.

Vapour check
A single layer of BS 747 felt type 2B will suffice to form a vapour check on timber. This should be nailed with the insulation fully bonded to it in hot bitumen. An alternative method is to mechanically fix the insulation through a loose-laid layer of BS 747 felt type 2B or 3B.

Metal decks

Vapour barrier
The majority of flat roofs on high humidity buildings are likely to be of metal deck construction, as these are comprised of materials particularly suited to high humidity conditions. The vapour barrier is applied direct to the metal and will span the open troughs of the deck. It is not recommended that any single layer system be used for an efficient vapour barrier on metal decks. The bonding of the laps will be unreliable as they are not continuously supported, and it would be unreasonable to expect roofers to make a satisfactory vapour barrier of a one layer system working under conditions that are often difficult in terms of wind, dampness, dust and temperature.

The minimum reasonable option for an efficient vapour barrier on metal deck is to apply a first layer of reinforced or high performance material. This should be bonded in bitumen to the top flats of the deck. A second layer of roofing is then applied by bonding in hot bitumen on top of the first layer. This gives good lap protection in the normal traditions of built-up roofing. The choice of material for this layer is unimportant, and a layer of BS 747 felt type 3B is satisfactory.

It is not advisable to introduce any mechanical fixings which might penetrate the vapour barrier. It is sometimes considered that mechanical fixings are self-sealing through a bitumen membrane of reasonable thickness; but little work has been done on the subject and it is best to avoid such penetration even though this will prove difficult on occasions.

An alternative specification for roofs exposed to high winds and requiring mechanical fixings is given in the Wind Attachment Design Guide Section 1.7, page 50.

Vapour check
A single layer vapour check on metal decking can normally only be regarded as of low efficiency. It is extremely difficult to ensure that it is fully sealed and undamaged, and the calculation methods of analysis use a reduced permeability for a single layer vapour check on metal decks.

Even though penetrations through a vapour check are acceptable, it would not be desirable to risk substantial damage to the membrane during application, and a high performance single layer system on metal decks will be the best choice. BS 747 type 1F hessian reinforced roofing is often specified as a vapour check, but this material is not as strong as one would wish and can be superseded in specifications now that better high performance materials are available.

The vapour check may be continuously bonded to each top flat of the decking. The insulation is fully bonded to the vapour check with supplementary mechanical fixings if necessary. Alternatively, the vapour check and insulation may be laid loose and the attachment formed solely by mechanical fixings through the insulation and vapour check into the deck.

Paper based underlays, or vapour checks, are available as an alternative to bitumen roofing and may be used on metal decks where they will reduce the combustible content of the materials below the insulation. These paper underlays may be foil faced to provide reasonable impermeability to water vapour, and glass reinforced for strength. Although generally effective as underlays or vapour checks, these materials are necessarily applied in a single layer system and are penetrated by the large number of mechanical fixings required for attachment. They should not be regarded as an efficient vapour barrier on metal decks. Laps may be bonded with proprietary adhesive, or sealed with tape, but this is unlikely to make a significant improvement bearing in mind the penetration from fixings, the possibility of damage and the difficulties of ensuring an effective seal on a deck which does not provide a continuous firm support for the material.

TABLE 1.13 VAPOUR BARRIERS AND VAPOUR CHECKS: SUMMARY OF RECOMMENDATIONS

DECK	VAPOUR BARRIER Side and end laps fully sealed	VAPOUR CHECK End and side lap sealing not necessary
Metal deck	**First layer:** High performance roofing fully bonded to deck flats **Second layer:** Type 2B or 3B felt fully bonded	Reinforced or high performance roofing bonded to deck flats
In-situ cast concrete	**First layer:** Type 2B or 3B felt fully bonded **Second layer:** High performance roofing fully bonded	Type 2B or 3B felt fully bonded
Precast concrete units, woodwool slabs	**First layer:** Type 3G felt partially bonded **Second layer:** High performance roofing fully bonded	Type 2B or 3B felt fully bonded
Plywood	**First layer:** Type 3G felt partially bonded **Second layer:** High performance roofing fully bonded	Type 2B or 3B felt fully bonded
Timber tongued and grooved	**First layer:** High performance roofing nailed **Second layer:** Type 2B or 3B felt fully bonded	Type 2B felt nailed

1.4 TRAPPED MOISTURE

INTRODUCTION

In addition to designing for the control of moisture vapour generated within the building by occupants and processes, precautions must be taken to prevent excessive moisture becoming trapped during construction and also to allow for the release of vapour pressures which may build up under the waterproof covering as a result. Failure to allow for trapped moisture may lead to blistering of the waterproofing or to residual dampness on the underside of the roof.

Construction water
Roofing components such as a poured concrete slab or a wet screed contain quite large quantities of water. For example a dense concrete slab 150mm thick will give up about 10 litres of water per square metre as it dries. The amount of water in the slab will be increased by subsequent wet screeding and also by rainfall if left exposed before the waterproofing is applied. Precast concrete units and other preformed deck and insulation materials may also absorb considerable amounts of water if left exposed to the rain.

Water within the structure is best drained away through temporary weep holes formed at the lowest points of the slab. These can be formed by casting-in blocks of expanded polystyrene for subsequent burning out, by conical timber plugs or cardboard formers or by drilling into the slab from below. The holes are then left open until seepage stops when they can be filled with sand cement mortar before applying the soffit finish.

Once the waterproof covering has been laid, drying out of a structural slab will largely take place from the underside of the deck. Brick box ventilators about 600mm square on plan provide an improved chance of drying out as do certain other proprietary drying ventilators, but it is not likely that these will act fast enough to make more than a modest contribution to drying.

Any trapped water within the roof will feed moisture vapour into all the air spaces and permeable open cells in the construction until the air in the system is saturated with moisture vapour. Increased temperatures will increase the pressure of the air and moisture vapour mixture above atmospheric and will also increase the capacity of the air to take in moisture with the result that the vapour pressures will increase over and above the normal increases from thermal expansion alone.

In the case of a wet deck construction trapped air and moisture vapour under the waterproof covering will, therefore, form pressures which may lead to blistering of the waterproofing unless some form of vapour pressure release is provided. Dry deck constructions should not contain construction water but certain apparently dry substrate materials absorb or transmit water vapour. These are treated as if they were a wet surface and pressure release is provided. In this category are timber, and cement based slabs or screeds. Pressure release is also normally provided above polyurethane foam.

Vapour pressure release
In the majority of cases when the waterproof covering is applied direct to a wet deck or screed, a part bonded attachment for the first layer will give sufficient vapour pressure release and will prevent cycles of air and water pressure from building up under the waterproofing.

The specific provision of breather vents to the external air is not usually necessary with part bonding because the construction will normally have joints and cracks or permeable sections which will allow sufficient vapour pressure release. Indeed it would be extremely difficult to seal a part bonded membrane round the edges of a roof so that it could hold a significant pressure, but some designers prefer to make quite sure by the addition of breather vents. These should then be installed between the part bonded layer and the deck, taking care that a free passage of air and water vapour is available from the part bonded interface, into and through the breather vent.

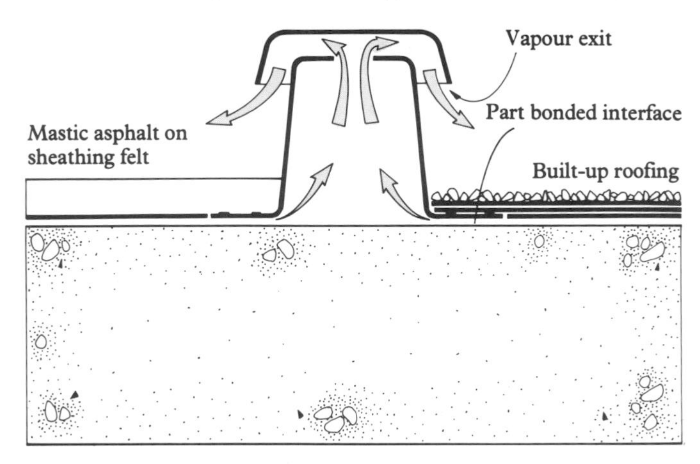

It would be wrong to rely upon small breather vents to provide a useful contribution to the drying of a screed or deck. Drying out will normally take place from the under surface of the deck and the addition of breather vents will make only the smallest contribution. They disrupt the continuity of the membrane and any damage or displacement by kicks or knocks can lead to leakage problems.

BLISTERING

The mechanism of blistering arises from pockets of trapped air and moisture which expand in the sudden heat of the sun and displace the membrane to form a small blister. The blister will only be formed when dense substrate materials allow air and moisture to pass through them at a slow rate of flow. Under these circumstances, temperatures from the heat of the sun can rise too quickly for trapped air and water vapour to escape back through the substrate. Pressures will then develop in the air pockets and cause a displacement and stretching of the waterproofing producing an increase in the size of the air pocket.

If the waterproofing has suffered an irreversible stretch, subsequent cooling will not cause the air pocket to return to the original size and a partial vacuum occurs in the now partly developed blister. Air and water vapour may be drawn slowly through the substrate to re-fill the original pocket of air, now slightly increased in size and ready to start another cycle of development of the blister when the sun appears again. The amount of expansion possible in a single cycle is calculable and is in the order of a 35% increase in volume, for a temperature change of 50°C.

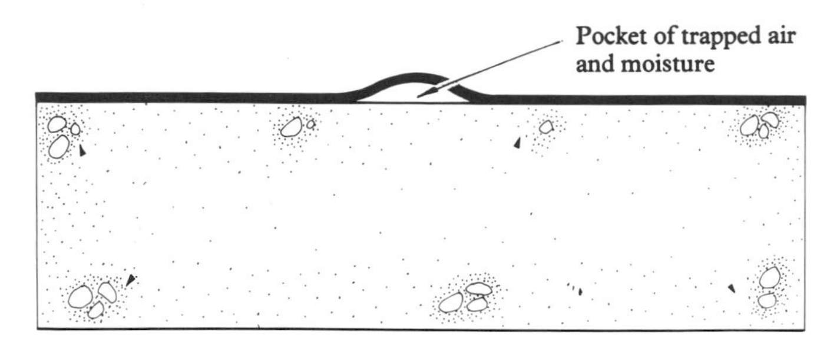

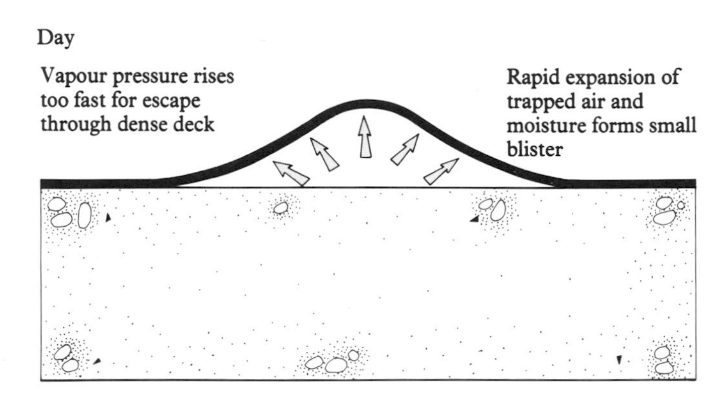

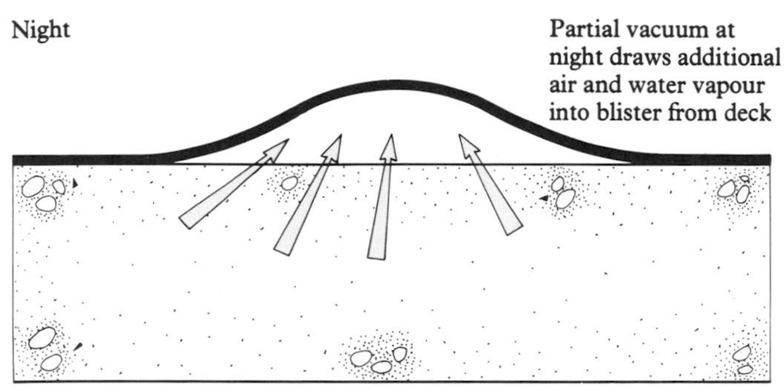

Cycle repeats to
increase size of blister

Disruption of the waterproof covering by vapour pressures takes many forms and it is important that the nature and causes of blisters and similar defects are understood so that built-up roofing and mastic asphalt specifications can be designed and laid to minimise their occurrence.

BUILT-UP ROOFING

Blistering associated with built-up roofing takes three basic forms: full membrane blistering, inter-layer blistering and top pitting.

Full membrane blistering

This refers to the blistering of the total waterproof covering from the surface it is fixed to without separation between the individual layers of roofing.

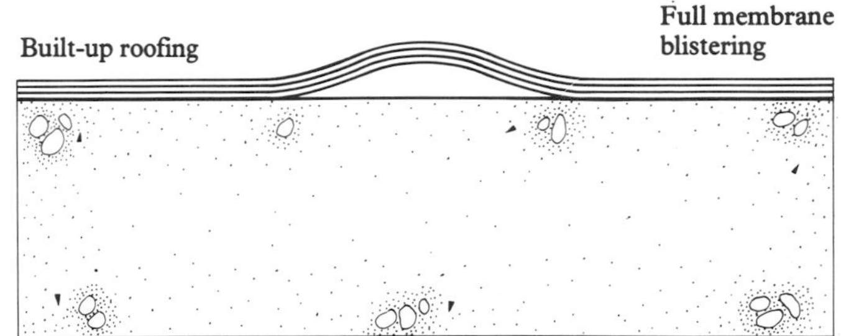

Full membrane blistering starts from the expansion of trapped pockets of air and moisture as described before. With normal bitumens and roofing felts, it is not usually possible to achieve such a good bond to a wet deck that blistering can be entirely prevented by adhesion. Specification design therefore relies on part bonding to prevent full membrane blistering occurring.

Certain torch-on materials with modified bitumen coatings achieve a bond which is an improvement on that obtained using unmodified bitumen and these materials exhibit an increased resistance to flow or stretching. All of these properties help the formation of specifications which are resistant to blistering pressures.

An application of chippings helps to cool the membrane and the weight of the chippings helps to hold down the membrane against blistering.

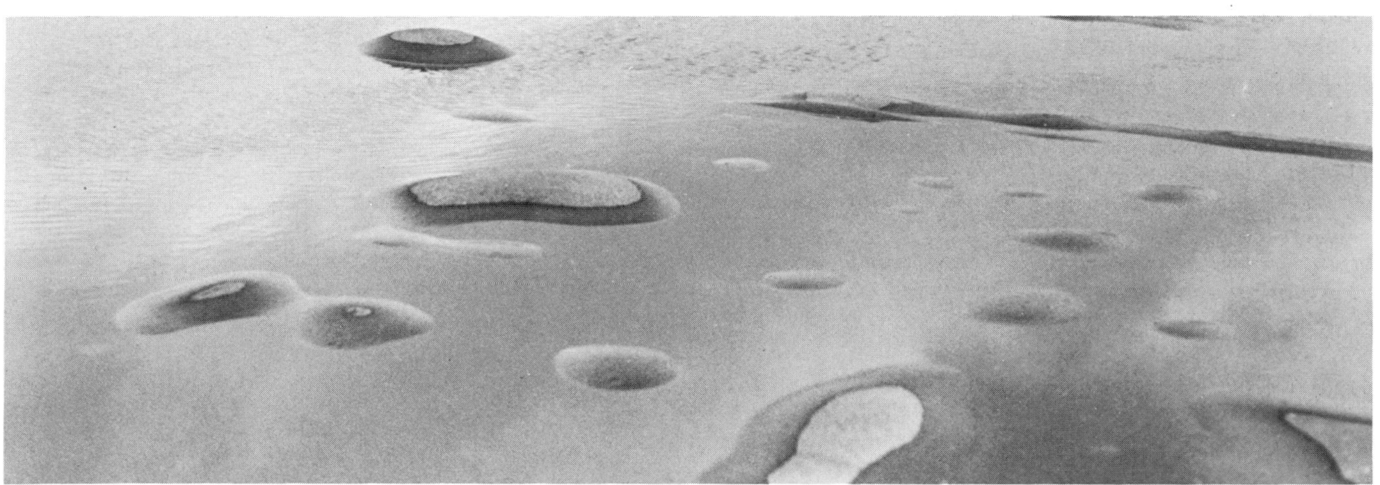

Full membrane blistering

Inter-Layer Blistering

Inter-layer blistering usually takes the form of a blister formed under the cap sheet of the membrane.

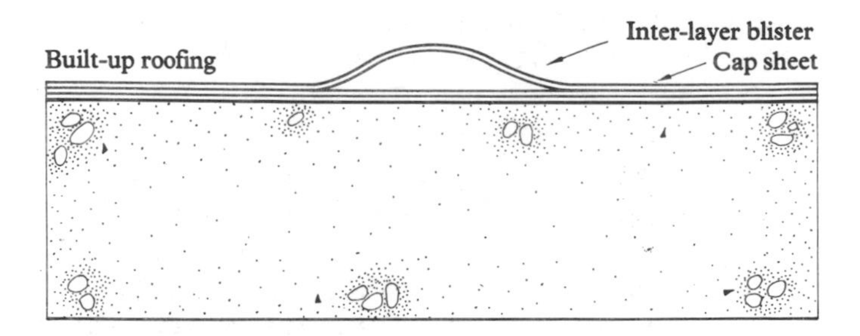

The traditional concept of inter-layer blistering assumes that the blisters are caused by air and moisture trapped between the layers.during application. In reality, substantial blisters cannot form without a topping-up process to add more air and moisture between the layers and this can only arise by the passage of air and moisture through the cap sheet.

To understand the process, it is necessary to consider the mechanism of air and moisture movements in a little more detail. The blister will start at a pocket of trapped air, and there may be many such pockets as it is impossible to achieve total continuity of bond. In practice only thin layers of air can be trapped during the course of application of roofing materials. Taking a simple case where a 1mm thickness of dry air is trapped in a cavity between two impermeable layers at 0°C (273° absolute). If the temperature rises on a hot day to 80°C (353° absolute), the volume of air would increase by the ratio 353:273 or about 30% provided there is no force to restrain the expansion. The 1mm layer of air cannot therefore increase to more than 1.3mm and would not form a noticeable blister.

If air and water are trapped together the water will keep the air 100% saturated with moisture vapour and the increased vapour pressure will cause additional expansion. When the temperature rises to 80°C, the increased air and vapour pressure will expand the cavity and if the pressure is allowed to expand the cavity freely, the volume increase will be 140%.

If the temperatures rise were only 40°C, the volume increase would be 23%. It should be noted that if the temperature returns to 0°C the blistering action will disappear and the felts will return to their original position, provided they have not been permanently elongated by the expansion.

Large blisters are, therefore, not explained by the expansion of trapped air and water alone. It should be appreciated that the mechanism depends on a feed of additional air and water vapour through the top layer of the felt. This is possible because of the vacuum effect which will develop inside the space formed by blistering action during the cooling process when the walls of the blister are stiff enough to resist returning to their original position.

The vacuum effect which draws air and moisture through the cap sheet into the blister can take place over a long period of time, but when exposed to sunshine the temperature will increase rapidly causing a rapid expansion of the air and moisture which will not have time to return through the cap sheet. This will cause an increase in pressure and further movement of the felt.

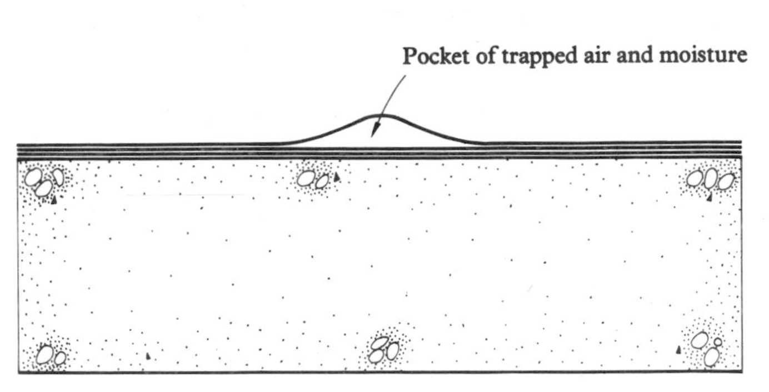

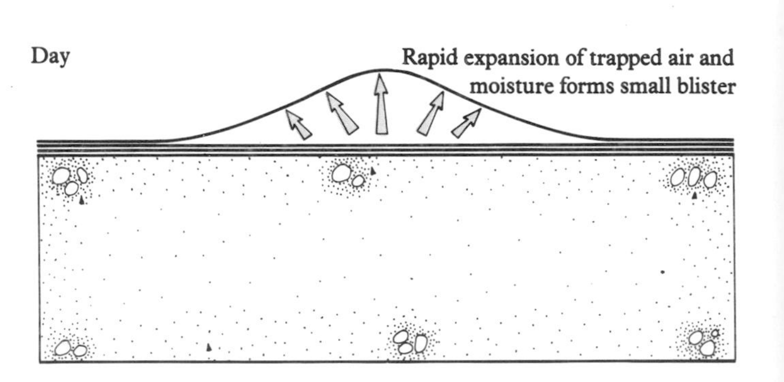

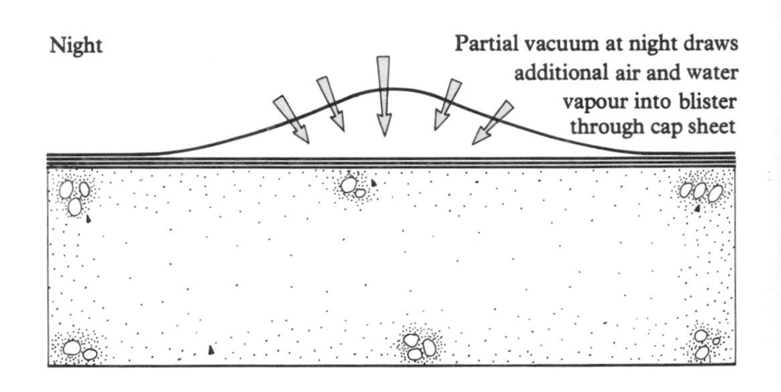

Observations of inter-layer blistering on site confirm that trapped water is not necessary to initiate the mechanism. Such blisters will occur on roofs which have been installed in entirely dry weather but the scale and frequency of blistering will tend to increase with roofs which have been installed in wet weather. Working in dry and warm weather is clearly desirable but the ideal conditions are not often experienced in the UK climate.

BS 747 type 1 fibre base felts often suffer from inter-layer blistering in service. Type 2 asbestos base felts exhibit blistering but to a lesser degree, and type 3 glass fibre felts are extremely resistant to blistering. Experience with polyester felts, although limited, indicates that they are resistant to blistering but possibly not to the same degree as glass fibre felts.

The majority of inter-layer blistering occurs with unsurfaced or mineral surfaced felts on vertical or sloping work or at upstands, but the incidence of this type of blister can now be reduced or avoided by using high performance membranes.

On flat roofs, blisters are almost entirely prevented by the application of a layer of 10mm stone chipings: smaller chippings are less effective. In recent years many specifiers have turned away from the use of chippings on flat roofs; some because of fears of blocked outlets, and others because of the difficulty of tracing leakage when a roof is covered with bonded chippings which are difficult to remove. In place of chippings, mineral surfaced roofing or a variety of liquid applied surface coatings are specified, and where these are applied on glass-base felts the danger of separation of the top layer by blistering is not severe. However, on most other felts the danger of blistering is increased and there is little doubt that the roofing industry will suffer increased problems of separation from inter-layer blistering with specifications which do not include a surfacing of chippings.

Inter-layer blisters can be almost entirely prevented by application of a layer of stone chippings

The cap sheet of a multi-layer roofing using BS 747 felts can to some extent be regarded as a sacrificial layer and there is likely to be some separation from a blistering mechanism in service and some degradation due to weathering and abrasion. In the event of inter layer blistering it is usually wise to leave the blisters in place. Aesthetically this may be unfortunate but blistering has always proved a characteristic of BS 747 felts and designers should expect the possibility that blisters might form. It is extremely unlikely that leakage will result from such blisters.

Top pitting
Certain felts can exhibit a miniature surface blistering of the coating bitumen and this is most marked with BS 747 type 1E and 2E mineral surfaced felts. In severe cases these blisters, known as top pitting, can be so numerous that a surface crust develops and separates from the membrane causing a loss of mineral surfacing. Top pitting is composed of small blisters in the order of 1mm to 3mm and is unsightly if widespread.

The small blisters can be formed by air and moisture trapped in the material during the course of manufacture; they can also be caused by the incompatibility of the saturating bitumen with the coating bitumen.

The incompatibility of the bitumens, referred to as Oliensis, causes oils from the oxidised coating bitumen to separate as a result of contact with an incompatible saturant bitumen.

An oily layer may develop at the interface and rise to the surface through pin holes in the coating to cause a spread of oily material on the surface, or miniature blisters in hot sunshine.

As with other blistering, there is no need to fear for the efficiency of the waterproofing and no remedial treatment is recommended for top pitting.

Cockling
Fibre and asbestos base felts expand when their moisture content is increased, and this expansion after application can cause cockles or rounded ridging to occur which usually runs in line with the length of the material, as the expansion is greatest across the width of the material. Efforts to reduce this by unrolling and weathering felts before use have not proved effective or necessary in practice.

Coated felts are less vulnerable to cockling, particularly those with a glass or polyester base. Fibre and asbestos felts would have a tendency to cockle if not fully bonded in bitumen and it is for this reason that these felts are not recommended for the partial bonded layer in a specification. Asbestos is frequently used as a nailed first layer on timber roofs and here the possibility of cockling is accepted, as it is minimised by nailing at close centres. Any remaining minor cockling will do no harm and is not expected to significantly reduce the life of the waterproofing, although there may be an increased risk of damage from traffic.

Some polyester felts are rather heat sensitive and cockles can be produced from the heat of the bonding bitumen during the application process, although they are generally not extensive and take the form of minor wrinkling. The action of the formation of the cockling in this case is from thermal expansion of the base as opposed to moisture expansion and such wrinkling often disappears at the material settles in warm weather.

MASTIC ASPHALT

Asphalt is relatively stiff and heavy compared to built-up roofing. It is less prone to blistering, but does pose special problems from trapped moisture and air which must be allowed for in the application of the asphalt.

Blowing
The action of trapped moisture and air during the initial application of mastic asphalt is dramatic. The extreme heat of the asphalt expands the air quickly and raises the temperature of moisture droplets above boiling point to cause steam pressure. The result is the formation of blows or bubbles on the surface as the gas forces its way through the asphalt. Blowing also increases the difficulty of achieving a satisfactory adhesion of the asphalt to a surface that is too wet or to a dusty surface which contains too much air.

Blowing is primarily associated with vertical work where the asphalt is bonded directly to the substrate. For the main horizontal roof areas, the mastic asphalt is applied on top of sheathing felt or sometimes glass tissue separating membrane. This allows the air and water vapour pressures to escape laterally beneath the asphalt as it is applied. Any lack of adhesion or formation of blows are noted by the asphalter and the situation is rectified as it occurs by piercing to release the pressure and working over the asphalt to prevent voids.

To contain the problem, it is necessary to prepare the surface before application of the asphalt, usually with a high bond primer in order to secure a satisfactory adhesion and minimise blowing. If the surface preparation is not effective it may be necessary to allow the surface to dry if it is too damp, or it may be necessary to damp down the surface it it is too dry.

The main problems of blowing are faced during the application of the first coat and although it can be expected that imperfections of bond or permanent air pockets will occur, they should be small in scale. The first coat will form an ideal dry smooth surface for the second coat which will be largely free of problems during application and free of significant imperfections.

It is important that the second coat is applied as soon as possible after the first coat and certainly on the same day,. in order to minimise deposits of dirt and dust or contamination from foot traffic which could prevent the formation of a full adhesion between the coats. These deposits could also cause the entrapment of sufficient air and moisture to give rise to a degree of blowing which would be hard for the asphalter to control.

Whilst imperfections in the first coat will seldom affect the performance of the finished asphalt, significant imperfections in the second coat could lead to visible defects. The majority of blows are visible and easily corrected but some small scale blows can stop short of showing on the surface of the asphalt but leave a small hidden void. This may not be immediately visible after the asphalt has cooled but gradual settlement of the unsupported asphalt above will take place to leave a depression on the surface, known in the trade as a 'sinker'. Sinkers are not a defect likely to lead to leakage but remedial action will normally be taken to make good the surface of the asphalt if the incidence is widespread.

Full membrane blistering
Sheathing felt is a fully efficient separating layer for the flat areas of a roof and air and vapour pressures in service will be released by lateral escape under the membrane. Asphalt which is bonded to concrete or sand and cement rendering or screeds, however, gives no escape routes for trapped air and moisture other than through the concrete itself. In these cases the cyclic formation of blistering can be set up with the same mechanism of formation previously described for full membrane blistering of built-up roofing on a wet deck.

As a full adhesion will only be formed on details or vertical or sloping surfaces, it is only in these situations where the mastic asphalt is at risk of blistering. Detail work is usually near the edge of the asphalt and it is likely that air and vapour pressures will disperse through the concrete sufficiently fast to prevent a build-up of enough pressure to cause the asphalt to blister. Concrete surfaces should be wood float finished to provide a coarse texture which allows a measure of pressure release. Smooth surfaces should be avoided.

The sides of gutters or large areas of vertical or sloping work are more likely to contain local pockets of increased pressure and these situations are rather more likely to blister.

As with any blistering the process must start from a small nucleus of unbonded material and the incidence of blistering will be reduced by a positive and continuous adhesion of the asphalt to the concrete. Unfortunately building construction can never be so accurate that 100% adhesion is achieved and even though blistering is not very likely with asphalt, it is always possible for a certain amount of blistering to take place on large bonded areas which are exposed to the sun.

Inter-Layer Blistering
Inter-layer blistering is not a common occurrence on mastic asphalt roofs but it can take place between coats under exceptional circumstances. Again the blister must start at a nucleus of unbonded material and this may occur where the second coat has not fully adhered, usually due to dirt or dust on the first coat. If this results in a blow, a cavity will be formed during the application of the asphalt but may remain unobserved. The second coat above the cavity will be thinned and conditions may occur where air and moisture can penetrate slowly into the cavity to feed the cyclic growth of the blister until a significant blister is formed. Finally, the blister erupts or cracks to the extent that it will no longer hold a pressure. At this stage no further growth is possible.

1.5 MOVEMENT AND MEMBRANES

INTRODUCTION

All flat roofs comprise a number of elements which expand, contract or move in relation to each other and therefore subject the waterproofing element to stress. Failure of roofs due to the action of movement has been a principal cause of roofing problems in the past and in almost all cases has been the result of not isolating the waterproofing specification from movement.

Movement in roofs is primarily caused by thermal expansion and contraction of the roof structure or insulation, or in the case of hygroscopic materials, their expansion and contraction as a result of wetting and drying.

Thermal movement cycles may occur daily or even hourly as the sun is obscured by clouds or as rainfall causes sudden drops in temperature. On the other hand, moisture movement is usually infrequent and occurs in relation to longer spells of wet or dry weather, although a more rapid cycle of moisture movement can occur as a result of the changing conditions inside a building which may involve large amounts of water vapour generated over short periods.

These cyclic movements must be allowed for in the design of the structural deck and in the selection and form of attachment of the insulation and waterproofing.

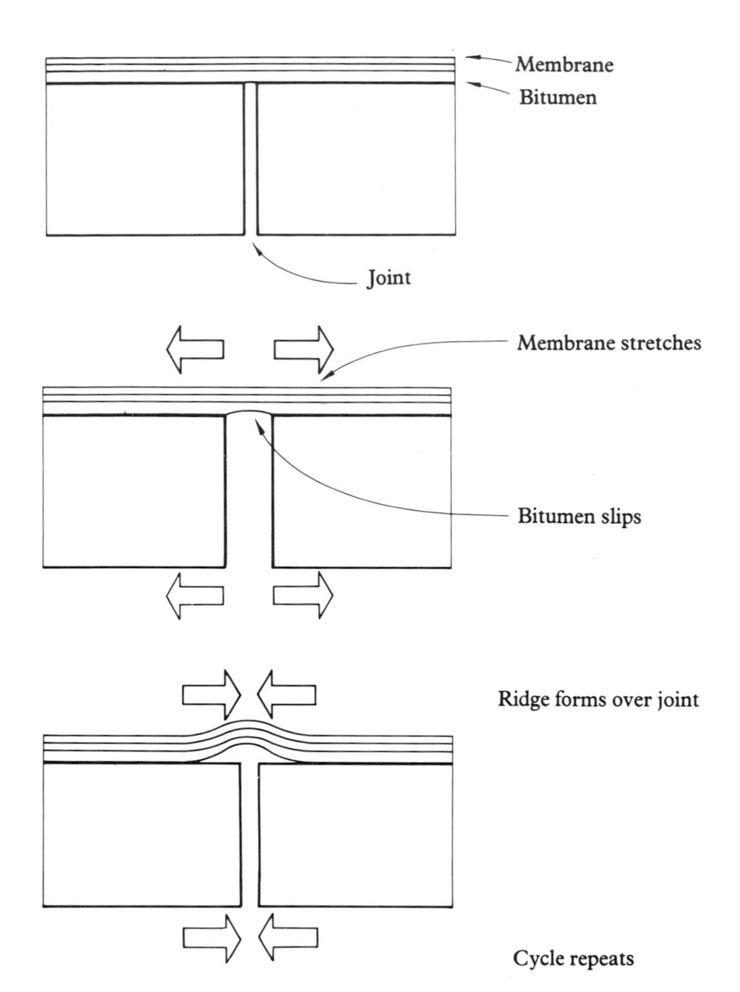

BUILT-UP ROOFING

Splitting problems are likely to occur in built-up roofing when there is not enough allowance in the design for movement either in the deck joints or the insulation joints. It is useful for the designer to have an understanding of the effect of joint movement in order to select specifications which reduce the likelihood of movement being transmitted to the waterproofing covering.

Ridging and splitting

The properties of bitumen can allow a measure of slip under load which does not fully recover when the load is released. In the case of built-up roofing bonded to a deck or insulation with joints which move significantly, the slip of the bitumen adhesive and the stretch of the membrane can build up from cycles of movement to form loose material above the joint in the form of a ridge.

The build-up of loose material arises as a result of the non-recovery of the slip or stretch in what is termed a ratchet effect, and is generated by temperature cycling or moisture cycling. For example, ridges over joints between precast concrete units or polystyrene insulation slabs are likely to be due to temperature cycling, whilst the ridges over joints between wood fibre insulating board or chipboard decking are usually due to moisture cycling.

After a ridge has formed, continual joint movement will flex the ridge and can lead to a split due to flexural fatigue.

Splits may also occur which are not accompanied by local ridging of the membrane. This is usually when a membrane is fully bonded on tightly butted joints where cyclic movement is small. The membrane does not become significantly stretched and the bond does not break down or slip locally to the joint, but a fine hair crack appears which is often hardly visible. The mechanism is again one of fatigue.

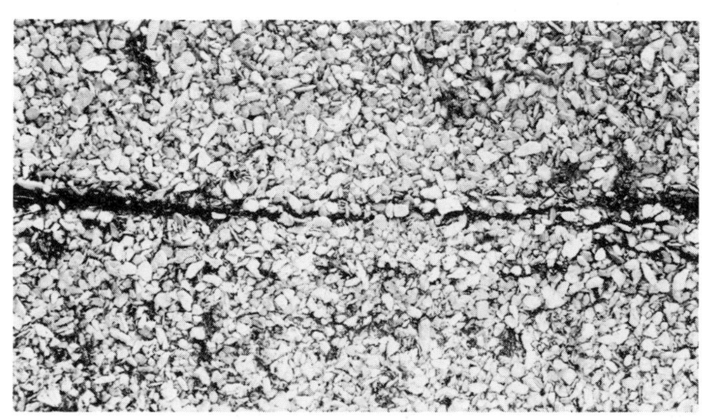

Continual joint movement can lead to fatigue failure

Ridging most often occurs with no question of fatigue arising and without leading to any splitting. Built-up roofing membranes are fairly tolerant of movement, and it can be expected that the surface of a membrane will show some disturbance in the form of ridging as the membrane takes up the natural movement of the substrate.

It is therefore the severe cases that must be avoided, and the design of the specification must make allowance for substrate movement by choosing a suitable method of attachment and a roofing material with appropriate properties of strength and elongation.

Deck movements

When built-up roofing is applied direct to the structural deck, movement of the deck will be transmitted directly into the membrane. This can be a single event, such as the initial curing and shrinkage of concrete, or repeated movements from the thermal or moisture cycling of prefabricated deck units. Differential movements may also occur as a result of a change in direction of deck or at a change of deck material and from differential deflection between adjacent units.

Most prefabricated deck materials will move too much to allow a fully bonded specification. There are two methods employed to isolate the movement of the deck from the membrane. The first method is a partial bonded first layer and the second and more certain method is to introduce an isolating insulation layer between the deck and the waterproofing.

The most common part bonded system is by the use of BS 747 type 3G perforated and gritted venting felt. The perforations produce a controlled part bond and the gritted underface prevents self-adhesion and ensures that no seal is formed by heat from the sun in service.

A partial bitumen bond is used to attach built-up roofing to substrates of precast concrete, lightweight concrete screeds, sand/cement screeds, woodwool slabs and timber based products such as plywood and chipboard.

Partial bonding is also usual when laying built-up roofing direct to an in-situ cast concrete slab, even though there is little movement. In this case, however, the primary purpose is to provide a release zone for vapour pressures which may build up as a result of trapped air and moisture beneath the waterproofing and may lead to blistering.

The alternative and often preferred method of isolating movement is to use an insulation layer which must break joint with the deck and will prove wholly effective in isolating the membrane from deck joint movements. If an underlay or vapour check is required, this would be bonded to the deck, before application of the insulation.

Timber boarded decks present a large number of closely spaced joints, and the traditional method of accommodating the deck movement is to nail the first layer of roofing.

Insulation movement

The movement characteristics of insulation materials vary considerably and determine whether the waterproofing membrane is to be fully bonded, partially bonded or protected from movement by using a more stable insulation as an overlay to the main insulation board before applying the waterproofing.

Fibrous insulating materials such as wood fibreboard, cork, glass fibre, mineral wool, or perlite board suffer little movement from temperature and it is usual to fully bond the membrane to these insulants as fatigue failure is unlikely to occur. Some ridging of the membrane above the joints in the insulation may be evident, but this should not be regarded as a cause for concern.

Partial bonding using 3G perforated and gritted venting felt

The high efficiency foam insulants, however, have high coefficients of expansion and rapid response to temperature change. They will absorb deck movements, but will also impose their own range of movements which must be taken into account when designing the specification.

Expanded polystyrene moves substantially as a result of temperature changes and new material has a tendency to shrink permanently during curing which is in addition to any thermal movement. An overlay of a more stable insulating material is required to prevent movement of the expanded polystyrene being transmitted directly into the membrane.

The movement of expanded polystyrene insulation is restrained by fully bonding it to the roof deck in bitumen, but attachment of the board by mechanical fasteners does little to restrain movement. Field tests by BRE show that a full bitumen bond reduces movement by about one third. Similarly, a full bond of the waterproofing to the insulation reduces movement by about one third, but in this case the restraint is a result of stresses in the waterproofing and is not recommended. The tests demonstrate that for all foam insulations it is desirable to have the insulation fully bonded to the deck but avoid a full bond between the waterproofing and the insulation.

Unfaced polyurethane foam has similar expansion characteristics to expanded polystyrene, but it is restrained considerably by facings provided these are not themselves sensitive to temperature or moisture movement. A glass fibre tissue base is adopted by most UK manufacturers, and the facings reduce the movement of the polyurethane foam to approximately one-third of the movement of unfaced foam. This restraint by the facings is similar, as one would expect, to the restraint imposed on expanded polystyrene when fully bonded to the deck and with a fully bonded waterproofing. The important difference is that the restraint on polyurethane is achieved without using the waterproofing in restraint. The facing takes the stress and provides the opportunity for part bonding, so that the membrane is substantially freed from stresses which could otherwise be induced by the movement of the insulation.

In addition to thermal movement, there is a tendency for polyurethane to expand from the absoprtion of air and moisture vapour into the cells in the first few years of service, and this is quite separate from the expansion and contraction that can arise through temperature changes. The expansion from absorption is permanent, and tends to close or tighten the joints between the insulation slabs, with a consequent reduction in joint movement from subsequent temperature changes.

Movement at details

Over the years splits in the waterproofing at skirtings have proved a source of trouble. The splits can arise from the joint movement mechanism previously described, but there are other reasons and it is wise to bear in mind the different movements which can cause splitting at skirtings.

Most common is differential movement or differential deflection between the roof structure and the wall. Typically movement occurs at the junction of timber, woodwool, or metal deck roofs and independent brick walls. However, the junction between metal deck and vertical metal cladding is not at risk of joint movement because the structural curb is normally formed of pressed metal sections connected firmly to the deck and to the structural frame.

Differential movement can also arise because the roof and walls act as separate single plates. In this case, the splits are normally towards the end of the skirting and may be accompanied by shear deformations in the waterproofing upstand. These show as slanting ruckles and are a sure sign of differential movement between the roof structure and the wall.

A sure sign of differential movement between the roof structure and the wall

It is usually possible to predict movements at skirtings, and effective precautions in the form of independent or reinforced upstands are easy to arrange.

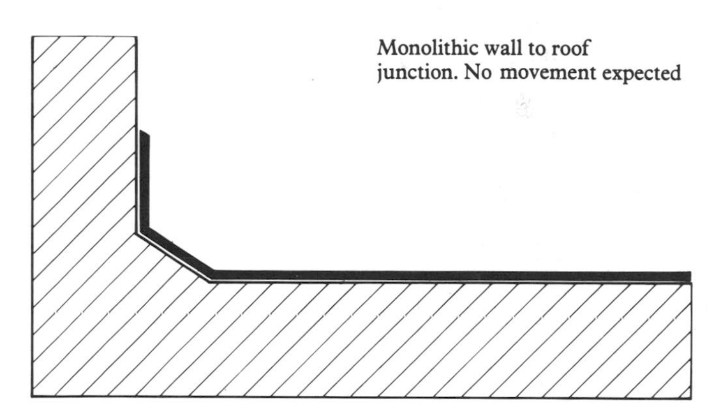

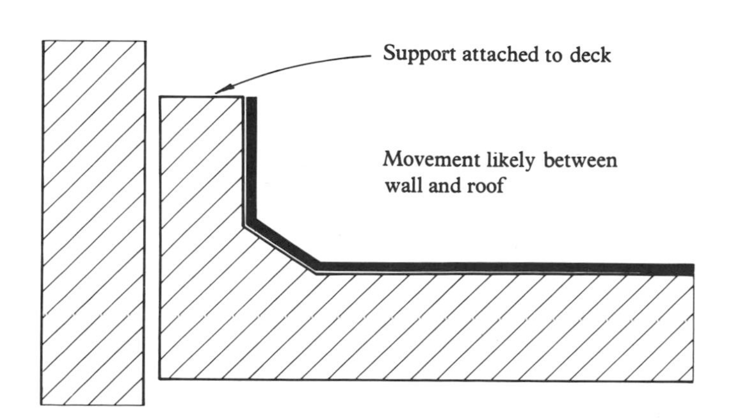

It is unfortunate that the need for independent upstands is so frequently ignored. They make an effective expansion joint at the junction between roof and wall when differential movement is expected, and the detail generally proves entirely effective. The reason for the omission is probably the simple acceptance of a risk and the knowledge that tracing leakage and repairing the membrane at the upstand may not be too onerous a task. Traditionally it is the custom to add tilting fillets at skirtings and these can reduce the effect of differential movement by removing the felt from the vulnerable intersection,. This is not always the case, however, and timber fillets nailed to a timber deck may merely alter the positioning of the split line to the top of the fillet. There is no doubt that the introduction of high performance membranes will substantially reduce splitting problems, but it will be best to continue with independent upstands if the risk of split skirtings is to be minimised. The detail design of independent upstands is fully described in Section 5.

Measurement of joint movement
As splits due to fatigue form a high percentage of leakage problems, various fatigue test programmes on roofing materials and specifications have been carried out, including tests by the Building Research Establishment. These tests simulate roof movement under controlled conditions and record the number of movement cycles which can be withstood by a membrane before failure occurs.

The tests demonstrate that classic fatigue failure as experienced on site can be reproduced in the laboratory showing that the form of failure is normally by formation of a ridge and subsequent flexural fatigue at the base of the ridge.

The results of the tests can be expressed in a number of ways, but they relate to the gap movement and the number of cycles.

A difficulty of comparing the best materials with the worst arises because of the inherent weakness of traditional British Standard roofings complying with BS 747. These only withstand small movement in the order of 1mm for a limited number of cycles. The best high performance roofings cannot be taken to failure with only a 1mm gap opening, even after many thousands of cycles and these materials are therefore tested at a greater gap opening, commonly 3mm or more.

The results of the fatigue testing relate to the form of apparatus used, the speed and form of movement cycling and the test temperature. No doubt standardised procedures will be agreed in the future, but at this stage it is not normally possible to correlate the work of one laboratory with another. The most useful common yardstick is the performance of traditional BS 747 felts, particularly a three-layer specification comprising type 3B glass base roofing and it is suggested that this specification is always tested and used as the base for comparison of results with other specifications. It is necessary to carry out tests in two directions, down the length of the roll and across the roll, as the behaviour often differs substantially.

The fatigue tests may show one waterproofing to be better than another, but this is unlikely to suggest that one will succeed and the other fail. If the tests show that the specification will accept movement in the order of 1mm for some thousands of cycles, or 3mm of movement for some hundreds of cycles, the material is likely to prove satisfactory in service with regard to fatigue.

Tests to date indicate that the majority of high performance roofings are so good that it is clear they will not suffer fatigue failure as long as the original properties of the material are retained in service and it is this aspect which will prove critical to the long term fatigue resistance. Favourable ageing characteristics are more likely to influence the life of the waterproofing than the fatigue test performance.

A method of obtaining an indication of the long term durability is to heat a sample of the roofing in an oven for 56 days at 80°C. Samples thus treated were used for the fatigue test figures given in the table below, which shows that a ranking order can be established between various known roofings. This is the subject of work carried out at the Princes Risborough Laboratory and was covered in a paper presented at the International Symposium of Roofs and Roofing in Brighton in September 1981.

	Fatigue resistance x 100	
	New	Aged
BS 747 felt	2	1
Polyester (oxidised bitumen)	20	10
Polyester (modified bitumen)	1500	200
Elastomeric	1500	700

Strength and elongation
Whilst fatigue testing gives information on the ability of a specification to resist splitting in service, it does not give information on the ultimate strength of the membrane, the elasticity or the elongation.

Membranes can be broadly described as either plastomeric or elastomeric. Most felts are in the plastomeric category. They extend in length before failure, but do not return to their original length.

The BS 747 felts are relatively weak and fail in tension at an elongation in the order of 2%. Polyester based roofings are considerably stronger and fail in tension at an elongation in the order of 40%. It is therefore hardly surprising that polyester based roofings should show a greatly increased order of fatigue resistance over BS 747 felts.

Some felts, notably the bitumen polymers or pitch polymers, are largely elastomeric in behaviour and show good recovery after elongation. They are generally used in conjunction with traditional felt underlays, however, and the completed system shows reduced elastic recovery.

Strengths and elongations of British Standard and proprietary roofings are available from most manufacturers, but the following properties are typical:

	Tensile strength in machine direction (N/50mm)	Elongation at break in machine direction (%)
BS 747 felt	200	2
Polyester (oxidised bitumen)	600	40
Polyester (modified bitumen)	600	60
Elastomeric	550	100

MASTIC ASPHALT

When designing with mastic asphalt, it is necessary to consider the movement of the membrane as a separate structure as well as the movements which are imposed on it by movement of the substrate.

Movement of the asphalt membrane

Mastic asphalt has no continuous reinforcement to control the natural movement of the bitumen other than the fine and coarse aggregates which form 85% of its weight and add stiffness and hardness. The properties of the asphalt are therefore primarily influenced by the rheological properties of the bitumen and the grading and proportions of the aggregate.

The ideal formulation for asphalt is one which is not too soft to take traffic at the highest temperatures and not so hard at low temperatures that it would crack from self-induced stresses during contraction or become easily damaged by knocks and strains. This requires a plastic performance over a temperature range of -20°C to 80°C which are the likely extremes of temperature for an exposed roof surface. This is a huge range and it is not possible to formulate asphalt to be plastic over the entire range.

In practice the transfer from the plastic state to a hardness and behaviour more related to solid state occurs at a temperature around 5°C. Below this temperature, significant self-induced stresses can be produced by a sharp drop in temperature, known as thermal shock, and in extreme cases the relief of the stress may take the form of cracking of the asphalt rather than plastic flow. Fortunately the weather in the UK is temperate; few problems from thermal shock arise and the expansion and contraction of the asphalt is accommodated by the method of application.

The thermal expansion of asphalt in the horizontal direction is effectively restrained and the major thermal expansion takes place in the vertical direction.

The sun heats the top surface of the asphalt whilst the underside remains cooler and the hotter more mobile mastic at the top surface is free to expand upwards. During the cooling of the asphalt, however, the temperature gradient reverses.The upper face of the asphalt will be at a lower temperature than the underside and being stiffer will take up an effectively solid form before the layers below. The solid upper face is not fully restrained by the more plastic lower layers and there is a tendency for the upper layer to contract in the horizontal plane. The contraction movement is non-reversible as it takes place only on the cooling cycle and the tendency to contract, if unrestrained, would continue progressively.

The self-induced stresses during contraction are generally not greater than the strength of the mastic asphalt. The tendency to contract merely results in the slight stressing of the asphalt and this is the key to the design and formation of detail work. It is essential that all edge details are securely bonded into position to restrain the forces of contraction. Skirtings can be pulled away from the vertical if not properly bonded and the asphalt can be drawn back from handrails and outlets if a satisfactory bond is not formed. If a skirting is badly bonded along its entire length, there will be nothing to restrain continued movement, and cases have been observed where the movement is so great that the fillets have disappeared and the upstand has been dragged away, until very little upstand is left.

Crazing

A secondary effect caused by thermal expansion and contraction is the phenomenon of crazing. This occurs to a greater or lesser degree on all exposed roofing grade asphalt, generally appearing in the first spring or summer after application. It is a surface effect only and should not prove a cause for concern other than on grounds of appearance.

When asphalt is laid, a surface skin rich in bitumen is formed which is sensitive to photo-oxidation and it is the formation of a thin but hard oxidised skin that results in crazing of the surface. With the continual thermal movement of the main body of the asphalt, the hard skin gradually breaks into a mass of pieces which are extremely small. These shrink and curl to give the typical crocodile skin effect associated with crazing.

Crazing is only a surface effect

Crazing can give the appearance of a series of splits right through the asphalt, but it is only a surface effect and there is no danger of complete splitting nor of the waterproofing being impaired. The crack is seldom deeper than the height of the ridge in which it is formed and the effective thickness of the mastic asphalt is not normally reduced by crazing. The raised and crazed portion can often be totally removed by rubbing the surface of the asphalt with a coarse abrasive, for example by rubbing with a brick.

Crazing on new work can be controlled at the time of laying the hot asphalt by rubbing clean coarse sand onto the surface with a float. This breaks the bitumen skin by combining the sand with the bitumen, reduces the richness of the surface and reduces the differential expansion between the surface and the main body of the asphalt.

The incidence of crazing is reduced by sand rubbing

Clearly surface crazing is a function of the surface temperature and asphalt applied on top of an insulation will be more prone to crazing unless further protected. At the higher temperatures unprotected mastic asphalt is also likely to prove soft and, as a general rule, all asphalt which is installed on an insulation must be protected by a stone chipping finish or a reflective coating to keep surface temperatures down.

With natural rock asphalt and asphalt made with lake asphalt, the bitumen at the exposed surface breaks down rapidly to expose the aggregate. This erosion has no significant effect on thickness but leaves a characteristic light grey appearance which increases the reflective properties of the asphalt and consequently reduces the surface temperatures of the roof in hot weather to less than would be the case if the surface remained dark. Crazing tends to come away with the erosion, with the result that natural rock asphalt seldom shows crazing to any great extent, as the crazing erodes as fast as it forms.

A further advantage of the rough texture which the erosion leaves is the increased skid resistance of the material.

Effects of movement

In warm weather the action of joint movement of a substrate onto which asphalt is bonded is to create an undulation of the surface of the asphalt in a series of wrinkles. These can cause minor changes of thickness but although unsightly the wrinkles will not normally lead to a rupturing or splitting of the membrane.

In severe cold weather, the opening of the substrate joint can cause a tear in the undersurface of the asphalt membrane. Repeated movement can gradually open up the tear from the underside until it reaches so far up into the main body of the material that the thinned asphalt at the surface is finally overstressed and cracks right through.

The extent of movement causing the tear is generally small and it is only the relatively small temperature changes in cold weather which contribute to the tear. The formation can take a long time and it may be many years before failure occurs but the action is remorseless and it is essential to isolate the membrane from substrate movement to ensure a satisfactory performance of the asphalt.

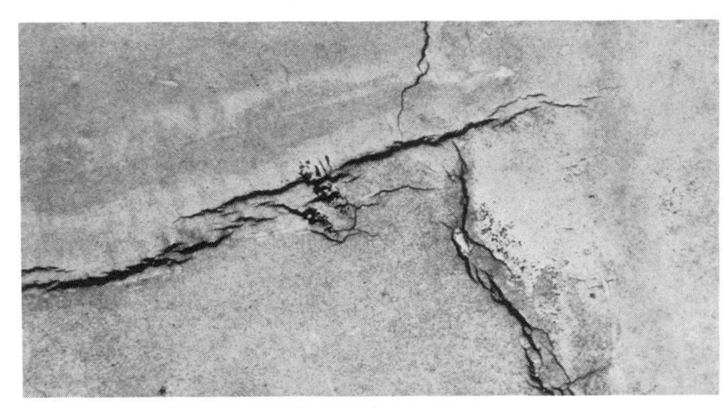

Failure to isolate asphalt from substrate movement can cause cracking

Allowing for movement

Horizontal surfaces

On horizontal areas a complete separation of asphalt from the roof deck can be formed by the use of sheathing felt. Many alternative separating layers have been tried but sheathing felt has particular qualities which are not matched by the alternatives. Being a loose mat of fibres bound together with bitumen of similar properties to that of the asphalt, the felt introduces no stresses or strains to the system. It effectively prevents the bond of the asphalt to the deck but at the same time presents a close, rough contact resulting in a fairly high level of friction. The sheathing therefore provides a measure of lateral restraint to the asphalt. This is an advantage in cold weather when the friction acts to restrain the contraction of the asphalt, but at the same time it will allow differential joint movement in the substrate without transmitting sufficient stress or strain to cause wrinkling or cracking.

Sheathing felt also acts as a form of reinforcement to the asphalt particularly at the lap joints which are welded together by the heat of the asphalt. This provides a continuity of reinforcement which prevents a concentration of stresses or strains in the asphalt above the lap.

Sheathing felt is normally applied to isolate the asphalt from substrate movement

If building paper is used as a separating layer, the overlapping joints of the paper do not adhere and will concentrate thermal movement at these joints. These repeated movements cause cavitation and thinning to the underside of the asphalt which eventually results in a fracture line.

The sheathing will only be fully effective against joint movement of the deck when applied to a reasonably level deck surface. Irregularities in the deck surface will form a key to the underside of the asphalt and impose local restraints which can cause local stresses from substrate movement or from the cold weather contraction of the asphalt itself. Changes of level between concrete slabs and between insulations of different thicknesses are examples of localised restraint which can break the efficiency of the separating layer and impose local stresses sufficient to cause cracking of the mastic asphalt.

It is also possible to cause restraint by damaging the deck or insulation surface before application of the asphalt. If depressions are formed by such damage, this will lead to an extra thickness over the area of the depression which develops a key to the deck. A restraint will then be formed which can again cause cracking, most probably around the edges of the depression. In practice it is the isolated points of restraint which are more likely to give rise to problems. An overall keying from a uniformly rough or indented surface is not likely to cause cracking, as a uniform restraint is imposed on the asphalt.

Vertical surfaces

Having assured adequate separation of the membrane from the flat area by means of sheathing felt, it is also necessary to restrain the natural contraction movement of the mastic asphalt in cold weather. The frictional resistance of the sheathing provides a measure of lateral restraint but it is necessary to ensure good edge restraint at the perimeter of the asphalt by forming a strong bond to vertical surfaces or a suitable turndown anchorage at eaves and outlets. This prohibits the use of separating layers on vertical areas but fortunately the common forms of building construction for vertical surfaces, other than timber, do not involve panel joints. Direct applications of asphalt are likely to be to brick or in-situ concrete where there is no great risk of joint movement but if brickwork or concrete cracks, it is likely to crack the mastic asphalt bonded to it.

In the case of vertical timber, not only must a secure attachment be formed but it is also necessary to separate the asphalt from direct contact with the timber, because of its moisture content and movement, both of which would act against the asphalt if it was fully bonded. The double duty of overall uniform mechanical attachment with appropriate interfacial separation between timber and asphalt is achieved by the use of expanded metal lathing on sheathing felt. The lathing is nailed at 150mm centres to keep it as tightly attached to the timber as possible and the first coat of asphalt is worked thoroughly into the lathing.

It is sometimes necessary to form vertical work on steel upstands particularly when the roof is of metal decking. In this case, timber is attached to the metal upstands and the asphalt detail is completed as normal for timber using sheathing felt and expanded metal lathing. In some cases, the expanded metal lathing can be welded direct to the steel upstand but only if the runs are short and the metal upstands are continuous with no joints which might give rise to movement. Small rooflight curbs are an example of the possible application of expanded metal lathing direct to steel.

Precast concrete wall, eaves or gutter units can sometimes cause problems. The movement at the joints of the pre-cast units is normally small but can be expected to cause wrinkling of the asphalt surface. There is a slight risk of cracks developing from the movement, however, and it is best to avoid the direct application of asphalt to precast units or to lock the units together to prevent joint movement. When circumstances permit, an application of sheathing felt and expanded metal lathing will alleviate the situation.

The effects of joint movement on asphalt applied direct to a precast concrete parapet

Joint movement at skirtings
Asphalt is bonded tightly to the wall or kerb face and is therefore vulnerable to splitting at the base of the skirtings, if there is differential movement between the wall face and the roof. Independent skirtings must be used where the wall and roof are not of an integral construction or not held together at the junction.

All timber, plywood and similar roof decks butting to brick or concrete walls are likely to suffer sufficient joint movement between the wall and roof to cause splitting of the skirting. Independent upstands are always included for these constructions. Woodwool is less prone to joint movement and whilst independent upstands are always desirable, it is common practice to dispense with them provided the woodwool is firmly fixed to supports which are themselves firmly fixed to the wall face. Metal decking against brick or concrete walls will always call for independent upstands.

As with built-up roofing, internal gutters form a line on which movement can take place and they are best avoided. In the case of mastic asphalt the movement of the material is constrained in the sole of the gutter and severe wrinkling is common. This may not lead to leakage but it is unsightly and shows that the asphalt is acting under too much constraint.

An asphalt gutter also forms a heat trap if exposed to continuous sunshine and the sides of a gutter are thus prone to slumping and blistering as well as to substrate movements. If gutters are incorporated in a roof, designers should ensure that joint movements are restrained as much as possible and that the asphalt is firmly supported.

Sloping surfaces
Sheathing felt can be used to provide separation on uninsulated roofs on slopes up to about 10° and on insulated roofs on slopes of up to about 5°. Slopes less than these can be treated as normal flat roof areas but for steeper slopes the asphalt will be difficult to apply on sheathing felt and slump and creep of the finished work will begin to cause problems. It is therefore necessary to carry out the work as if it were vertical. A full bond to concrete is necessary and for this purpose the concrete should be lightly tamped or wood float finished. On sloping timber decks, a firm attachment is achieved by applying the asphalt to expanded metal lathing, securely fixed to the timber at 150mm centres and over a layer of sheathing felt. Asphalt must be avoided on sloping roofs if joint movement of the substrate is likely.

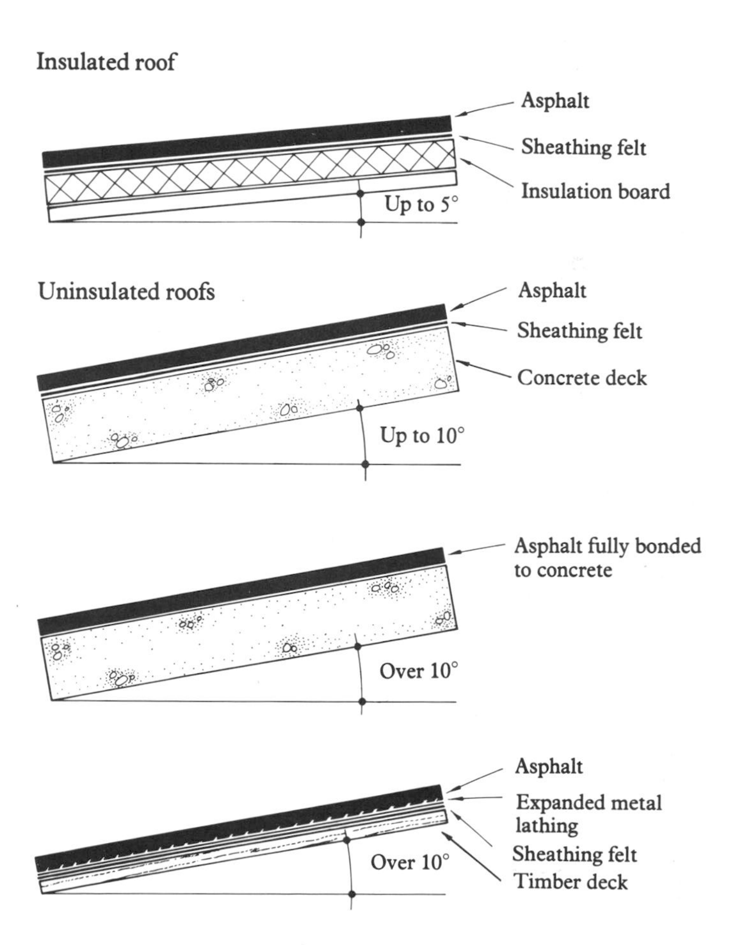

The figures of 5° and 10° are arbitrary and, as so often happens in the building industry, have no absolute significance. Long slopes in a heat trap situation may well be restricted to lesser angles of slope but a slightly increased slope might be in order on short and shaded slopes.

1.6 WIND

INTRODUCTION

It is not generally appreciated that a significant number of roof failures in the UK are due to gale damage, and that protection against wind forces should be one of the fundamental principles behind good roofing design.

When wind strikes a building, it is deflected to generate a positive pressure on the windward face, and it accelerates round the side of the building and over the roof, leading to a reduced, or negative pressure over the roof and in the lee of the building.

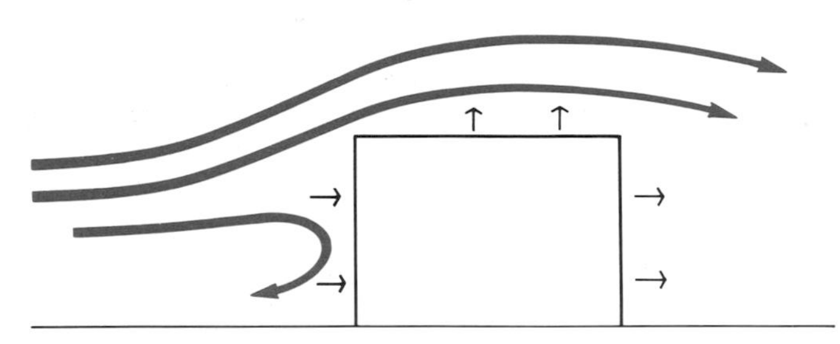

Distribution of pressure over the roof is far from uniform, even for the simple box type structure illustrated. The wind does not normally strike square to the face of a building. When, as is more usual, it strikes at an oblique angle, the air deflected up and over the roof is at the same time moving along the face of the building, creating vortices along the roof edges. The greatest wind pressures are experienced at the windward corners and edges of the roof, where the negative pressure can be several times that experienced in the central areas.

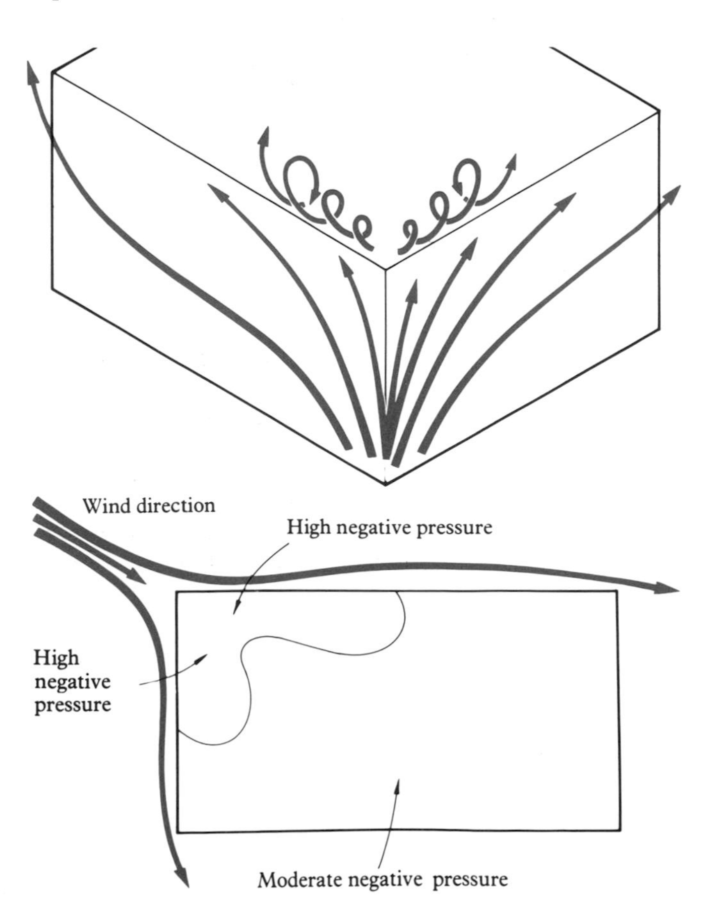

THE CRITICAL LAYER

When there is no wind, a roof system will be subject to atmospheric pressure on both the upper surface and the underside. Atmospheric pressure can be taken as roughly 100kN/m^2, or 10 tonnes/m^2 and is a huge pressure acting in all directions so that, in the case of a roof, the upward pressure is balanced by the downward pressure.

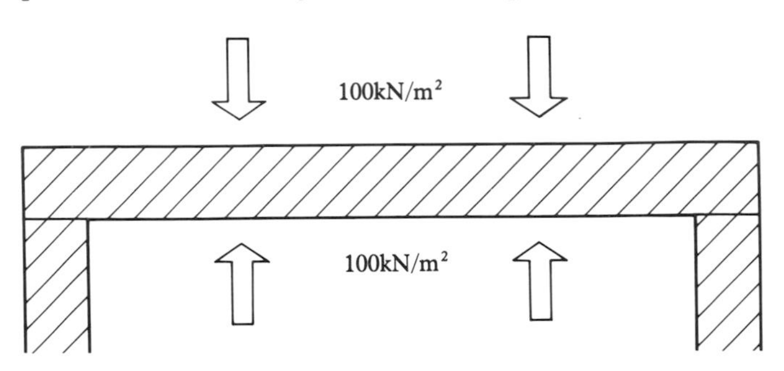

Wind changes this equilibrium by reducing the atmospheric pressure on top of the roof. If we take a typical negative wind pressure of 2kN/m^2 to illustrate the situation, it means that there is a reduction of atmospheric pressure on the surface of the roof by 2kN/m^2, so that the downward push is now 98kN/m^2. The underside of the roof will remain at 100kN/m^2 although this may be increased if the windows or doors are open on the windward side of the building.

The resulting pressure is 2kN/m^2 from the underside, generating an upward thrust from below, not in any sense a wind grab from above.

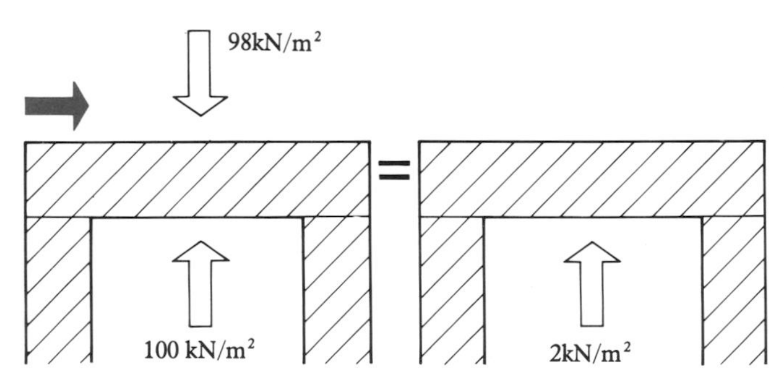

The importance of understanding this basic concept lies in the need to provide secure attachment at the correct position to withstand this upward thrust wherever it occurs. The thrust will be exerted on the first air impermeable layer in the roof which will stop air from flowing further into the system. In most roof constructions there is one layer which provides the dominant barrier against the upward thrusting flow of air, and this is referred to as the critical layer. The attachment of this layer is referred to as the critical attachment.

A flat roof is composed of many layers of material and it is essential to decide which of these is the first air impermeable barrier to form the critical layer and require the critical attachment. The critical layer will sometimes be the roof deck itself, sometimes a vapour barrier or thermal insulation and sometimes the waterproof covering. This is best illustrated by considering a few typical roof constructions.

1. Concrete deck with insulation and waterproofing with or without an underlay

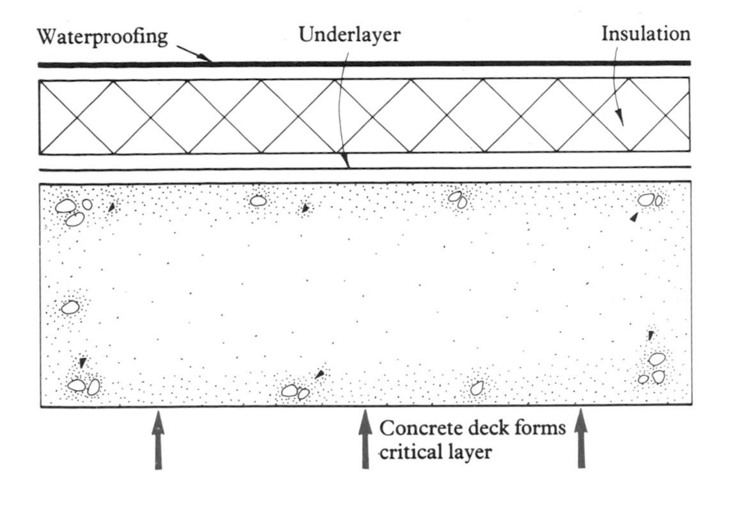

The first air impermeable layer is clearly the concrete deck itself. This will take the resultant wind thrust direct, and as there is no critical layer of attachment above the deck, the roofing specification is at minimum risk.

2. Metal decking with an underlay, insulation and waterproof covering

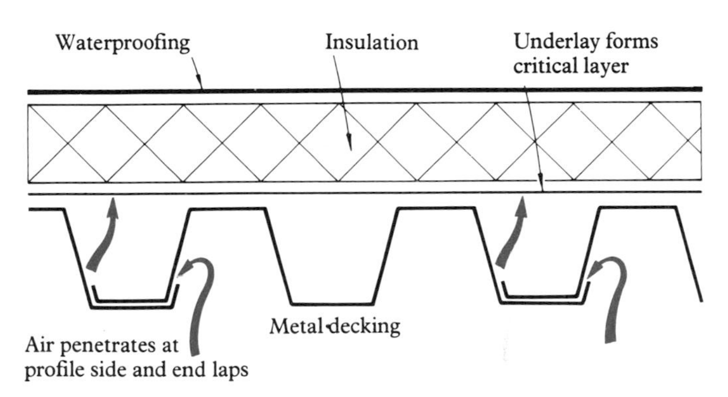

The metal deck itself can be taken as permeable to air through the profile side and end laps. The first impermeable layer in this case is the underlay and it is the attachment of the underlay to the top flats of the deck which is the critical attachment.

3. Metal decking with insulation and waterproof covering but no underlay

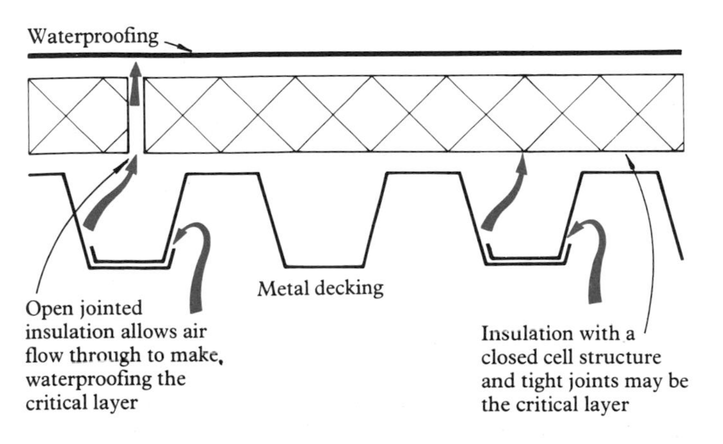

In the absence of an underlay there are two contenders for the critical layer. The first air impermeable layer will be the insulation if the joints are reasonably tight and the insulation is of a closed cell structure. If the insulation is open jointed and allows air through the joints to the underside of the membrane, the waterproof covering will then be the first impermeable layer. If in addition the waterproofing is partially bonded, a flow of air beneath the membrane will be allowed and will exert a high pressure on the underside of the entire membrane.

In practice the insulation will probably restrict the flow of air to the extent that it will take most of the air pressure. Enough air may, however, pass through the joints to make the membrane an equally critical layer and unless there is some over-riding factor it would be wise to treat the attachment of both insulation and waterproofing as critical.

INTERMEDIATE LAYERS OF ATTACHMENT

Having identified the critical layer it must be firmly attached so that wind loads are transferred back to the roof deck where necessary. Consideration then needs to be given to the attachment of layers of the roof construction above the critical layer. Waterproofing specifications will either be loose laid such as mastic asphalt or partially or fully bonded as for built-up roofing and most insulation layers. There is only a high risk of wind damage at intermediate layers if air can flow to the underside of that layer. This flow of air must be prevented by closing off the edges of the roof by the suitable formation of edge details and edge bonding.

Under some circumstances, certain specifications allow for only minimal forms of attachment or for completely unattached waterproofing. The prime example is mastic asphalt which is applied over a loose-laid sheathing felt, leaving no form of attachment to the deck. The self-weight of asphalt waterproofing is approximately 50kg/m^2 and yet it does not blow off even when wind forces considerably exceed the self-weight. This is due to a suction effect below the asphalt which depends on the provision of a wind impermeable deck to hold the suction.

To explain the mechanism, it must be appreciated that there is a thin layer of air between the mastic asphalt and the deck. When wind reduces the atmospheric pressure above the roof, this thin layer of air will expand until its pressure is reduced to equal the pressure of the air above the membrane. At this stage an effective suction is formed between the asphalt and the deck, and lift-off is prevented. If, on an open-jointed deck, the air can reach the underside of the asphalt fast enough to break the suction effect, damage could result. A butt-jointed timber deck where shrinkage has taken place will allow a free flow of air through the open joints and the suction effect will not develop. Fortunately the weight and the stiffness of the asphalt add a further protection and it is only roofs which are exposed to very high winds on an open jointed deck which would be at significant risk.

Lightweight flexible membranes
The negative pressures which develop over the roof will often be quite small in area and may traverse the roof or parts of the roof in the form of eddies. With lightweight flexible waterproofing, it only requires a small flow of air to the underside of the waterproofing to allow air to collect locally under an eddy formation and form a wave which will follow the eddy as it passes over the roof surface.

If the waterproofing is fully bonded to a good stable surface, the wind forces involved will not be sufficient to break down the bond, and will not allow a wave to form. Extreme wind forces can, however, be sufficient to break down a partially bonded waterproofing specification or even a fully bonded specification to an insulation board with a loose or friable surface. If a wave-form develops, the amount of air under the waterproof covering can sometimes increase rapidly, and failure can occur by simple lift-off. Alternatively, splitting of the waterproofing can occur from stressing as the wave hits up against an edge or at the junction between full bonded and part-bonded areas. It is also possible for the waterproofing to be lifted up and down or dragged out of position until air can find an entry at an edge condition and cause further damage.

The stiffness of the waterproof covering and the application of loading coats will do much to reduce the possibility of this form of wind damage, and the stronger the membrane, the less likely it is to tear in the wind. It should be appreciated that once wind damage has started, the stronger membranes are more likely to develop into a sail by catching the wind and this can lead to large scale blow-offs. Modern high performance felts and membranes are more likely to form sails than traditional BS 747 roofing felts which tear more easily, and extra care needs to be taken with the design and formation of details.

LOADING COATS

In the design and testing of a membrane, it is normal to assume that the wind exerts a continuous static force. It should be remembered, however, that CP 3: Chapter V: Part 2 calculations relate only to three-second gusts which snatch at small areas of the waterproofing. In practice, a layer of 10mm gravel provides sufficient weight and inertia to prevent damage to most partially bonded specifications, provided the roof edges are closed off to prevent entry of air. Unfortunately, the present tendency to exclude stone chipping surfacings from flat roofs is taking away the loading coat which has so often prevented widespread damage.

Occasionally a loading coat of chippings may be unacceptable, for example when process dust may settle on the roof and regular sweeping is necessary. A mineral surfaced roofing will be easier to clean but the specification will be lighter and more vulnerable to the wind. If it is installed on an insulation which has a low laminar strength, the risk of delamination will be increased. Under these circumstances, a first layer of high performance roofing mechanically fixed through the insulant and into the deck may be necessary to present a stable base for the rest of the waterproofing.

PROTECTED MEMBRANE SPECIFICATION

With a protected membrane specification the insulation is laid loose, and the security of the entire system depends on the loading coat which may be 50mm of gravel or 50mm concrete paving.

There is not yet enough evidence in the UK to know the circumstances of exposure to wind which will call for an extra weight of loading coat to prevent wind uplift for a protected membrane roof. In this respect, it should be noted that experience in other parts of Europe may not be a safe guide. The UK is situated on the exposed edge of Europe, and the west coast of the UK experiences the highest winds in Europe other than exceptional winds which occur in mountainous areas. It will be prudent for designers to discuss extra precautions with the manufacturers of the insulation when conditions of extreme exposure are encountered.

EDGE DETAILING

Surveys after widespread gales indicate that the majority of damage caused to flat roofs starts at the exposed windward corners and edges of the roof. Oversails, fascias, cappings, trims and drips take the brunt of the wind forces and these details or their grounds are usually the first components to fail.

A calculation of the actual forces on these roof edge details is usually difficult, but the cost of fixing is cheap, and there is no reason why these details should not be amply fixed. As a rough guide, all metal cappings and trims should be fixed at 300mm centres, with extra fixings added under conditions of extreme exposure. Cappings which have both edges exposed will need at least two lines of fixing at 300mm centres.The grounds to which the details are secured must themselves be firmly attached to the structure. Wall plates which support a roof oversail must be firmly anchored to the supporting walls with straps and fixings specially designed to withstand the calculated wind forces.

All details should be designed to reduce the free entry of air beneath the membrane. In particular, profiled sidewall cladding and overhangs can be permeable to air and may allow additional wind pressures into the system and additional bonding or fixing may be required.

DESIGN WIND LOADS

Tables 1.14 to 1.17 give design wind loads for flat roofs, based on calculation methods given in CP 3: Chapter V: Part 2. To use these tables, select the basic wind speed from the wind map, choose the relevant table according to the ground roughness (S_2) and read off the wind load for the appropriate height of the building.

In extremely exposed, or fully sheltered locations, adjust the wind load given in the table by the topography factor (S_1) and allow for the altered pressure coefficient (Cp) if there is likely to be a dominant opening, see below.

Topography factor S_1

Topography	Value of S_1
a All cases except those in b and c below	1.0
b Very exposed hill slopes and crest where acceleration of the wind is known to occur. Valleys shaped to produce a funnelling of wind	1.1
c Steep sided, enclosed valleys sheltered from all winds	0.9

The wind loads shown in the tables are based on an S_1 factor of 1. If a factor of 1.1 is required the loads given should be multiplied by 1.21, similarly for a factor of 0.9 the loads should be multiplied by 0.81.

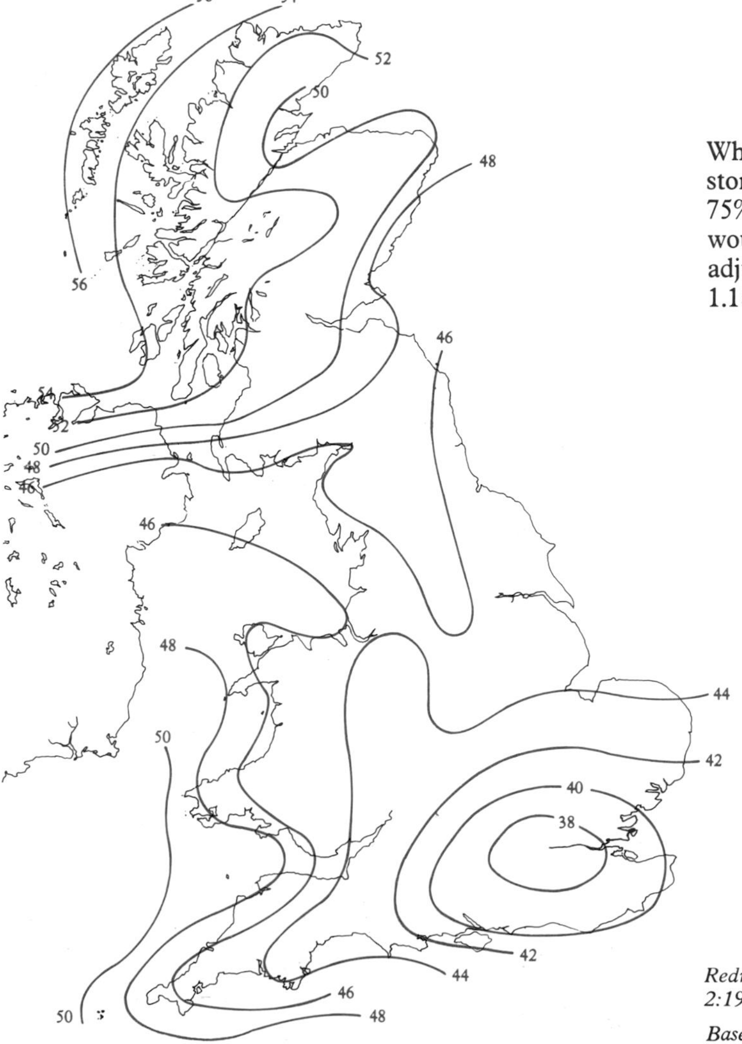

Ground roughness, building size and height above ground factor S_2 Class A

Table 1.14 Open country with no obstructions
Table 1.15 Open country with scattered wind breaks
Table 1.16 Country with many wind breaks, small towns, outskirts of large cities
Table 1.17 Surface with large and frequent obstructions eg city centres.

Statistical factor S_3

The S_3 factor is taken as 1.0 which is almost invariably used for roof constructions, and represents a return period of 50 years.

Pressure coefficients Cp
The variations of wind pressure over the roof are taken into account by the application of pressure coefficients. In Tables 1.14 to 1.17 the external pressure coefficient (Cpe) has been taken as -2.0 for local or edge areas and -1.0 for the general or central areas of the roof. The internal pressure coefficient (Cpi) is taken as +0.2 and represents the case when there is only a negligible probability of a dominant opening occuring during a severe storm.

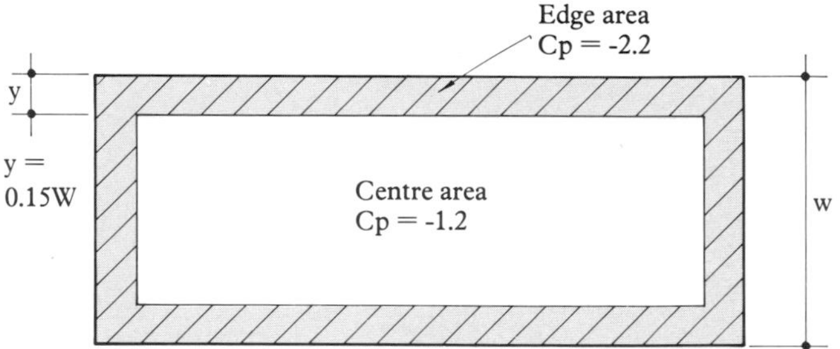

Where a dominant opening is likely to occur during a storm, the Code of Practice suggests that a Cpi figure of 75% of the Cpe outside the opening is appropriate and this would give a Cpi of +0.6 in the worst case. In order to adjust for this, multiply the loads given for edge areas by 1.18 and for centre areas by 1.33

Redrawn from British Standards Code of Practice CP3:Chapter V:Part 2:1972

Based upon the Ordnance Survey map with the permission of the Controller of Her Majesty's Stationery Office, Crown copyright reserved.

Basic wind speeds (m/s). Maximum gust speed likely to be exceeded on average only once in 50 years at 10m above ground in open level country.

WIND LOADS ON FLAT ROOFS (kN/m²)

TABLE 1.14 OPEN COUNTRY WITH NO OBSTRUCTIONS

Wind speed m/sec	40		42		44		46		48		50		52		54	
Height m	Centre	Edge	Centre	Edge	Centre	Edge	Centre	Edge	Centre	Edge	Centre	Edge	Centre	Edge	Centre	Edge
3	0.81	1.49	0.89	1.64	0.98	1.80	1.07	1.97	1.17	2.14	1.27	2.32	1.37	2.51	1.48	2.71
5	0.91	1.67	1.00	1.84	1.10	2.02	1.21	2.21	1.31	2.41	1.42	2.61	1.54	2.82	1.66	3.05
10	1.18	2.16	1.30	2.38	1.42	2.61	1.56	2.85	1.69	3.11	1.84	3.37	1.99	3.65	2.15	3.93
15	1.25	2.29	1.38	2.52	1.51	2.77	1.65	3.03	1.80	3.30	1.95	3.58	2.11	3.87	2.28	4.17
20	1.32	2.42	1.46	2.67	1.60	2.93	1.75	3.21	1.90	3.49	2.07	3.79	2.23	4.10	2.41	4.42
30	1.40	2.56	1.54	2.83	1.69	3.10	1.85	3.39	2.01	3.69	2.18	4.01	2.36	4.33	2.55	4.67
40	1.48	2.71	1.63	2.98	1.79	3.28	1.95	3.58	2.13	3.90	2.31	4.23	2.50	4.57	2.69	4.93
50	1.53	2.80	1.69	3.09	1.85	3.39	2.02	3.71	2.20	4.04	2.39	4.38	2.58	4.74	2.79	5.11

TABLE 1.15 OPEN COUNTRY WITH SCATTERED WIND BREAKS

Wind speed m/sec	40		42		44		46		48		50		52		54	
Height m	Centre	Edge	Centre	Edge	Centre	Edge	Centre	Edge	Centre	Edge	Centre	Edge	Centre	Edge	Centre	Edge
3	0.61	1.12	0.67	1.23	0.74	1.35	0.81	1.48	0.88	1.61	0.95	1.75	1.03	1.89	1.11	2.04
5	0.73	1.35	0.81	1.48	0.89	1.63	0.97	1.78	1.06	1.94	1.15	2.10	1.24	2.28	1.34	2.45
10	1.02	1.87	1.12	2.06	1.23	2.26	1.35	2.47	1.47	2.69	1.59	2.92	1.72	3.15	1.86	3.40
15	1.18	2.16	1.30	2.38	1.42	2.61	1.56	2.85	1.69	3.11	1.84	3.37	1.99	3.65	2.15	3.93
20	1.25	2.29	1.38	2.52	1.51	2.77	1.65	3.03	1.80	3.30	1.95	3.58	2.11	3.87	2.28	4.17
30	1.35	2.47	1.49	2.72	1.63	2.99	1.78	3.27	1.94	3.56	2.11	3.86	2.28	4.18	2.46	4.50
40	1.42	2.61	1.57	2.88	1.72	3.16	1.88	3.45	2.05	3.76	2.23	4.08	2.41	4.41	2.60	4.76
50	1.48	2.71	1.63	2.98	1.79	3.28	1.95	3.58	2.13	3.90	2.31	4.23	2.50	4.57	2.69	4.93

TABLE 1.16 COUNTRY WITH MANY WIND BREAKS, SMALL TOWNS, OUTSKIRTS OF LARGE CITIES

Wind speed m/sec	40		42		44		46		48		50		52		54	
Height m	Centre	Edge	Centre	Edge	Centre	Edge	Centre	Edge	Centre	Edge	Centre	Edge	Centre	Edge	Centre	Edge
3	0.48	0.88	0.53	0.97	0.58	1.07	0.64	1.17	0.69	1.27	0.75	1.38	0.81	1.49	0.88	1.61
5	0.58	1.06	0.64	1.17	0.70	1.28	0.76	1.40	0.83	1.52	0.90	1.65	0.97	1.79	1.05	1.93
10	0.72	1.31	0.79	1.45	0.87	1.59	0.95	1.74	1.03	1.89	1.12	2.05	1.21	2.22	1.31	2.39
15	0.91	1.67	1.00	1.84	1.10	2.02	1.21	2.21	1.31	2.41	1.42	2.61	1.54	2.82	1.66	3.05
20	1.06	1.95	1.17	2.15	1.29	2.36	1.40	2.58	1.53	2.80	1.66	3.04	1.80	3.29	1.94	3.55
30	1.20	2.20	1.32	2.43	1.45	2.66	1.59	2.91	1.73	3.17	1.88	3.44	2.03	3.72	2.19	4.01
40	1.30	2.38	1.43	2.62	1.57	2.88	1.72	3.15	1.87	3.43	2.03	3.72	2.19	4.02	2.36	4.34
50	1.37	2.52	1.51	2.77	1.66	3.05	1.82	3.33	1.98	3.62	2.15	3.93	2.32	4.25	2.50	4.59

TABLE 1.17 SURFACE WITH LARGE AND FREQUENT OBSTRUCTIONS, EG CITY CENTRES

Wind speed m/sec	40		42		44		46		48		50		52		54	
Height m	Centre	Edge	Centre	Edge	Centre	Edge	Centre	Edge	Centre	Edge	Centre	Edge	Centre	Edge	Centre	Edge
3	0.37	0.68	0.41	0.75	0.45	0.82	0.49	0.89	0.53	0.97	0.58	1.06	0.62	1.14	0.67	1.23
5	0.42	0.78	0.47	0.86	0.51	0.94	0.56	1.03	0.61	1.12	0.66	1.21	0.72	1.31	0.77	1.42
10	0.53	0.97	0.58	1.07	0.64	1.17	0.70	1.28	0.76	1.39	0.83	1.51	0.89	1.64	0.96	1.77
15	0.64	1.18	0.71	1.30	0.78	1.43	0.85	1.56	0.93	1.70	1.01	1.85	1.09	2.00	1.17	2.15
20	0.73	1.35	0.81	1.48	0.89	1.63	0.97	1.78	1.06	1.94	1.15	2.10	1.24	2.28	1.34	2.45
30	0.95	1.75	1.05	1.93	1.15	2.11	1.26	2.31	1.37	2.52	1.49	2.73	1.61	2.95	1.74	3.19
40	1.11	2.03	1.22	2.24	1.34	2.46	1.46	2.68	1.59	2.92	1.73	3.17	1.87	3.43	2.02	3.70
50	1.22	2.24	1.35	2.48	1.48	2.72	1.62	2.97	1.76	3.23	1.91	3.51	2.07	3.79	2.23	4.09

1.7 WIND ATTACHMENT DESIGN GUIDE

INTRODUCTION

Once the design wind loads have been determined for a roof, the appropriate method of attachment of the roofing specification can be considered.

Mastic asphalt is laid unbonded to the substrate, and only the attachment of the insulation and underlays need be considered.

For built-up roofing, the attachment of the waterproofing layer, the insulation and underlay all need to be designed to take into account the wind loads.

The efficiency of a bitumen bond, a partial bitumen bond or the holding power of mechanical fixings can only be forecast within approximate limits. It is usual to carry out formal calculations for the attachment of insulation layers to metal deck, but the attachment of insulation to other decks and the attachment of waterproofing to insulations is normally based on customs which have developed in the industry. No quantitative guidance is available but it is generally accepted that some combinations of insulation and waterproofing are not suitable for exposure to high wind forces.

In order to provide guidance for designers, the following section proposes limitations on the application of flat roofing specifications for three categories of wind risk.

The first category relates to nett wind loads up to $2.4kN/m^2$, the second category wind loads up to $3.6kN/m^2$ and the third category wind loads over $3.6kN/m^2$.

The nett wind load is the design wind load less the self weight of the waterproofing specification.

These categories are notional and have been selected from experience to represent broad bands of measurement for the severity of the wind. Suggested forms of attachment and the influence of loading coats are indicated for each category of wind force, taking into account the type of roof deck and the laminar strength of the insulation.

In some cases specific product data and test results may be available to give more accurate forecasts of performance.

DELAMINATION OF INSULATION

Insulation materials exhibit different levels of resistance to delamination and it is useful to consider them in two general groups: those insulations which from experience exhibit relatively high resistance to delamination and those which exhibit a low resistance to delamination. This grouping provides an easy method of analysis and is referred to in the tables which follow.

Typical high resistance materials are cork, wood fibreboard, polyurethane foam board, and paper-faced glass fibre board. Materials with a low resistance to delamination include mineral wool slabs, unfaced glass fibre board and perlite board. Isocyanurate foam boards may tend to a glass-like cell structure which could be broken down by continuous traffic after the roofing has been completed. This may weaken the bond between the foam and the facing, and if continuous traffic conditions are likely the board can be strengthened by the addition of an overlay of a stable insulation such as wood fibreboard.

Low resistance boards should be used with caution, and it will sometimes prove necessary to hold them against delamination by securing them to the deck with mechanical fixings and large washers. As a generalisation, insulation boards with a low resistance to delamination should only be applied on wind impermeable decks, or with an underlay which will act as the critical layer and relieve the insulation from much of the wind load. Built-up roofing specifications on these insulants should also be provided with a loading coat of stone chippings to reduce the possibility of disturbance of the waterproof covering by the wind.

The strength of bitumen bond to the aluminium foil or plastic facings of some insulation boards offers little peel resistance and additional fixings or loading coats may be required when these materials are used.

TIMBER BOARDED ROOFS

Built-up roofing direct to deck

Tongued and grooved timber boarding may be taken as an air impermeable deck and the small amount of air leakage to the underside of the membrane is not of great significance. The traditional nailing patterns have proved satisfactory and clearly make sufficient allowance for the air leakage through the deck. Butt jointed timber decks should not be used with built-up roofing because of warping and shrinkage.

BS 747 type 2B asbestos base roofing is frequently used as the first layer of built-up roofing applied direct to a timber deck. This should be nailed at 150mm centres in both directions. The deck should be close fitting to provide sufficient suction to back-up the strength of the nailing. A stone chipping finish is recommended to provide a loading coat.

Table 1.18 Built-up roofing direct to timber deck

Nett wind load	Attachment of first layer
Up to 2.4kN/m²	BS 747 type 2B roofing nailed at 150mm centres with 10mm layer chippings as a loading coat or High performance layer nailed at 200mm centres.
Up to 3.6kN/m²	High performance layer nailed at 200mm centres.
Over 3.6kN/m²	High performance layer nailed at 150mm centres, with 10mm layer of chippings as a loading coat.

If there are fears that the suction effect will not be created on the underside of the waterproofing, and if loading coats are not to be used, it will be necessary to use a high performance felt for the first layer. The frequency of nailing of this layer will depend on the nail-holding power of the felt used. Assuming that the holding power of the nail into the timber is the weakest part of the system, nails at approximately 200mm centres will prove satisfactory on the majority of sites.

In cases of extreme exposure to wind, the high performance first layer should be nailed at 150mm centres and a loading coat of chippings or equivalent weight of paving should be applied.

Table 1.18 gives suggested forms of attachment suitable for given wind loads.

Built-up roofing and insulation to deck

When an insulation is applied to timber boarding, a vapour check or underlay will normally be required. The underlay will be nailed to the deck at 150mm centres, and the insulation will be bonded in bitumen on top of it. The attachment of the waterproofing layer will depend on the type of insulation used and on the provision of loading coats as shown in table 1.19.

The insulation may also be mechanically fixed through a loose laid underlay. Proprietary fixings which are threaded from point to head continuously are preferred, as these give maximum holding power. The same number of fixings should be used as indicated for mechanically fixing to metal decking in tables 1.24 to 1.26 provided the timber is of minimum 19mm thickness.

Table 1.19 Built-up roofing to insulation on timber deck

	High laminar strength insulation board		Low laminar strength insulation board	
Nett wind load	Partially bonded waterproofing	Fully bonded waterproofing	Partially bonded waterproofing	Fully bonded waterproofing
Up to 2.4kN/m²	No special precautions	No special precautions	Not recommended	10mm layer of chippings as a loading coat or First layer of high performance roofing mechanically fixed through insulation into deck.
Up to 3.6kN/m²	10mm layer of chippings as a loading coat	No special precautions	Not recommended	Increase to 25mm loading coat or First layer of high performance roofing mechanically fixed through insulation into deck
Over 3.6kN/m²	Increase to 25mm loading coat or Change to full bond. A change to full bond may be effected by overlaying with an insulation such as cork or wood fibreboard to which a full bond is usual.	No special precautions	Not recommended	Increase to 25mm loading coat or 10mm layer of chippings as a loading coat to a specification with a first layer of high performance roofing mechanically fixed through insulation into deck.

CONCRETE, WOODWOOL, PLYWOOD AND CHIPBOARD DECKS

Built up roofing direct to deck

In-situ concrete decks, precast concrete units and pre-screeded woodwool slabs are generally considered to be air impermeable. The joints should be sufficiently close to prevent the free flow of air and any open joints should be filled or taped, otherwise the deck should be considered open jointed.

Plywood and chipboard decks may also be considered air impermeable provided the edges of the panels are supported on joists or noggings and those not closed off by the support system are taped. Otherwise these decks should be considered as open jointed and the attachment designed accordingly, see table 1.20.

Direct applied membranes are normally part bonded to concrete, woodwool, plywood and chipboard decks, using BS 747 type 3G perforated roofing or other part bonding systems. A bitumen based priming coat should be applied to concrete or screeded surfaces if they are dusty or damp.

Built-up roofing and insulation to deck

An underlay or vapour check will be fully bonded on these decks and a vapour barrier will be fully or partially bonded as indicated in the Vapour Barrier Design Guide. The insulation board is usually fully bonded to the preceding layer and this will suffice for all conditions of exposure. The attachment strength of the waterproofing to the insulation will depend on the laminar strength of the insulation, the method of attaching the waterproofing to the insulation, and the provision of loading coats, as shown in Table 1.21.

Table 1.20 Built-up roofing direct to deck

	Concrete decks, woodwool, plywood and chipboard decks with closed and taped joints	Woodwool, plywood and chipboard decks with open joints
Nett wind load	Attachment of first layer	Attachment of first layer
Up to 2.4kN/m²	Partial bond	Partial bond and minimum 10mm layer of chippings as a loading coat
Up to 3.6kN/m²	Partial bond and 10mm layer of chippings as a loading coat	Partial bond and increase to 25mm loading coat or Change to full bond. A change to full bond may be effected by overlaying the deck with an insulation such as cork or wood fibreboard to which a full bond is usual
Over 3.6kN/m²	Partial bond and increase to 25mm loading coat or Change to full bond. A change to full bond may be effected by overlaying the deck with an insulation such as cork or wood fibreboard to which a full bond is usual	Partial bond and increase to 25mm loading coat or Change to full bond. A change to full bond may be effected by overlaying the deck with an insulation such as cork or wood fibreboard to which a full bond is usual

Table 1.21 Built-up roofing to insulation: Concrete, woodwool, plywood and chipboard decks

	High laminar strength insulation board		Low laminar strength insulation board	
Nett wind load	Partially bonded waterproofing	Fully bonded waterproofing	Partially bonded waterproofing	Fully bonded waterproofing
Up to 2.4kN/m²	No special precautions	No special precautions	Not recommended	10mm layer of chippings as a loading coat or First layer of high performance roofing mechanically fixed through insulation into deck
Up to 3.6kN/m²	10mm layer of chippings as a loading coat	No special precautions	Not recommended	Increase to 25mm loading coat or First layer of high performance roofing mechanically fixed through insulation into deck
Over 3.6kN/m²	Increase to 25mm loading coat or Change to a full bond. A change to a full bond may be effected by overlaying with an insulation such as cork or wood fibreboard to which a full bond is usual	No special precautions	Not recommended	Increase to 25mm loading coat or 10mm layer of chippings as a loading coat to a specification with first layer of high performance roofing mechanically fixed through insulation into deck.

METAL DECKING

Built-up roofing to insulation

Metal decks can be considered as being open-jointed and when there is no underlay both the insulation and the membrane should be considered as critical layers for attachment. The normal method of attaching the first layer of waterproofing to the insulation will be a partial or a full bitumen bond, depending on the type of insulation and its laminar strength as indicated in table 1.22.

An underlay or vapour barrier beneath the insulation will act as a critical layer and the waterproofing will then be fully or partially bitumen bonded as indicated in table 1.23 according to the ability of the selected insulation board to resist wind delamination, and the influence of loading coats.

When lightweight flexibile membranes are to be applied to low strength insulation and additional loading coats are not admissible, the conditions of exposure may dictate the need for the first layer waterproofing to be mechanically fastened, together with the insulation, to the deck. In this case, a high performance first layer waterproofing will be required.

Table 1.22 Built-up roofing to insulation with no underlay

	High laminar strength insulation board		Low laminar strength insulation board	
Nett wind load	Partially bonded waterproofing	Fully bonded waterproofing	Partially bonded waterproofing	Fully bonded waterproofing
Up to 2.4kN/m^2	10mm layer of chippings as a loading coat	No special precautions	Not recommended	Add underlay (see table 1.23)
Up to 3.6kN/m^2	Add underlay (see table 1.23)	10mm layer of chippings as a loading coat	Not recommended	Add underlay (see table 1.23)
Over 3.6kN/m^2	Add underlay (see table 1.23)	Add underlay (see table 1.23)	Not recommended	Add underlay (see table 1.23)

Table 1.23 Built-up roofing to insulation with underlay

	High laminar strength insulation board		Low laminar strength insulation board	
Nett wind load	Partially bonded waterproofing	Fully bonded waterproofing	Partially bonded waterproofing	Fully bonded waterproofing
Up to 2.4kN/m^2	No special precautions	No special precautions	Not recommended	10mm layer of chippings as a loading coat or First layer of high performance roofing mechanically fixed through insulation into deck
Up to 3.6kN/m^2	10mm layer of chippings as a loading coat	No special precautions	Not recommended	Increase to 25mm loading coat or First layer of high performance roofing mechanically fixed through insulation into deck
Over 3.6kN/m^2	Increase to 25mm loading coat or Change to a full bond. A change to a full bond may be effected by overlaying with an insulation such as cork or wood fibreboard to which a full bond is usual.	No special precautions	Not recommended	Increase to 25mm loading coat or 10mm layer of chippings as a loading coat to a specification with first layer of high performance roofing mechanically fixed through insulation into deck.

ATTACHMENT OF INSULATION TO METAL DECKING

Bitumen bonding of the insulation provides the most stable base for the waterproofing membrane, and this can be supplemented by mechanical fixings if necessary. Mechanical fixings are efficient when properly selected and applied and may also prove economic compared with bitumen bonding, but the fixings mar the appearance of the underside of the deck.

Sole reliance on mechanical fixings may be dictated by specific fire requirements and under these circumstances the entire specification, including underlays and insulation, must also be chosen to satisfy these requirements.

Bitumen bonding
The strength of a bitumen bond to the top flats of a metal decking profile is commonly taken as 2.4kN/m^2. Profiles used for metal decking are available with a range of trough spacings and widths of top flat for bitumen bonding. The allowance of 2.4kN/m^2, however, will give a good indication of performance on the majority of decks available in the UK. For wind forces exceeding 2.4kN/m^2, it will be necessary to add mechanical fixings to supplement the bitumen bond.

A difficulty arises when the wind uplift exceeds 2.4kN/m^2 and an unpenetrated vapour barrier is required. If the wind force is not too much greater than 2.4kN/m^2, it is possible to take special precautions in the application of the bonding bitumen and increase the design load to about 3kN/m^2, on the basis that the deck will be fully primed, a generous application of bitumen will be applied and no work is allowed to proceed in cold weather, or when conditions leave a bloom of dampness on the deck.

A satisfactory alternative when a vapour barrier is essential is to add an extra loading coat, such as 25mm of paving slabs or chippings in place of the usual 10mm layer of chippings, and this will be satisfactory under most conditions of exposure in the UK. Nett design uplift in the order of 3.6kN/m^2 should be appropriate.

A better alternative may be to design for the vapour barrier to be installed at an intermediate layer in the insulation. This allows a thin base layer of insulation to be mechanically fixed. The vapour barrier is fully bonded to this base layer and the main insulation is bonded to the vapour barrier.

The thickness of each layer of insulation must be chosen to ensure that the dew point is located above the vapour barrier.

Mechanical fixings
Several types of fastener systems are available for mechanically fixing insulation boards to metal decking. All use washers or discs of varying shapes but approximately the same size, in the order of 50-70mm diameter, to provide the necessary holding power.

Self-tapping, self-piercing screws are normally used into steel and aluminium decks. Fine screw threads strip easily in the thin metal of the decking and specially designed coarse thread screws are essential.

For convenience of design, it is usually best to work to one standard design strength per fixing. A conservative value of 0.4kN per fixing is recommended by the Felt Roofing Contractors Advisory Board.

A minimum four fixings per board placed with one fixing near each corner will ensure that every board is firmly and independently secured. When more than four fixings per board are necessary, the exact pattern is of little importance provided a regular distribution is achieved.

Fixings should be inserted approximately 75mm from the edges of the insulation boards. The practice of inserting fixings at the joints between boards should be avoided as this results in a weakening of the board around the fixing. The practice can also lead to confusion about the number of fixings allowed per board as illustrated.

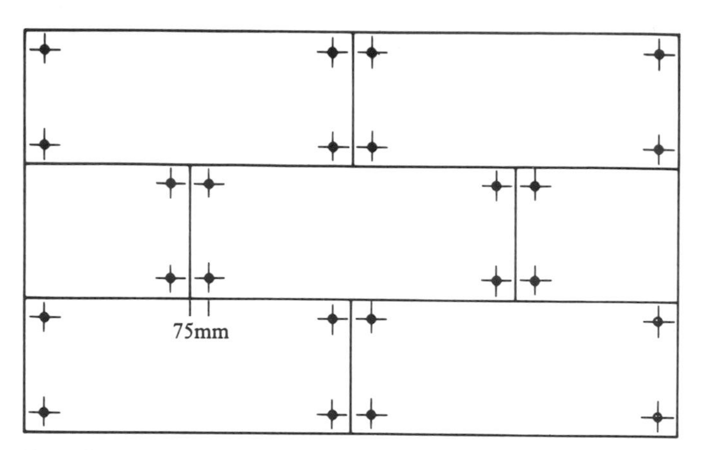

Four fixings per board

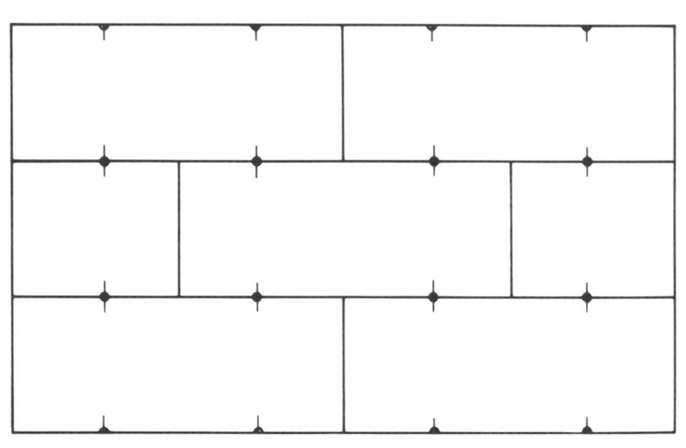

Numerical distribution two fixings per board, even though four fixings are touching each board.

The number of mechanical fixings required for securing insulation boards to metal decking for commonly occuring wind speeds are given in Tables 1.24, 1.25 and 1.26, assuming a design strength of 0.4kN per mechanical fixing and classifying building location, height and wind speeds according to CP 3:Chapter V: Part 2. Values of S_1, S_3 and Cp are as described on page 44 and the numbers of fixings relate to the wind loads given in Tables 1.14 to 1.17

The tables give data for the three most common sizes of insulation board and indicate the number of fixings required at roof edges and roof centre areas, based on there being no bitumen bond.

TABLE 1.24 NUMBERS OF MECHANICAL FIXINGS REQUIRED - SIZE OF BOARD 900mm x 600mm

Exposure: open country with no obstructions

Wind speed m/sec	40		42		44		46		48		50		52		54	
Height m	Centre	Edge	Centre	Edge	Centre	Edge	Centre	Edge	Centre	Edge	Centre	Edge	Centre	Edge	Centre	Edge
3	4	4	4	4	4	4	4	4	4	4	4	4	4	4	4	4
5	4	4	4	4	4	4	4	4	4	4	4	4	4	4	4	4
10	4	4	4	4	4	4	4	4	4	4	4	5	4	5	4	5
15	4	4	4	4	4	4	4	4	4	5	4	5	4	5	4	6
20	4	4	4	4	4	4	4	4	4	5	4	5	4	6	4	6
30	4	4	4	4	4	4	4	5	4	5	4	6	4	6	4	6
40	4	4	4	4	4	5	4	5	4	5	4	6	4	6	4	7
50	4	4	4	4	4	5	4	5	4	6	4	6	4	7	4	7

Exposure: open country with scattered windbreaks

Wind speed m/sec	40		42		44		46		48		50		52		54	
Height m	Centre	Edge	Centre	Edge	Centre	Edge	Centre	Edge	Centre	Edge	Centre	Edge	Centre	Edge	Centre	Edge
3	4	4	4	4	4	4	4	4	4	4	4	4	4	4	4	4
5	4	4	4	4	4	4	4	4	4	4	4	4	4	4	4	4
10	4	4	4	4	4	4	4	4	4	4	4	4	4	4	4	5
15	4	4	4	4	4	4	4	4	4	4	4	5	4	5	4	5
20	4	4	4	4	4	4	4	4	4	4	4	5	4	5	4	6
30	4	4	4	4	4	4	4	5	4	5	4	5	4	6	4	6
40	4	4	4	4	4	4	4	5	4	5	4	6	4	6	4	7
50	4	4	4	4	4	5	4	5	4	5	4	6	4	6	4	7

Exposure: country with many windbreaks; small towns, outskirts of cities

Wind speed m/sec	40		42		44		46		48		50		52		54	
Height m	Centre	Edge	Centre	Edge	Centre	Edge	Centre	Edge	Centre	Edge	Centre	Edge	Centre	Edge	Centre	Edge
3	4	4	4	4	4	4	4	4	4	4	4	4	4	4	4	4
5	4	4	4	4	4	4	4	4	4	4	4	4	4	4	4	4
10	4	4	4	4	4	4	4	4	4	4	4	4	4	4	4	4
15	4	4	4	4	4	4	4	4	4	4	4	4	4	4	4	4
20	4	4	4	4	4	4	4	4	4	4	4	4	4	5	4	5
30	4	4	4	4	4	4	4	4	4	4	4	5	4	5	4	6
40	4	4	4	4	4	4	4	4	4	5	4	5	4	6	4	6
50	4	4	4	4	4	4	4	5	4	5	4	5	4	6	4	6

Exposure: surface with large and frequent obstructions

Wind speed m/sec	40		42		44		46		48		50		52		54	
Height m	Centre	Edge	Centre	Edge	Centre	Edge	Centre	Edge	Centre	Edge	Centre	Edge	Centre	Edge	Centre	Edge
3	4	4	4	4	4	4	4	4	4	4	4	4	4	4	4	4
5	4	4	4	4	4	4	4	4	4	4	4	4	4	4	4	4
10	4	4	4	4	4	4	4	4	4	4	4	4	4	4	4	4
15	4	4	4	4	4	4	4	4	4	4	4	4	4	4	4	4
20	4	4	4	4	4	4	4	4	4	4	4	4	4	4	4	4
30	4	4	4	4	4	4	4	4	4	4	4	4	4	4	4	4
40	4	4	4	4	4	4	4	4	4	4	4	4	4	5	4	5
50	4	4	4	4	4	4	4	4	4	5	4	5	4	5	4	6

TABLE 1.25 NUMBERS OF MECHANICAL FIXINGS REQUIRED - SIZE OF BOARD 1200mm x 600mm

Exposure: open country with no obstructions

Wind speed m/sec	40		42		44		46		48		50		52		54	
Height m	Centre	Edge	Centre	Edge	Centre	Edge	Centre	Edge	Centre	Edge	Centre	Edge	Centre	Edge	Centre	Edge
3	4	4	4	4	4	4	4	4	4	4	4	4	4	5	4	5
5	4	4	4	4	4	4	4	4	4	4	4	5	4	5	4	6
10	4	4	4	4	4	5	4	5	4	6	4	6	4	7	4	7
15	4	4	4	5	4	5	4	6	4	6	4	6	4	7	4	7
20	4	4	4	5	4	5	4	6	4	6	4	7	4	7	4	8
30	4	5	4	5	4	6	4	6	4	7	4	7	4	8	5	8
40	4	5	4	5	4	6	4	6	4	7	4	8	5	8	5	9
50	4	5	4	6	4	6	4	7	4	7	4	8	5	9	5	9

Exposure: open country with scattered windbreaks

Wind speed m/sec	40		42		44		46		48		50		52		54	
Height m	Centre	Edge	Centre	Edge	Centre	Edge	Centre	Edge	Centre	Edge	Centre	Edge	Centre	Edge	Centre	Edge
3	4	4	4	4	4	4	4	4	4	4	4	4	4	4	4	4
5	4	4	4	4	4	4	4	4	4	4	4	5	4	4	4	4
10	4	4	4	4	4	4	4	4	4	5	4	5	4	6	4	6
15	4	4	4	4	4	5	4	5	4	6	4	5	4	7	4	7
20	4	4	4	5	4	5	4	5	4	6	4	6	4	7	4	8
30	4	4	4	5	4	5	4	6	4	6	4	7	4	8	5	8
40	4	5	4	5	4	6	4	6	4	7	4	7	4	8	5	9
50	4	5	4	5	4	6	4	6	4	7	4	8	5	8	5	9

Exposure: country with many windbreaks; small towns, outskirts of cities

Wind speed m/sec	40		42		44		46		48		50		52		54	
Height m	Centre	Edge	Centre	Edge	Centre	Edge	Centre	Edge	Centre	Edge	Centre	Edge	Centre	Edge	Centre	Edge
3	4	4	4	4	4	4	4	4	4	4	4	4	4	4	4	4
5	4	4	4	4	4	4	4	4	4	4	4	4	4	4	4	4
10	4	4	4	4	4	4	4	4	4	4	4	4	4	4	4	4
15	4	4	4	4	4	4	4	4	4	4	4	5	4	5	4	6
20	4	4	4	4	4	4	4	5	4	5	4	6	4	6	4	6
30	4	4	4	4	4	5	4	5	4	6	4	6	4	7	4	7
40	4	4	4	5	4	5	4	6	4	6	4	7	4	7	4	8
50	4	5	4	5	4	6	4	6	4	7	4	7	4	8	5	8

Exposure: surface with large and frequent obstructions

Wind speed m/sec	40		42		44		46		48		50		52		54	
Height m	Centre	Edge	Centre	Edge	Centre	Edge	Centre	Edge	Centre	Edge	Centre	Edge	Centre	Edge	Centre	Edge
3	4	4	4	4	4	4	4	4	4	4	4	4	4	4	4	4
5	4	4	4	4	4	4	4	4	4	4	4	4	4	4	4	4
10	4	4	4	4	4	4	4	4	4	4	4	4	4	4	4	4
15	4	4	4	4	4	4	4	4	4	4	4	4	4	4	4	4
20	4	4	4	4	4	4	4	4	4	4	4	4	4	4	4	4
30	4	4	4	4	4	4	4	4	4	5	4	5	4	5	4	6
40	4	4	4	4	4	4	4	5	4	5	4	6	4	6	4	7
50	4	4	4	5	4	5	4	5	4	6	4	6	4	7	4	7

TABLE 1.26 NUMBERS OF MECHANICAL FIXINGS REQUIRED - SIZE OF BOARD 1200mm x 750mm

Exposure: open country with no obstructions

Wind speed m/sec	40		42		44		46		48		50		52		54	
Height m	Centre	Edge	Centre	Edge	Centre	Edge	Centre	Edge	Centre	Edge	Centre	Edge	Centre	Edge	Centre	Edge
3	4	4	4	4	4	4	4	4	4	5	4	5	4	6	4	6
5	4	4	4	4	4	4	4	5	4	5	4	6	4	6	4	7
10	4	5	4	5	4	6	4	6	4	7	4	8	4	8	5	9
15	4	5	4	6	4	6	4	7	4	7	4	8	5	9	5	9
20	4	5	4	6	4	7	4	7	4	8	5	8	5	9	5	10
30	4	6	4	6	4	7	4	8	4	8	5	9	5	10	6	10
40	4	6	4	7	4	7	4	8	5	9	5	9	6	10	6	11
50	4	6	4	7	4	8	4	8	5	9	5	10	6	11	6	11

Exposure: open country with scattered windbreaks

Wind speed m/sec	40		42		44		46		48		50		52		54	
Height m	Centre	Edge	Centre	Edge	Centre	Edge	Centre	Edge	Centre	Edge	Centre	Edge	Centre	Edge	Centre	Edge
3	4	4	4	4	4	4	4	4	4	4	4	4	4	4	4	5
5	4	4	4	4	4	4	4	4	4	4	4	5	4	5	4	5
10	4	4	4	5	4	5	4	5	4	6	4	7	4	7	4	8
15	4	5	4	5	4	6	4	6	4	7	4	8	4	8	5	8
20	4	5	4	6	4	6	4	7	4	7	4	8	5	9	5	9
30	4	5	4	6	4	7	4	7	4	8	5	8	5	9	5	10
40	4	6	4	6	4	7	4	8	5	8	5	9	5	10	6	11
50	4	6	4	7	4	7	4	8	5	9	5	9	6	10	6	11

Exposure: country with many windbreaks; small towns outskirts of cities

Wind speed m/sec	40		42		44		46		48		50		52		54	
Height m	Centre	Edge	Centre	Edge	Centre	Edge	Centre	Edge	Centre	Edge	Centre	Edge	Centre	Edge	Centre	Edge
3	4	4	4	4	4	4	4	4	4	4	4	4	4	4	4	4
5	4	4	4	4	4	4	4	4	4	4	4	4	4	4	4	4
10	4	4	4	4	4	4	4	4	4	4	4	5	4	5	4	5
15	4	4	4	4	4	4	4	5	4	5	4	6	4	6	4	7
20	4	4	4	5	4	5	4	6	4	6	4	7	4	7	4	8
30	4	5	4	5	4	6	4	7	4	7	4	8	5	8	5	9
40	4	5	4	6	4	6	4	7	4	8	5	8	5	9	5	10
50	4	6	4	6	4	7	4	7	4	8	5	9	5	10	5	10

Exposure: surface with large and frequent obstructions

Wind speed m/sec	40		42		44		46		48		50		52		54	
Height m	Centre	Edge	Centre	Edge	Centre	Edge	Centre	Edge	Centre	Edge	Centre	Edge	Centre	Edge	Centre	Edge
3	4	4	4	4	4	4	4	4	4	4	4	4	4	4	4	4
5	4	4	4	4	4	4	4	4	4	4	4	4	4	4	4	4
10	4	4	4	4	4	4	4	4	4	4	4	4	4	4	4	4
15	4	4	4	4	4	4	4	4	4	4	4	4	4	4	4	5
20	4	4	4	4	4	4	4	4	4	4	4	5	4	5	4	5
30	4	4	4	4	4	5	4	5	4	6	4	6	4	7	4	7
40	4	5	4	5	4	5	4	6	4	7	4	7	4	8	4	8
50	4	5	4	6	4	6	4	7	4	7	4	8	5	8	5	9

1.8 FIRE

INTRODUCTION

Experience of fire, large scale fire tests and laboratory fire tests demonstrate the enormity of fire once it has developed and the power of a fire to search out weaknesses in the construction.

Any new building must comply with the fire protection requirements of the relevant Building Regulations, Building Standards and Local Authority Bye Laws.

In essence, these performance requirements are concerned with the preservation of life, rather than the preservation of property or goods. The material loss of the building and its contents is the concern of insurance companies, and certain insurance bodies have adopted approved roof constructions based on their own tests and experience. In addition, some industrial concerns have themselves adopted standards which may be in excess of the statutory requirements, and it is therefore important to establish whether there are any special fire performance requirements, and consult with the relevant insurance body for advice and approval at the planning stage.

Opinions on the subject of fire are divided and little authoritative guidance has been issued. It is easy for designers to make decisions that are misguided or at least controversial and may be at the expense of overall roof performance. The best approach will be to consult with and take advice from the fire authorities who will have the task of attending the fire if one arises.

The various statutory regulations consider the performance of a flat roof in respect of external and internal fire conditions, and are based on performance tests set out in British Standard 476 'Fire Tests on Building Materials and Structures'.

EXTERNAL FIRE

A fire may create a hazard to neighbouring property and the chief danger to a nearby roof arises from radiant heat, combined with burning embers thrown off from the fire.

The performance requirements for a roof construction exposed to external fire conditions relate to the ability of the roof to act as a protective barrier against penetration by fire and the spread of flame over the roof surface, both of which are tested in Part 3 of BS 476.

BS 476: Part 3: 1958. External Fire Exposure Roof Test

In the test procedures, samples of the roof construction are subjected to radiant heat on the upper surface and measurements are made of the possibility of fire penetration during a 60 minute heating period. A test flame is applied after 5 minutes to simulate the fall of a burning brand and the spread of flame is observed. A preliminary test is also made in which the specimen is subjected to a flame in the absence of radiant heat to identify highly flammable coverings. The two criteria of performance are penetration time and distance of spread of flame along the external surface and the performance of the total roof construction is represented by the following letter system, with an AA designation indicating the best performance that can be obtained.

Roof fire test

First Letter - Penetration Classifications

A Specimens not penetrated within 1 hour
B Specimens penetrated in not less than ½ hour
C Specimens penetrated in less than ½ hour
D Specimens penetrated in the preliminary flame test.

Second Letter - Spread of Flame Classifications

A Specimens with no spread of flame
B Specimens with not more than 533mm spread of flame
C Specimens with more than 533mm spread of flame
D Specimens which continue to burn for 5 minutes after the withdrawal of the test flame or spread more than 381mm in the preliminary test.

Attention is also drawn to dripping from the underside of the specimen, any mechanical failure or the development of any hole, by the addition of a suffix 'X' to the designation. This suffix, however, carries no restriction in Building Regulations. A typical roof designation would be presented as:

EXT.F.AA, where

EXT = External, F = Flat, AA = Achieved designation.

An A classification for spread of flame is most desirable as it is only too easy for flames fanned by a steady wind to spread across the roof and enter the building through openings such as ventilators and rooflights. The potential spread of flame on the roof is therefore important. It is regrettable that some roof treatments do not perform well in this respect and that the orginal spread of flame requirement is often ignored during re-roofing or maintenance.

BS 476: Part 3 was revised in 1975 with slight variations in the testing procedure and with the results expressed by a new method. The Building Regulations do not make reference to the new standard however and it is likely that the 1958 version will remain in general use until a further revision has been issued.

In BS 476: Part 3: 1975 the surface spread of flame element of the test has been dropped and replaced by a measurement of surface ignition, made at the same time as the penetration test and using the same level of radiation intensity.

The previous method required a test flame to be applied only once during the early part of exposure. With the new test, the flame is applied at intervals throughout the test. Designations in the revised standard are expressed by the letters 'X' and 'P' followed by the time in minutes for the sample to be penetrated by fire.

X indicates that in the preliminary ignition test, the duration of flaming of the specimen exceeded 5 minutes or that the maximum distance of flaming exceeded 370mm.

P indicates that the above conditions were not exceeded.

The table below shows the basic relationship between the old and new designations.

BS 476 Part 3 1958	BS 476 Part 3 1975
AA, AB, AC	P60
BA, BB, BC	P30
AD, BD, CA, CB, CC, CD	P15
UNCLASSIFIABLE	P5

The standard allows the test period to be extended to 90 minutes if required so that the highest designation obtainable would be P90. The incidence of dripping from the underside, hole formation or mechanical breakdown, (which were previously referred to by the suffix X) are now considered within the new overall classification.

Rooflights of plastic materials cannot be subjected to the BS 476 tests due to the relatively low softening point of the materials. Performance tests designed specifically for plastics are described in BS 2782 'Methods of testing plastics'.

Boundary distance

The designations achieved in BS 476: Part 3: 1958 are used in Building Regulations to define acceptable roof constructions in relation to their distance from a possible external fire source.

As the distance of the roof from the boundary increases, a relaxation is made for the ability of the roof to resist ignition and penetration as shown in Table 1.27.

Table 1.27

Designation	Minimum distance from boundary*	Maximum size	Minimum spacing**
AA, AB, AC	No restriction	No restriction	No restriction
BA, BB, BC	6m	No restriction	No restriction
AD, BD, CA, CB, CC, CD	12m	No restriction	No restriction
AD, BD, CA, CB, CC, CD	6m	3m^2	1.5m
DA, DB, DC, DD	22m	3m^2	1.5m

* A boundary is defined in the Building Regulations as the boundary of land belonging to the building up to and including the centre line of any abutting street, canal, or river.
** Minimum spacing refers to the requirement for designated roofs indicated to be separated from similarly designated roof areas by a minimum spacing of 1.5m, covered by non-combustible material.

Although designations AA, AB and AC are accepted by Building Regulations for most roofs, certain authorities may insist on the designation AA.

Notional Designations of Roof Coverings

Schedule 9 of the Building Regulations, England and Wales and Northern Ireland, and Schedule 9, Table 7 of the Scottish Building Regulations give notional designations for roof coverings to BS 476: Part 3: 1958. Mastic asphalt is deemed to provide an AA designation over deckings of timber, woodwool, plywood, chipboard, concrete, steel, aluminium or asbestos cement and this designation is achieved without a surface dressing of chippings. A flat roof covering of bitumen felt on these decks (irrespective of the felt specification) is also deemed to be of AA designation, provided that the roofing has a surface finish of bitumen bedded stone chippings covering the whole surface to a depth of not less than 12.5mm, or non-combustible tiles.

If the roof is required to support or stabilise the loadbearing walls, or if the roof surface forms part of a fire escape route, there may be a requirement for the roof to provide fire resistance as defined in BS 476:Part 7:1972 'Test methods and criteria for the fire resistance of elements of building construction'.
Regulation E9 gives requirements for the treatment of the top of compartment walls.

Fire on the underside of a concrete deck is unlikely to lead to structural collapse or cause ignition of the overlying roofing materials, or cause them to give off combustible gases which might enter the building.

Metal decking and woodwool decks would generally maintain their structural integrity with some distortion and deflections, although the steel frame could collapse or distort to the extent that the structural deck is no longer supported and may tilt or fall towards the fire. At this stage the combustible components may add fuel to the main fire but that section of the building will be a total loss and the additional fuel may not prove significant.

Aluminium deck is an exception as this will usually fail before the supporting steel frame. Typically a hole will burn through the roof above the fire and opinions have sometimes been expressed after such fires that the ventilation effect of the penetration of the roof by fire has proved beneficial and has helped to avoid the spread of fire further into the building.

As the development of the fire increases, roof temperatures may be reached which support a heat transfer through the deck to generate combustible gases from the insulation or roofing materials above. These gases will not burn until in contact with air containing normal proportions of oxygen to support the combustion, but the gases may pass down through the joints in the roof deck to burn on the underside of the deck if a suitable air supply is available.

The channels of metal decking will conduct the gases along them but it would be unusual for flames to travel along the channels as there is not usually a sufficient supply of air to support the combustion. The detailing of the roof usually closes off the ends of the channel at the edges of the building.

Fire in cavities
Experience of fire and fire tests indicates that the greatest danger is the spread of fire through cavities, particularly the space between the ceiling or lining and the underside of the roof deck. It was with the object of restricting this unseen spread of fire that the Building Regulation requiring cavity barriers and fire stops was introduced. Even if the ceiling has good fire performance on test, it is possible for fire to enter the space at a badly fixed ceiling panel or a panel which has been removed. A designer may rely on the integrity of a ceiling in fire and the fire performance which is established by test, but this counts for nothing if panels are loosely fixed or missing. The building owner or those concerned with maintenance and repair should appreciate the continued need for the integrity of the ceiling and panels should always be replaced and firmly fixed as soon as possible after removal.

SECTION 2 DECKS, SCREEDS AND INSULATIONS

2.2 SCREEDS

INTRODUCTION

Screeds provide a suitable surface to receive the waterproofing, and can also be used to achieve falls and cross falls when concrete slabs and precast concrete units are installed flat. In addition some screeds can provide a level of thermal insulation and contribute to the U-value of the roof.

Sand and cement screeds have a high thermal conductivity and do not significantly contribute to the overall thermal transmittance value of the roof. Aerated screeds and lightweight aggregate screeds bound by cement or bitumen have a lower thermal conductivity and will make a useful contribution to the U-value when dry. Additional insulation where required will normally be added above the screed.

Sand and cement screeds are usually laid by the general contractor, but other types of screeds are normally laid by specialist contractors and their advice and recommendations should be followed. These screeds should also be protected from damage by other trades.

Wet screeds which contain large quantities of water cannot be covered by the waterproofing membrane on the same day. A period of time is necessary to allow for curing so that the top surface is suitable to accept the waterproofing. If the screed cannot be protected from rain, drainage holes should be formed in the deck.

SAND AND CEMENT SCREEDS

Sand and cement screeds are normally mixed in the ratio 4:1 and the surface should be finished with a wood float.

The screed should be laid direct to the deck to obtain a good key. It should not be laid continuously but in areas not exceeding $10m^2$, to reduce the incidence of cracking due to drying and shrinkage. The screed contains considerable amounts of water and the surface should be adequately cured and dry before the roofing specification is applied.

AERATED SCREEDS

Portland cement, water and a foaming emulsion are combined to produce a cellular material which offers a hard surface when dry.

The k-value of aerated screeds varies between 0.10 and 0.54W/m°C. The insulation value should only be taken into account when it is reasonable to assume that the screed will be efficiently drained and dry. Roof top ventilators may assist the drying out. Certain proprietary screed systems include drainage and ventilation channels in the depths of the screed to provide the best opportunity for achieving dry conditions in service.

LIGHTWEIGHT AGGREGATE SCREEDS: CEMENT BONDED

Suitable lightweight aggregates are formed from expanded clay or sintered pulverised fuel ash, bonded with a cement binder.

The material must be laid soon after mixing otherwise the cement binder may dry and not bond the aggregate together. A 13mm sand and cement topping is necessary to give a smooth level surface for the roofing specification. Walkboards must be used when applying the topping to prevent displacement of the aggregate.

The k-value of cement bonded lightweight aggregate screeds is in the order of 0.29W/m°C. The insulation value should only be taken into account when it is reasonable to assume that the screed will be efficiently drained and dry.

LIGHTWEIGHT AGGREGATE SCREEDS: BITUMEN EMULSION BONDED

The lightweight aggregate consists of expanded clay bonded together with a bitumen binder. This screed does not require a topping, only the passage of a light roller before the application of the waterproofing; this is not to compact the screed but to form a level surface.

Only as much screed should be laid as can be waterproofed the same day. It is recommended that this type of screed is laid with an underlay of bitumen roofing. This will allow a temporary seal to be formed between the underlay and the waterproofing to fully protect the screed from overnight rain. The k-value of this type of screed is in the order of 0.144W/m°C when dry.

PERLITE/BITUMEN SCREEDS

A 50/50 mix by weight of perlite and bitumen binder is mixed hot on site, laid as a loose fill to prepared gauges or screeding bars, and then compacted in the order of 30% by rolling. There are no delays for drying and curing and the waterproofing can follow immediately.

Perlite bitumen screeds laid hot and dry allow the waterproofing to follow immediately

Under normal conditions of internal humidity, this type of screed should be laid without an underlay. If a vapour barrier or vapour check is required by the humidity conditions, a smooth even finish to the roof slab is needed to prevent the formation of voids under the vapour barrier and ensure successful compaction of the screed by roller. Perlite is an efficient insulation and perlite bitumen screeds have good resistance to moisture and a relatively low k-value in the order of 0.076W/m°C.

2.3 INSULATION BOARDS

INTRODUCTION

All insulants rely for their thermal efficiency on trapped air or gas to provide a resistance to the flow of heat.

Insulation foams trap the air or gas within a cellular structure. Fibrous materials trap air between the fibres and their insulation efficiency depends on the size and orientation of the fibres.

Insulation board materials can be divided into three main groups according to their origin.

Vegetable: Wood fibre
Cork

Mineral: Glass fibre
Mineral wool
Cellular glass
Perlite

Plastic foams: Polystyrene (bead)
Polystyrene (extruded)
Polyurethane
Polyisocyanurate

A wide range of insulation boards is available which incorporate proprietary refinements to the basic material or are manufactured with facings of plastic, paper, metal foil, glass fibre tissue or bituminous roofing felt. In addition, composite insulation boards are now more common, such as woodfibre/polystyrene or perlite/polyurethane laminates.

The world of roofing insulation is therefore becoming one of proprietary products and it becomes increasingly difficult to provide general guidance on the performance characteristics of the basic insulation materials. But, at the same time, it is important that all the factors influencing the selection of an insulation board are understood.

A board insulant should not be chosen solely on the basis of thermal efficiency. An equally important function of the insulation is to provide firm support for the waterproofing. Soft compressive boards do not provide suitable support for BS 747 roofing felts or for mastic asphalt. High performance roofings may be suitable over such boards provided they are not subjected to continuous traffic.

The long term resilience of an insulation is important and of growing significance because of the introduction of cellular foams, some of which have relatively brittle foam cell walls which can be broken by traffic over the roof. Similarly mineral fibre insulation cannot accept unlimited traffic and can break down to the extent that the waterproofing is not properly supported and becomes easily damaged. The selection of an insulation should therefore take account of expected traffic. Those insulations which break down easily should only be used on roofs subject to minimal traffic in service, or should be overlaid with a stronger board.

It is advisable to choose moisture resistant materials for roof insulation. In practice it is inevitable that occasional leakage will allow rainwater to make contact with the insulation, and the amount of damage suffered by the roofing system as a result of leakage will depend on the ability of the insulation to resist degradation from water.

The thermal and moisture movement characteristics of the insulation must be understood, and the waterproofing partially bonded or fully bonded accordingly.

Satisfactory site handling qualities should not be ignored. Unless a material can be satisfactorily handled and fixed by the contractor, the system may be put at risk. The desirable qualities looked for by a roofing contractor are ease of cutting, robust working surfaces which are not dusty or friable or abrasive to the touch, corners which are not too susceptible to breakage and materials which do not require special techniques of application.

The following pages provide a generalised guide to the performance characteristics of the various types of insulation board available, together with application techniques for warm roofs.

WOOD FIBREBOARD

Wood fibreboard was the first type of board extensively used for roof insulation and is still widely used when lower levels of thermal insulation are acceptable. Boards are manufactured by a felting process from an aqueous suspension of wood and other vegetable fibres, compressed to form a rigid insulation board.

BS 1192:Part 1 gives a water absorption test for a 2-hour period. Part 3 of this standard describes wood fibre insulating boards but gives no moisture absorption requirement for roof insulation. Tests show that certain untreated boards can have a moisture absorption of as much as 200% by weight during the test period. It is recommended that board used for roofing should have a moisture absorption limited to 50%, and leading suppliers make such a board available. Bitumen impregnated boards generally show a further reduction in the rate of moisture absorption but the bitumen will not act as a fungicide.

Moisture absorption can cause significant movement, and may cause ridging of the waterproofing over the joints, but the movement is extremely slow and will not normally lead to splitting from fatigue.

Boards will require dry site storage, and when applied to decks of wet construction, concrete and screeded surfaces, should always be used in conjunction with an underlay to act as a temporary damp proof course while the deck dries out. If the building conditions are such that a vapour barrier is required or a vapour check, it is better to choose a more moisture resistant insulation than fibreboard.

The treatment of wood fibreboard with a flame retardant may make the board weaker and more absorbent. The treatment may also cause adverse reactions with other roofing materials, particularly metal decking, and should be isolated from direct contact.

Wood fibreboards have good compressive and laminar strength and a very low coefficient of thermal expansion. They provide a good base for mastic asphalt waterproofing or fully bonded built-up roofing. The dimensional stability of this board makes it useful as an overlay to insulants more susceptible to thermal movement such as expanded polystyrene.

Wood fibreboards present no application problems to the roofing contractor. They are easily handled on site without excessive damage and are easy to lay with traditional bitumen bonding or mechanical fixing methods.

CORK

A well established and proven insulation material formed from pure granulated cork, compressed, steam baked and held together by the natural cork gum.

Cork is resistant to moisture and decay and is particularly useful for high humidity applications. It provides a substrate of good laminar and compressive strength suitable for asphalt roofing and with its low coefficient of thermal expansion will accept a full bonded built-up roofing specification.

Although cork is somewhat friable during handling, it is firm under foot traffic and is a suitable insulation to use below membranes which are subject to continual foot traffic. For general handling strength, a minimum 25mm board is recommended. An underlay may be required to provide support over the troughs of metal decking depending on the thickness of the cork and the trough opening of the deck.

PERLITE

Perlite is a mineral of volcanic origin. It contains a small amount of water which causes powdered perlite to expand some 20 times when heated. During manufacture, the perlite is combined with mineral fibres and binders to produce a roofboard with a thermal insulation value similar to wood fibreboard but with a greatly reduced combustible content.

One surface of perlite board is usually treated during manufacture with a bitumen emulsion to increase resistance to bitumen absorption and bind the surface.

Perlite boards generally have a good compressive strength but low laminar strength. They are also rather brittle and a minimum 25mm thickness is recommended. Perlite boards resist decay but the absorption of moisture can cause significant loss of strength. They have a very low coefficient of thermal expansion and will accept fully bonded waterproofing specifications. As with wood fibreboard the dimensional stability and heat shielding qualities of perlite are often used to improve the performance of other insulants either in the form of a composite with polystyrene or polyurethane or as an on-site overlay to these materials.

MINERAL WOOL AND GLASS FIBRE

Mineral wool is manufactured from volcanic rock, melted at extremely high temperatures. The molten rock is directed onto a series of rotating wheels where it is converted into thin fibres and during this process a small amount of resin is added to act as a binding agent. The fibres are gathered together to form a mat, which is then cured and compressed to form a rigid insulation slab.

Glass fibre boards are manufactured by a similar process but with molten glass spun into fine fibres, which are bonded together with resin and cured to form a rigid slab.

Several types of board and slab are available. Glass fibre insulation board is supplied with a facing designed to improve the surface texture and to provide a firm support for the membrane.

Mineral wool slabs and glass fibre roofboards provide a good level of thermal insulation and a dimensionally stable substrate. Both materials provide a board of low compressive strength and may also present a surface that will absorb an appreciable amount of bonding bitumen.

CELLULAR GLASS

Cellular glass is made from pure glass, expanded during manufacture and formed into slabs which are inorganic, will not rot or decay and are non-combustible. Slabs are available in constant thicknesses or tapered to provide falls.

The slabs are almost impermeable to water vapour, will not gain moisture by vapour diffusion and can normally be laid without a vapour barrier, provided that the joints are sealed in bitumen according to the manufacturer's instructions.

Cellular glass provides a stable base for mastic asphalt and fully bonded built-up roofing. Special laying techniques are required to seal the slab surfaces and joints, with liberal quantities of bitumen, to ensure physical stability and continuous support to reduce the incidence of breakage.

When mastic asphalt is applied, two layers of loose-laid non-bituminised paper are required between the slabs and the sheathing felt isolating membrane of the asphalt to prevent adhesion.

EXPANDED POLYSTYRENE

Expanded polystyrene boards, generally known as bead boards are formed by fusing together beads of pre-expanded polystyrene.

Roofboards are available in Grades HD (High Duty) and EHD (Extra High Duty). Each grade is identified by a coloured strip across the board edge: black for HD grade and green for EHD. Boards are available with a pre-felted upper surface and may be of a constant thickness or tapered to provide falls. Pre-felted HD grade board, overlaid with fibreboard or perlite board is normally used for roofing applications. Expanded polystyrene boards are also available as composites with wood fibreboard or perlite board laminated to the upper surface.

Expanded polystyrene boards exhibit large thermal movement and are heat sensitive. The boards cannot tolerate the application temperatures of hot bitumen or asphalt. Indirect laying techniques have to be adopted by applying a coat of hot bitumen to the substrate, and allowing it to cool to a tacky condition before the expanded polystyrene board is laid.

A wood fibreboard, perlite board or corkboard overlay performs the dual function of shielding the expanded polystyrene from hot bitumen and asphalt and absorbing the thermal movement suffcently to permit a fully bonded built-up roofing system. The overlay should be laid to break joint with the polystyrene although with factory applied overlays this is of course, not possible.

EXTRUDED POLYSTYRENE

Extruded polystyrene boards have similar thermal movement and heat sensitivity characteristics to expanded polystyrene but have an improved thermal conductivity. The boards are also exceptionally resistant to water absorption and are therefore normally used in the protected membrane roof system.

When selecting a board thickness to achieve a specific U-value for a protected membrane roof, it is necessary to allow for the loss in efficiency due to the effect of rainwater draining through the insulation. One method to take this into account is to add 20% to the thickness of the board. Another method is to adjust the calculated U-value by a factor depending on the amount of thermal resistance below the level of the waterproofing, see Appendix A.

Application in the protected membrane roof system consists of laying the boards loose above the waterproofing with anchorage against wind uplift and protection from ultra-violet degradation provided by a loading coat of gravel or paving slabs.

A loose-laid underlay may also be applied to even out surface irregularities or, if the waterproofing has a rough or sharp surface, to prevent abrasion to the underside of the board. The manufacturer of the insulation will advise on a suitable underlay material.

The insulation can be covered with rounded gravel with the minimum size in the order of 20mm. Standard gravel gradings allow a tolerance for quarries and a proportion of fine material may be present unless expressly excluded by agreement. Fine granules can work their way through the joints of the insulation and can cause damage and instability if they accumulate on the underside of the insulation. If it is expected that significant quantities of fine gravel will be present, it will be necessary to add a filter layer on top of the insulation to act as a sieve and prevent the passage of fine material through the joints. Again the manufacturers of the insulation will advise on a suitable material.

POLYURETHANE AND POLYISOCYANURATE FOAMS

Polyurethane and polyisocyanurate boards are foamed with an inert gas of high molecular weight. This is locked into the cells and gives the boards better insulation characteristics than foams which have air-filled cells.

The cell walls are slightly permeable to air and, over a period of several years, the air will diffuse into the cells until it reaches atmospheric pressure. The thermal efficiency of the foam decreases over this period and this is allowed for in the figures published for thermal conductivity.

The correct choice of density facing and form of membrane attachment are crucial to the success of specifications using polyurethane insulation.

The British Urethane Foam Manufacturers Association and the Felt Roofing Contractors Advisory Board have agreed specifications for the foam and facing, to define a board which is suitable to receive built-up felt roofings. It is recommended that only boards which comply with these specifications are used.

Built-up roofing systems used with polyurethane and polyisocyanurate boards should be partially bonded. Partial bonding is not suitable for sloping roofs and a wood fibreboard overlay is normally included to allow a fully bonded system.

Mastic asphalt retains a high temperature for long periods after application and insulation boards designed for use under mastic asphalt must be specifically formulated to maintain dimensional stability during the application and cooling of the asphalt. Polyisocyanurate foams accept higher operating temperatures than polyurethane and suitable formulations of polyisocyanurate are available for use under mastic asphalt.

SECTION 3 BUILT-UP ROOFING

3.1 MATERIALS

INTRODUCTION

Built-up roofing is formed on-site from two or more layers of roll roofing. The vast majority of materials available today, including those referred to as high performance roofings, consist of a reinforcement base, coated with bitumen. The continued development and popularity of roll roofings is due to the waterproofing and adhesive properties of bitumen, but it is the nature of the reinforcing base which dominates the strength, weathering and ageing characteristics of the membrane.

Rag or wood fibre was originally used to form the base reinforcement for all built-up roofing. This base was saturated with a penetrating bitumen. No other coating was added and the resulting felt was not waterproof in its own right. The waterproofing was achieved by the application of hot bitumen on-site, with the saturated felts providing strength and reinforcement. Such felts are still widely used in the USA but in Europe it has been the custom for many years to add a coating of filled oxidised bitumen during manufacture. This adds a waterproof coating to the felt, and the principle of waterproofing changes from using saturated felts to reinforce the hot applied bitumen to using felts as the waterproofing and bitumen primarily as an adhesive.

Both approaches have been successful but the fibre base, being composed of hollow, cellulosic fibres will rot in time through the absorption of moisture and the ageing process leads to weaker and weaker reinforcement, which in the end is insufficient to hold the waterproofing together. Asbestos fibres are more resistant to rotting but the absorption of moisture will again break down the asbestos base as a cohesive reinforcing layer and the finished roofing will suffer a similar weakening process.

Glass tissue reinforcement was introduced during the 1950's. Being composed of solid mineral fibres held together with an adhesive this base requires only a coating of oxidised bitumen. No breakdown of the glass through weathering has been observed in service and whilst the adhesive can break down, the bitumen coating holds the fibres together and the ageing and weathering process has little effect on the strength of the reinforcement provided by the glass tissue. However, this form of roofing has only limited strength and fatigue resistance in the first place.

High performance felts were developed to provide increased performance using stronger, more fatigue-resistant base materials, and by adding polymers to modify the bitumen to make it more flexible in winter, firmer in summer and more fatigue-resistant over a wide range of temperatures.

Little is known about the long term ageing and weathering characteristics of the new materials, and it would be unreasonable to suppose that the improved properties will be achieved for all time. Short term improved performance is assured and there is a reasonable expectation of improved performance in the long term. A number of high performance materials have been used for 10 years or more and have behaved extremely well.

The manufacturers of membranes will provide information on properties such as tensile strength, elongation at break, and perhaps an indication of fatigue resistance. This is useful in comparing one material with another but the roofer will make his judgement from the way the material handles, how easily it forms and stays in position, how easily it is damaged, how easily it tears in the hand, and how easily it ruptures. The completed membrane should also accept the knocks and wear which will occur during construction from following trades and over many years of service. There is no satisfactory laboratory test to simulate this long term requirement for ruggedness and toughness.

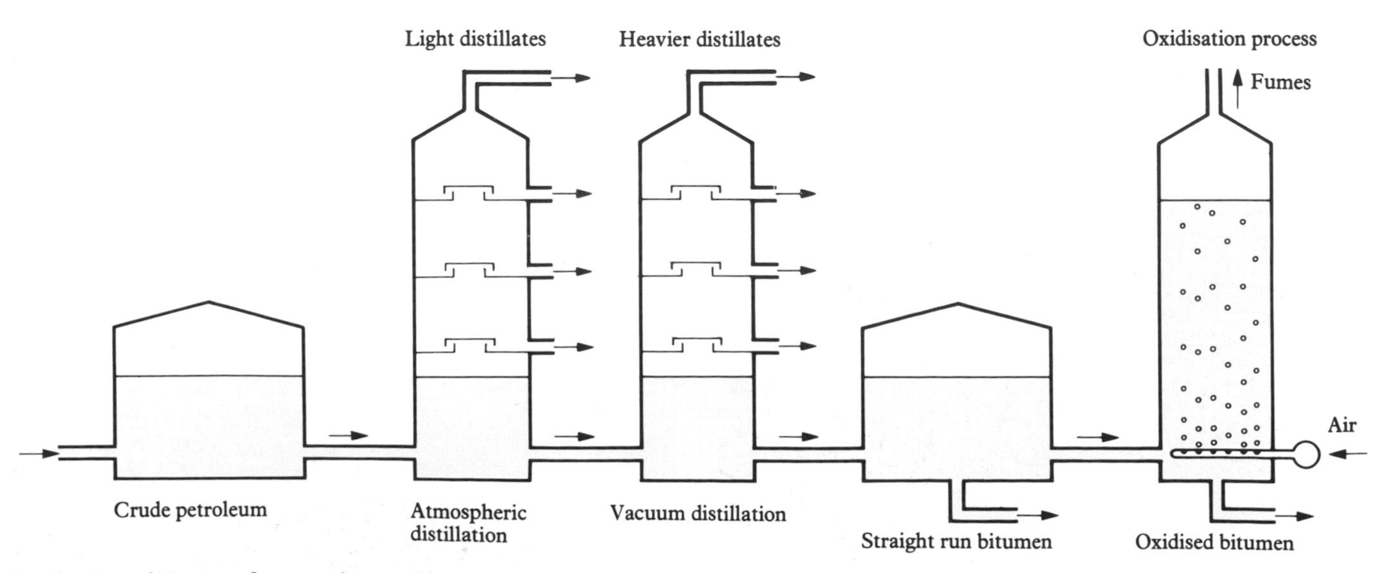

Production of bitumen from crude petroleum oil

BITUMEN MANUFACTURE

Bitumen is a component of crude oil and is separated by the refinery process of fractional distillation. Heat is applied to the crude oil in a distillation column until the lighter constituents vaporise and rise to condense on a series of trays within the column, with only the lightest remaining as vapours. A temperature difference between the top and bottom of the column ensures that the trays contain successively lighter fractions of the oil.

At atmospheric pressure the high temperature needed to vaporise some constituents would degrade the products. The distillation process is therefore carried out in at least two stages, one at atmospheric pressure and the other at reduced pressure or vacuum to lower the boiling point of the heavier constituents.

Because bitumen cannot be distilled it remains as an end product of fractional distillation after other products such as naphthas, white spirit, kerosene, diesel oil, and light, medium and heavy lubricating oils have been driven off. Bitumens obtained by this method are generally called penetration bitumens. They flow easily at high temperatures and are susceptible to cracking at low temperature. For roofing work it is necessary to make the bitumen less temperature sensitive and more rubber-like in consistency. These types of bitumens are known as oxidised bitumens and are made by passing a regulated stream of air through a soft bitumen under controlled temperature conditions.

BONDING BITUMEN

Oxidised bitumens are used by the roofing trade as bonding and coating bitumens and are identified by a two figure system, for example 95/25 bitumen. The first figure refers to the softening point of the bitumen in degrees centigrade when subject to a ring and ball test. The second figure refers to the result of a penetration test.

To carry out the ring and ball test, melted bitumen is poured into a tapered brass ring and allowed to cool. A 3/8th inch steel ball is placed centrally on the bitumen and the apparatus is heated in a suitable liquid, the temperature of which is increased at a rate of 5°C per minute.

The softening point is the temperature at which the bitumen has softened sufficiently for the steel ball to fall through the ring for a distance of 25mm and is an indication of temperature susceptibility.

The penetration test measures the distance in hundredths of a centimetre to which a standard needle penetrates into the bitumen under a load usually of 100grammes for 5 seconds at 25°C.

Both figures in the classification give an assessment of hardness, one at 25°C and the other at the softening point, which will be at a much higher temperature. Together they provide a good indication of the performance of the bitumen.

Ring and ball test

Penetration test

The bonding bitumens in normal use are listed below:

Grade	Softening point °C	Penetration at 25°	
10/20	63 - 73	10 - 20	Penetration grade bitumen
85/25	80 - 90	20 - 30	Oxidised bitumen
95/25	90 - 100	20 - 30	
105/35	100 - 110	30 - 40	
115/15	110 - 120	10 - 20	

10/20 is a penetration grade bitumen, too soft to be supplied in block form, but occasionally used from bulk deliveries by tanker. 10/20 does not have the rubbery characteristics of oxidised bitumen and is only suitable for flat roofs to nominal falls.

85/25 and 95/25 are the conventional general purpose oxidised bitumens for bonding. These grades are similar and are generally considered to be interchangeable. They are normally supplied in block form to be re-melted on site in a heated boiler but may also be supplied hot for bulk delivery by tanker.

105/35 bitumen is used by most leading manufacturers as a coating bitumen for roll roofing, and has only recently been made available as a bonding bitumen. It is more rubbery than 95/25, acts as a more powerful adhesive, is more resistant to slip and creep and is more flexible in cold weather. 105/35 is an improved all-round bonding bitumen but in practice there are some difficulties in application which prevent widespread use. It is hard to break up, can be slow to re-melt and is less suitable for pumping to roof level or for use in rooftop pouring equipment.

115/15 is the hardest bitumen and a strong adhesive in hot conditions. It is used as a general purpose bonding bitumen for roofs with a slope of 20° or more and is sometimes used on lesser slopes when improved bonding and resistance to slipping are required.

THE MANUFACTURE OF ROOFING FELTS

The manufacture of roofing felt is a continuous process involving the impregnation of those bases which are susceptible to moisture absorption with a penetration grade bitumen and coating with a filled oxidised bitumen to provide the waterproofing. Bases which require saturation to protect against the inclusion of moisture in the final roofing include fibre, asbestos and polyester.

The saturated felt is then coated on both faces with a relatively thick layer of oxidised bitumen and filler, which provides the waterproofing layer. The filler is incorporated in the coating to stabilise the bitumen and the thickness of the bitumen is controlled by a pair of nip rollers.

Glass fibre materials do not require impregnation; they are manufactured by passing the base through the coating stage only.

Sand surfacing is applied to the roofing felt to prevent sticking within the roll. For mineral surfaced roofing a range of coloured mineral aggregates, is applied and correct control of the coating bitumen temperature is necessary to allow the mineral surfacing to adhere to the roofing felt.

The felt is then cooled by passing over cooling rollers so that the felt can be reeled and wrapped without blemishing or bleed-through of warm bitumen.

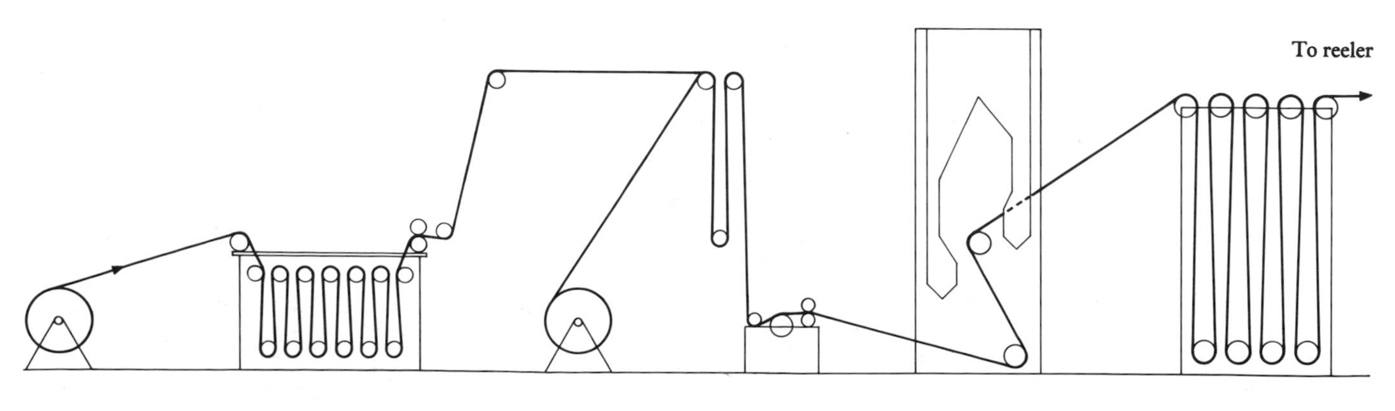

Manufacture of roofing felt

FELTS MANUFACTURED TO BS 747

BS 747 classifies felts according to base and surface finish by a number and letter system.
The number refers to the base material: Class 1 is fibre base, Class 2 is asbestos base and Class 3 is glass fibre base.
The letter refers to the type or finish of the felt. Type B is general purpose sand finished felt, Type E is mineral surface felt and Types G and H are perforated felts.
Each class of felt can be identified by the coloured sand strip which is applied to one edge during manufacture. An additional class, Class 4, includes black sheathing felt which is used as an underlay to isolate mastic asphalt from the substrate material.

Felt type	Surface finish	Weight/roll
Class 1 Fibre base (colour code white)		
1B	Fine granule surfaced	36kg/20m x 1m
1E	Mineral surfaced	38kg/10m x 1m
1F	Hessian reinforced base with fine granule surface	22kg/15m x 1m
Class 2 Asbestos base (colour code green)		
2B	Fine granule surfaced	36kg/20m x 1m
2E	Mineral surfaced	38kg/10m x 1m
Class 3 Glass base (colour code red)		
3B	Fine granule surfaced	36kg/20m x 1m
3E	Mineral surfaced	28kg/10m x 1m
3G (perforated)	Grit finished underside fine granule surfaced topside	32kg/10m x 1m
3H (perforated)	Fine granule surfaced	34kg/20m x 1m
Class 4 Sheathing felts		
4A (i)	Black sheathing felt (bitumen)	17kg/25m x 810mm

The problems associated with BS 747 felts which have beset the industry for many years are now much more clearly understood and can be avoided. It would be wrong to discount specifications comprised of BS 747 roofings, as a long life in service can be achieved by these felts when they are used under appropriate circumstances and with the correct specification. Some BS 747 felts are however so overtaken by improved alternatives that they have little use in roofing for permanent buildings.

Class 1 - Fibred based
These are the original felts used by the industry. They are still the cheapest roofings but they have failings which make them poor value for money for any purpose and it is suggested that these should no longer be included in built-up roofing.

Class 2 - Asbestos based
Some authorities require a layer of asbestos roofing in a multi-layer system in the belief that this will improve the fire performance. In some cases this may be achieved, although the addition of stone chippings gives an AA rating to most specifications. Nevertheless asbestos roofings may remain a useful component of some built-up systems as they maintain their integrity in fire better than most other roofings. Type 2B asbestos roofing is still used for the first nailed layer for some specifications on a timber deck. It is not a strong material but it can be used under these circumstances with confidence where the exposure to wind is not too severe. Asbestos roofing may also prove useful as an underlay or as a vapour check beneath insulation on concrete or screeded decks.
Asbestos felts are more expensive than glass base felts and a specification of three layers of asbestos type 2B material is rarely used for flat roofs. There is however still a use for asbestos felts on sloping roofs as they can be nailed and held at the top of the slope. A complete roof of asbestos base felts including a mineral surface type 2E cap sheet forms the minimum quality specification for sloping roofs although it is prone to blistering and high performance cap sheets are likely to prove much superior.

Class 3 - Glass fibre base
Glass base roofing can be expected to give the best waterproofing performance of the BS 747 felts and three layer glass base specifications will probably remain in use as the cheapest form of flat roof covering. This specification should give many years of good service if it is applied in accordance with good roofing practice on a firm substrate. Glass base roofings are not suitable for nailing as they do not have enough strength.

BS 747 type 3G is a perforated and gritted venting roofing that is used for the first layer in partially bonded specifications. BS 747 type 3H is also a perforated and sanded material but it is not widely used in the roofing industry. It can be used where a temporary separating layer is required, such as on bridge waterproofing, where the development of a full bond in time does not matter.

HIGH PERFORMANCE BITUMEN FELTS

Recent development work on bituminous built-up roofing felts has been concerned with improving two main areas of deficiency in BS 747 roofings; the age hardening of the bitumen and the long term strength or elongation properties of the felt base. Many other aspects can also be improved, such as handling characteristics, cold weather flexibility, hot weather firmness, tear resistance and puncture resistance. There are two separate lines of approach: improving the base, and improving and modifying the bitumen.

Improving the base
Most roofings are weakest in the machine direction and BS 747 felts always tear or split most easily in their length. Originally a low quality and weight of glass tissue was chosen for BS 747 Class 3 felts, but the revisions to the British Standard in 1977 required a heavier glass base with a modest strength test to control quality. As a result of the revision, the strength of the felt in the cross direction is almost doubled, and comes near to the strength in the length direction, which remains roughly as it was. The tissue still has a grain or natural line of cleavage, but there is no doubt that the specification is more resistant to splitting.

There is a trend towards heavier glass bases on grounds of additional strength. A heavy woven glass base is particularly strong in tension although the tear strength is not great.

The most popular line of development has been the use of polyester to provide a non-rotting base, with greater strength than glass, and greater elongation at break. Polyester bases appear to be stable and strong for a long life, and to be compatible with the bitumen. Elongation at break is in the order of 40% which is much higher than needed, as age hardened bitumen will fail long before this, resulting in a breakdown of waterproofing even though the base is still sound. The strength in the machine direction is in the order of 600N/50mm width, as opposed to the British Standard glass roofing at 200N/50mm width.

Improving and modifying the bitumen
The ageing and weathering characteristics of bitumen over a period of time are well-known. Bitumen will crack or craze in a crocodile skin formation from the combined effects of ultra-violet light, ozone and oxygen. This ageing process leads to increasing hardness which can be expected to take place over about 15 years. The rate of hardening is greatest in the early years with significant hardening taking place in the first three to five years. The fatigue resistance of bitumen decreases in this time and a bituminous membrane becomes much more vulnerable to splitting from fatigue as the ageing process continues.

The addition of a polymer to the bitumen improves its properties as a roofing material in almost all respects, and in particular the flexibility, strength and fatigue resistance. Polymer modified bitumens are usually applied to a base of polyester or glass.

The most commonly used modifying additives are SBS (styrene butadiene styrene) and APP (atactic polypropylene).

Roofings with SBS additives have the greatest elasticity and elongation and generally involve conventional hot bitumen bonding techniques.

Roofings with APP additives have improved high temperature performance and can have better weathering characteristics. They are not suitable for bonding with oxidised bitumens and are generally bonded by torching.

Modifications to the bitumen have also led to the development of roofings which are not strictly classifiable as felts in that they do not need a base to act as a carrier or reinforcement for the bitumen. Bitumen polymer and pitch polymer sheetings are examples of such materials with no base. They are manufactured by a cold mixing and calendering process. Both materials are strong, tough and flexibile. They are bonded with bitumen though high melting point bitumens are necessary to prevent lap separation and movement.

Metal foil surfaced felts
High performance materials with a polyester or woven glass base are available with aluminium or copper facings. Metal facings give an extremely effective protection to the membrane, as they totally exclude ultra violet light, oxygen and ozone - the factors which are most instrumental in the ageing and hardening of bitumens. Aluminium also has good reflective properties and effectively reduces the temperature of the bitumen in service.

Foil faced felts are suitable for nailing but on sloping roofs the metal facing can slip or slide at the interface with the bitumen backing. Most manufacturers indicate a maximum sheet length in the order of 2 to 2½ metres. Slope lengths which exceed this dimension must have nailing battens at intervals down the slope to allow nailing at the top of each length of material.

Aluminium faced roofing is frequently used for the formation of detail work and is also sometimes applied to flat areas of roof. In this case the drainage must be efficient to prevent corrosion from standing water and chemical pollution. Aluminium faced roofing should not be used for roofing or detail work near cement works or in similar alkaline atmospheres.

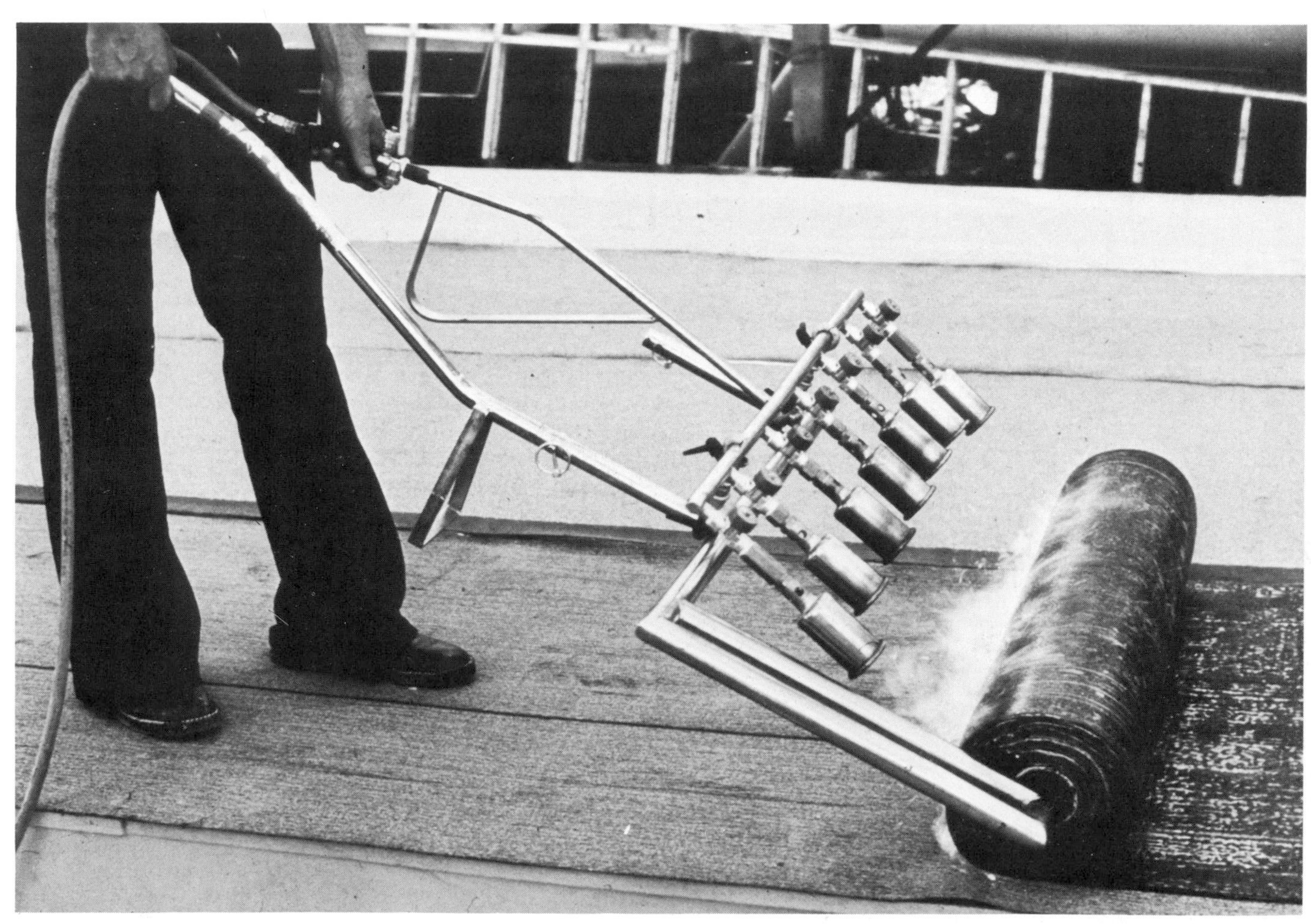

Roofings with APP additives are generally bonded by torching

3.2 APPLICATION TECHNIQUES

Bitumen preparation
Bitumen is heated in boilers to a temperature in the range of 200°C to 250°C; higher temperatures should be avoided due to the danger of fire. The boilers are usually placed at ground level but can also be placed on the roof, provided adequate fire precautions are taken and the roof area is suitably protected. The bitumen may be pumped from ground to roof level or transported in buckets, wheeled containers or dispensers.

The handling of hot bitumen requires rigorous safety procedures and the work must only be carried out by roofers trained to the necessary skills.

Priming
Metal decks, concrete decks and cement screeds can present dusty or damp surfaces which will increase the difficulty of forming a satisfactory bond.

If the surfaces are dry and free from dust, there should be no requirement for priming. Failure under test should be due to a breakdown of the bitumen rather than a release at the interface with the deck.

In damp or dusty conditions it is usually possible to carry out the work providing suitable priming is applied to bind any loose surface. Although bitumen will not bond to primer which is wet from rainfall or heavy dew, a reasonable bond will be achieved if there is merely a bloom of dampness of the sort which frequently occurs in winter.

Laying the roofing
Hot bitumen is poured in front of the roofing which is then unrolled into the bitumen, spreading it to give a continuous coating for the full width of the roll. This technique known as 'pour and roll', is also used for applying underlays, vapour checks and vapour barriers.

Alternatively the bitumen may be applied by mop, but most coated roofing materials will only be satisfactorily bonded by this method if the bitumen is applied to the back of the roofing and to the surface to which the roofing is to be applied. Mopping is generally only used for detail work and the main areas are usually applied by the pour and roll technique. Detail work may also be carried out by pour and roll using jugs to hold a relatively small quantity of hot bitumen and ensure a safe and accurate application.

Traditional pour and roll application of roofing

Most bitumen roofing is applied with 50mm side laps and 100mm end laps. On sloping roofs, the roofing is applied from ridge to eaves, but on flat roofs to falls the direction of application will not necessarily relate to the direction of fall. The formation of details, the economic use of materials and the need to leave a satisfactory edge for closing off at the end of a day's work will all be taken into account by the experienced roofer when deciding the direction and patterning of the roofing.

The laps of the differing layers will be staggered to ensure maximum lap security and minimum build up of thickness. This is achieved by starting the first layer as a narrow width against the eaves or skirting, using a one-third strip off the width of the roll, then continuing with full width rolls for the rest of the first layer. The second layer starts with a two-thirds strip of roofing, probably the off-cut from the previous cut roll, and with full width rolls thereafter. The third layer then starts and continues with full width rolls. Various patterns of laying are employed on site but the end result will always be to achieve staggered laps.

The first layer will be fully bonded or part bonded according to the nature of the substrate. Part bonding is most usually formed by the use of BS 747 type 3G perforated, gritted roofing. This is applied loose and the bond is achieved by rolling and pouring the second layer over the perforated layer, so that hot bitumen runs through the perforations to form a spot bond.

Care is taken to keep the perforated roofing flat and close to the substrate during application to prevent too much bitumen flowing under the layer. This would establish too great an area of bond and would not provide the efficiency of separation required.

It is undesirable to carry out large areas of first layer work as a temporary waterproofing. Building supervisors are often tempted to press the roofing contractors to adopt this practice, but the quality of finished work is bound to suffer by the entrapment of dampness, dust and dirt deposits. The best procedure is to complete all the layers of roofing as the work proceeds and to add the protective surfacing as soon as possible after application of the built-up roofing. An underlay or vapour check can be used to give a temporary waterproofing, as the consequences of small traces of entrapped damp or dirt at the interface of this layer and the insulation should not prove a cause for concern.

Sloping roofs

Specifications on sloping roofs of 10° and over should be nailed at the top of each cap sheet into timber nailing battens which must be included in the detailing of the ridge or top of the slope. Nailing is necessary at 50mm centres along the top of each sheet. This pattern of nailing is cheap and easy to apply, is of prime importance and should not be neglected by the designer and roofer.

When the specification is nailed to prevent slip, it is the shank of the nail which holds the roofing. The head of the nail does not retain a useful grip on the roofing and the number of fixings cannot be reduced by the incorporation of large washers. If screw fixings are used, these again need to be at 50mm centres.

115/15 bitumens should always be used when the slope exceeds 20° and sometimes as an extra precaution on lesser slopes which are exposed to heat trap conditions or when the slope length is considerable. The change of bitumen from the conventional 95/25 bitumen for flat roofing is in addition to nailing.

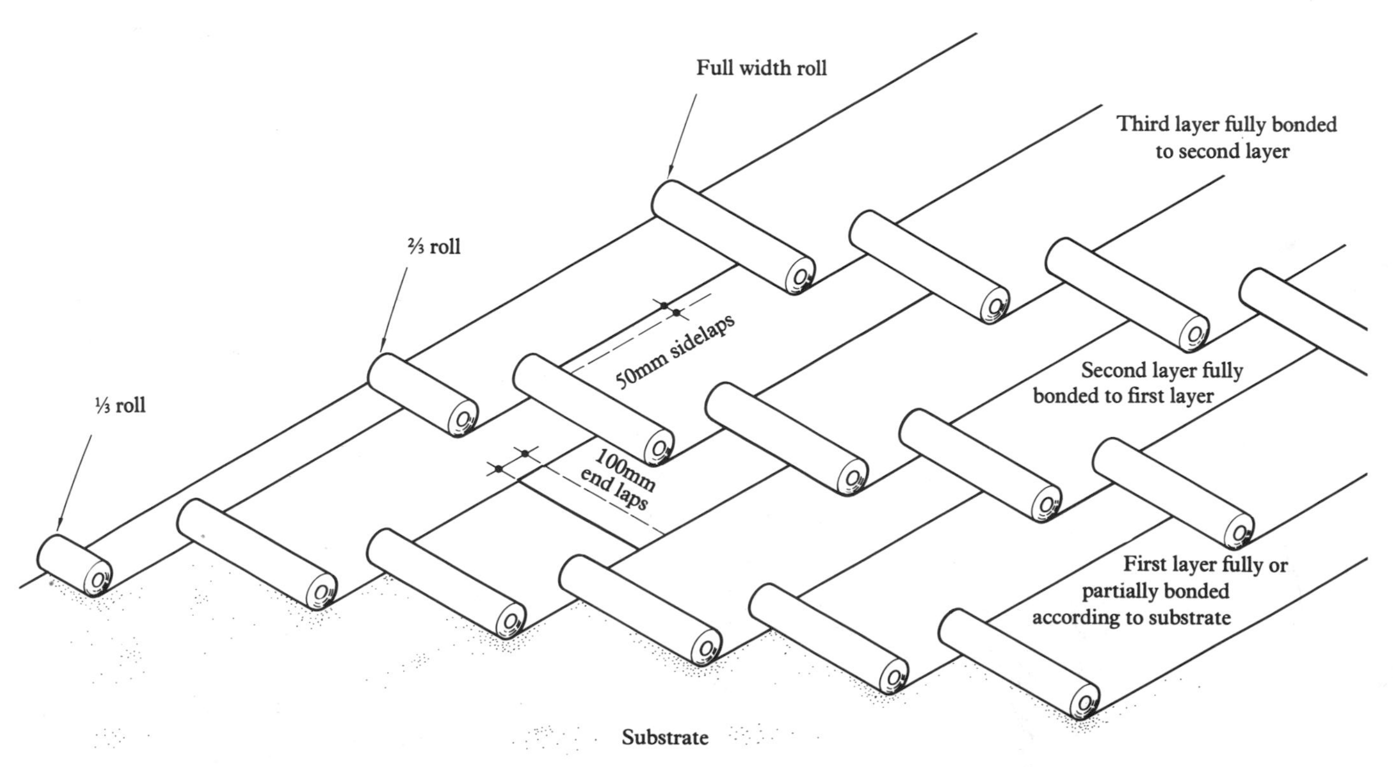

Formation of staggered side laps

Mineral surfaced roofing provides good protection for sloping roofs. Fibre base type 1E materials and asbestos base type 2E materials are prone to severe blistering and glass base felts or high performance alternatives are far superior. Mineral surfaced glass based type 3E materials perform well but they cannot be nailed and therefore are not suitable for vertical or sloping work unless the slope is shallow and of no great length. Opinions vary on the limiting slope and length but unless manufacturers advise to the contrary the maximum slope may be taken as approximately 10° coupled with a maximum slope length of about 5 metres. As glass fibre felts cannot be mechanically fixed they should be bonded in 115/15 bitumen for all sloping roofs.

Polyester felts are available with a mineral surface finish and most are suitable for mechanical fixings at the top of the slopes to prevent slip.

They may require fewer fixings than BS materials: follow the manufacturer's recommendations.

Metal faced membranes also provide good protection for sloping roofs. Care is needed in the application of these materials. They necessarily become hot during laying and are easily damaged in this condition as the bitumen ceases to provide firm support. Metal faced roofing cannot normally be applied without some creasing and deformation.

The laps of foil faced roofing must be pressed together firmly by roller or trowel during application. This will deform the foil locally and squeeze a little bitumen from under the lap. The black appearance may be concealed by the application of paint or metal powder, although powder is sometimes regarded as dangerous to work with. Copper facing has more spring than aluminium and may exhibit a greater tendency to open at the laps.

Applying the insulation

Most insulations are bonded in hot bitumen, applied by pouring or mopping onto the substrate or onto the back of the board or both as necessary. Straight joints must be avoided by staggering to minimise joint movement and a full half stagger on the longitudinal joints is to be preferred. The grain of fibres in roofing felts generally tends to be in the length direction of the roll leading to a natural tendency to tear down the length of the roll rather than across it. A diagonal pattern of laying in relation to the waterproofing has the advantage that all joints over which the waterproofing is applied are at 45° to the length of the roll and the weakest direction is avoided.

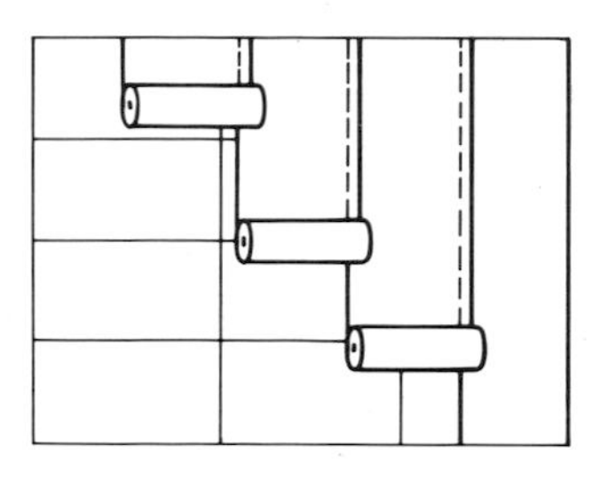

Not recommended

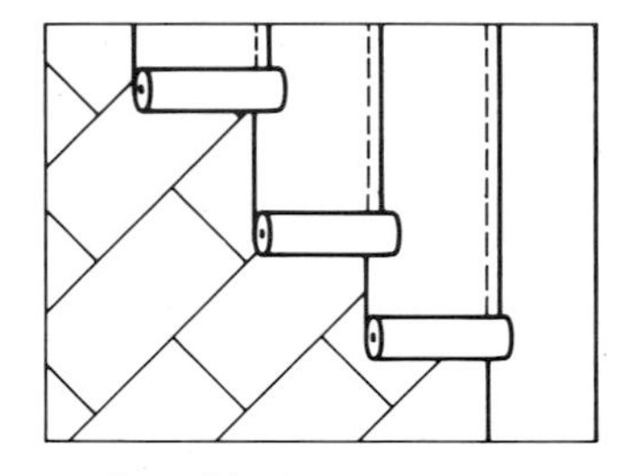

Preferred laying pattern

On metal decking, boards should only be placed in line with the troughs if the board size is a multiple of the trough centres so that all longitudinal joints are fully supported by the top flats of the decking. Unsupported longitudinal joints over the open troughs of the decking are undesirable and in certain cases can increase the risk of splitting or fracturing the waterproofing at this point. The diagonal pattern of application avoids the danger of unsupported board joints and improves the load distribution to the decking so that the finished roof feels firmer under foot. The worst structural condition for the board will be a T joint over an open trough, but in practice this does not lead to a significant lack of support for the waterproofing provided that the insulation is strong enough.

Partial bonding using BS 747 type 3G perforated gritted roofing

The edge of mineral surfaced roofing is usually wetted before the adjacent roll is applied. Excess bitumen from the lap will then be easy to remove while warm, to leave a clean finish.

Bonding of insulation to solid decks

As with many aspects of roofing work, there are a variety of application techniques for achieving a bitumen bond of insulation board to the deck. A continuous coat of hot bitumen can be applied to the back of the insulation board which is then laid onto the deck. The bitumen should be applied with sufficient thickness to ensure a satisfactory contact area.

The application of the insulation to a vapour barrier or underlay is usually by hot bitumen bonding applied by mop or by in-situ pouring to leave a pool of bitumen which is then spread with the edge of the insulation board which is about to be laid.

Bonding of insulation to metal decking

On metal decking an alternative method of bonding the insulation is with poured bitumen direct to the deck. The recommended method is to use dispensers which pour bitumen from spaced holes. The insulation is then applied direct to the bitumen and a good contact is obtained in strips which normally spread to widths between 25mm and 50mm.

This method is probably the most reliable and effective method for achieving a good bond but application of bitumen must not progress ahead of the insulation to the extent that the bitumen is cooling to form a skin. The new generation of metal deck profiles include stiffening ribs in the top flats and these form natural channels for the collection of bitumen. This thins the bitumen poured onto the top flats and can leave insufficient for a satisfactory bond. It is therefore necessary to pour a sufficient quantity of bitumen to fill the stiffening ribs and leave a surplus on top of the top flat for bonding.

Another method frequently adopted is to pour hot bitumen from a can in a serpentine movement across the deck. This method is wasteful of bitumen as much of it ends up in the open troughs of the deck, where it serves no useful purpose. Problems also arise from the temptation to make fast progress resulting in open loops of bitumen and too much discontinuity in the bond, in which case the nominal $2.4kN/m^2$ design load may not be justified.

Similar techniques to the above are used for the bitumen bonding of an underlay, vapour check or the first layer of a vapour barrier on metal deck and the aim is to apply bitumen continuously to the top flats of the deck and to roll the underlay in the hot bitumen.

Testing the bitumen bond

The bitumen bond of the insulation should be tested at random during the progress of the work and should withstand a strong pull using the normal strength of a man before the board releases. Examination after a test will usually reveal substantial patches of deck untouched by the bitumen but this is normal and the important criterion is the strength of bond overall. If the board releases easily when pulled, it is necessary to examine the form of failure to ensure that sufficient bitumen has been used. It may be found that a primer is necessary, or the failure could have occurred by delamination of the insulation. Under these circumstances the insulation may need to be rejected as having insufficient strength.

3.3 SURFACE PROTECTION

Flat roofs are particularly exposed to the effects of solar heat gain during the day and heat loss by radiation at night. When the roof covering is of a dark and non-reflective material such as a self-finished bitumen felt, a large amount of solar radiant heat can be absorbed which can raise the temperature of the surface considerably above that of the surrounding air. Conversely, radiation heat loss to a clear sky at night can cause the temperature of the surface to be about 5°C lower than the air temperature. It is not, therefore, unusual for a roof to be subjected to a temperature range from -10°C to +80°C.

The effects of solar heat gain and radiation heat loss are to increase the thermal movements in the roof structure and to subject the waterproofing material itself to cycles of softening and hardening. These effects can be reduced by applying a suitable reflective surface finish to the roof.

In addition to its influence on the temperature stability of the roof construction and waterproof covering, a surface protection layer is required to reduce the ageing effects that ultra-violet radiation and weathering have on most materials used in built-up roofing. A surfacing may also be required to protect the waterproof covering from damage by pedestrian traffic and provide an aesthetically pleasing finish to a roof that will be overlooked.

Stone chippings
The all-purpose surfacing preferred by most of the roofing industry is a layer of 10mm or 14mm stone chippings bonded in hot bitumen or a cold bitumen solution. This surfacing gives excellent protection from ultra-violet light due to the density of the stones. The chippings also act as a heat sink to slow the process of temperature change and reduce the extremes of temperature. Most stone chippings have reflective properties which significantly reduce the surface temperature in summer.

The lightest coloured chippings can have the same order of reflective efficiency as white reflective paint, but after some years they may become dirty and the efficiency of the reflection is reduced to that of medium grey chippings.

The weight of 10mm chippings coupled with the temperature control which they provide is sufficient to prevent blistering in the majority of cases and the weight alone helps to provide security against wind forces.

Although stone chippings give the necessary protection to a roofing membrane, there are disadvantages to be faced. Loose chippings can block rainwater outlets and tightly bonded chippings make inspection and fault tracing more difficult.

It has become common practice to try to achieve maximum adhesion of the chippings, mostly at the insistence of building owners or their site representatives. As a result it is extremely difficult to trace the position of a leak, sometimes with the unfortunate result that the roofing is renewed entirely as the only practical method of repair.

°C
60
Black surface
40
20
0
New white chippings
24 hours

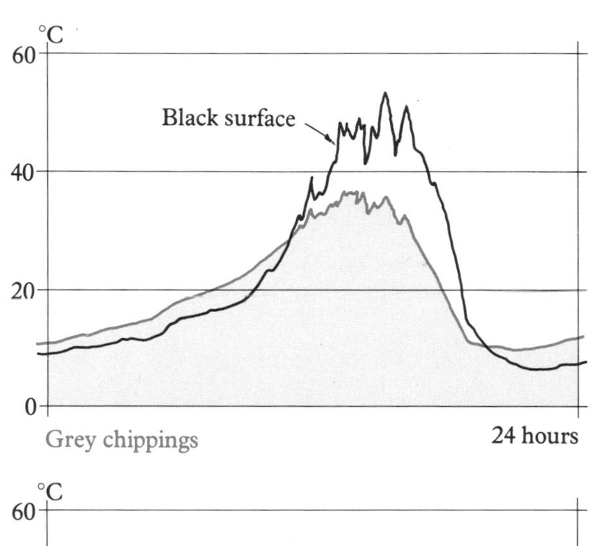

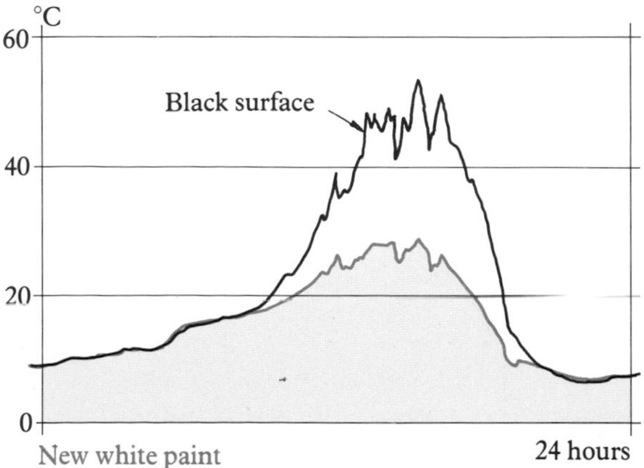

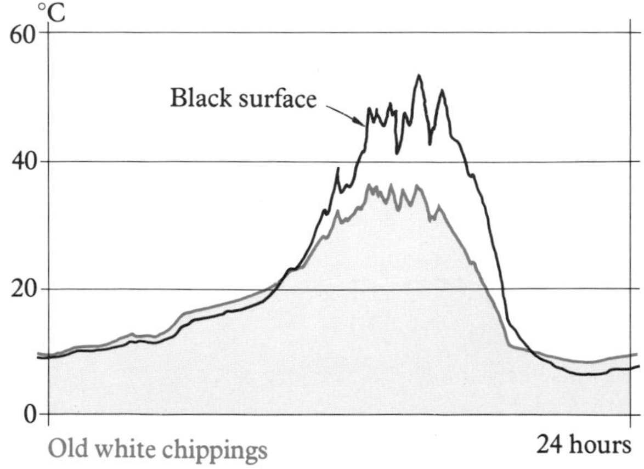

Temperature variation (°C) at waterproofing level for different surface finishes measured over a typical 24-hour period in summer.

The alternative is to aim at a key rather than a bond and although this may give a proportion of loose chippings this will be acceptable in the majority of cases. The outlets will not block provided that a satisfactory grating is positioned before the chippings are applied. It is sensible practice to set the chippings back from the outlets or form a firm bond in the immediate vicinity. For roofs subject to wind scour, a thorough bond of the chippings overall may be required.

The use of hot gritting solutions generally leads to an extremely solid bond and is not always to be recommended. Ordinary hot bitumen applied at the rate of approximately 1.5kg/m^2 will give a satisfactory bond for chippings provided they are dry and free of dust. An alternative bonding agent is cold applied bitumen solution which is more tolerant of damp chippings and will give a suitable degree of bond provided it is not applied too thickly.

Some roofing felts have a loose sand surfacing which prevents the bonding agent from making a wet contact with the membrane: the application of hot bitumen spread thinly with a squeegee will bind the surface and act as a primer for the cold applied adhesive.

Some designers are concerned about the possibility of the chippings penetrating a built-up roofing membrane as a result of foot traffic. In practice this is very rare after the surfacing is completed as the load is spread onto a large number of chippings. Hammer tests to represent heavy impact from traffic will demonstrate that the chippings do not penetrate the membrane.

Promenade surfacing

Where pedestrian traffic is anticipated, a surfacing of tiles or paving slabs can be used. All tiled or paved roofs must be drained efficiently with currents and falls as necessary. Small amounts of standing water or dampness will lead to mould growth and ice formation in cold weather, both of which make extremely dangerous conditions for pedestrians and can lead to the break up of the paving in severe weather. Falls of at least 1 in 80 are necessary to ensure adequate drainage. However, tiles and slabs can creep or slip when bonded in bitumen and unless precautions are taken to reduce the chance of slipping in hot weather, a maximum fall of 1 in 40 is advisable.

Asbestos cement or glass reinforced concrete tiles provide a lightweight promenade surfacing. They are applied in a thick coating of hot bitumen and the back of the tiles should be bituminised before placing to ensure a continuous bitumen bonded contact. It will also be necessary to treat the backs of the tiles with a bitumen primer if they are damp. Intermediate expansion gaps across the roof are not normally required as the tiles will be installed with a gap of a few millimetres at each joint to allow a tolerance for inaccuracies. They form a stable surface on three layer BS 747 glass base roofing as the laps of this specification do not stand proud. Many of the high performance materials however are extremely thick and form laps which stand proud causing tiles to rock over the lap. Attempts are sometimes made to overcome this problem by butt jointing the cap sheet but this is a departure from the proven principles of built-up roofing and is not recommended. It is preferable to choose specifications which use the thinner high performance roofings.

Concrete pavings, 25mm minimum thickness, provide a heavier and more stable promenade surface. They may be bedded in a thick layer of bitumen, but it is not possible to obtain a standard of flush finish to compare with that achieved when paving is bedded in sand or sand and cement. When concrete tiles or paving slabs are bedded in sand and cement, it will first be necessary to overlay the waterproofing with a building paper to allow differential movement between the waterproofing and surfacing. Polythene sheeting is sometimes applied to act as a slip-plane but this tends to cling tightly to the waterproofing and does not provide such a satisfactory separation. An allowance for expansion is necessary and it will usually be sufficient to set the tiles or slabs back 25mm from the vertical at the roof perimeter, and around major details, with intermediate joints at 3m centres.

Concrete tiles may be bedded on proprietary plastic corner supports. Irregularities of level can be made up by bitumen felt pads forming shims. The great advantages of these systems are the effective separation of the promenade surface from the waterproofing, the rapid drainage of surface water and the easy access to the waterproof covering should inspection and repair prove necessary. The ease of removal can however be a disadvantage on buildings where vandalism is likely.

Chippings in the immediate area of the outlet should be firmly bonded.

For roofs subject to wind scour, a thorough bond of the chippings overall will be required.

Certain specialist companies apply an in-situ sand and cement paving cut into a tile pattern before the material hardens. This paving must again be separated from the membrane by building paper to allow differential movement of the paving over the membrane. In-situ concrete pavings should only be undertaken by skilled and experienced specialist applicators.

Reflective coatings
In recent years reflective coatings have been used as treatments on roofing as a replacement for mineral surfaced roofing or stone chippings. The coating can give a measure of protection to the built-up roofing membrane while it remains in good condition but this is not likely to exceed a few years and significant bare patches may occur even after one or two years. Because most built-up roofing depends on a permanent protection, designers must be sure that if reflective coatings are to be specified, the building owner clearly understands the need to clean and re-coat the surface every few years as soon as significant bare patches appear. The permanent protection of stone chippings, paving, or factory applied finishes are much to be preferred.

It should be remembered that surface coatings may alter the external spread of flame rating to BS 476 (See Section 1.8), and this should be checked with the manufacturer to ensure that the coating does not reduce the rating to unacceptable levels.

Although certain coatings achieve an excellent rating for spread of flame it is recommended that no dependency is placed on the coating to improve the fire characteristics of the membrane. It can never be certain that the coating will be replaced with a coating of similar fire performance at a future date as the implications of the provision of fire protection are not likely to be appreciated or remembered by maintenance staff.

Roof gardens and terracing
The waterproofing under a roof garden or immovable terracing should be regarded as a buried membrane which is inaccessible for inspection and repair. Faults are extremely difficult to trace and expensive to repair. Flat roofs should only be used as a base for roof gardens or terraces when the roof design can comply with all requirements of good practice without compromise and also when the building owner is strongly financed and able to face up to costly repairs in the event of failure. A number of proprietary high-performance specifications are available and the advice of the manufacturers should be followed.

Asbestos cement or glass reinforced concrete tiles provide a lightweight promenade surfacing.

Proprietary support systems are available for concrete tiles.

3.4 TYPICAL SPECIFICATIONS

STANDARD SPECIFICATION

BS 747 Class 3 felts
Applications to concrete, cement screeds, woodwool, plywood and chipboard decks and polyurethane and polyisocyanurate insulation boards will require a partially bonded specification. The first layer will be BS 747 type 3G gritted perforated roofing, followed by two layers of type 3B roofing. The main virtue of type 3G in providing a part-bond is the control which is exercised on the extent of the bond and the certainty that the material will not develop a full bond through high sun temperatures in service. On timber boarded decks the method of achieving partial attachment will normally be by nailing a first layer of BS 747 type 2B asbestos base roofing.

Fully bonded specifications are suitable on wood fibreboard and cork board and the specification will be three layers of type 3B roofing, all fully bonded in bitumen.

Stone chippings are always recommended as they provide a protection which is likely to last, but where they are for some reason inadmissible, mineral surfaced roofing type 3E has given satisfactory results.

This specification has shown good resistance to blistering but the designer should not forget the limitations and design requirements which will be necessary to provide security against wind forces.

Glass fibre, mineral wool and perlite board are compressible or fragile and easily damaged. A specification on these insulations formed entirely from BS 747 glass base roofing would be too fragile, particularly as it will harden considerably over the years and become easily damaged as the ageing process continues. It is therefore recommended that an upgraded or high performance system is used when overlaying these insulants, as described in the following pages.

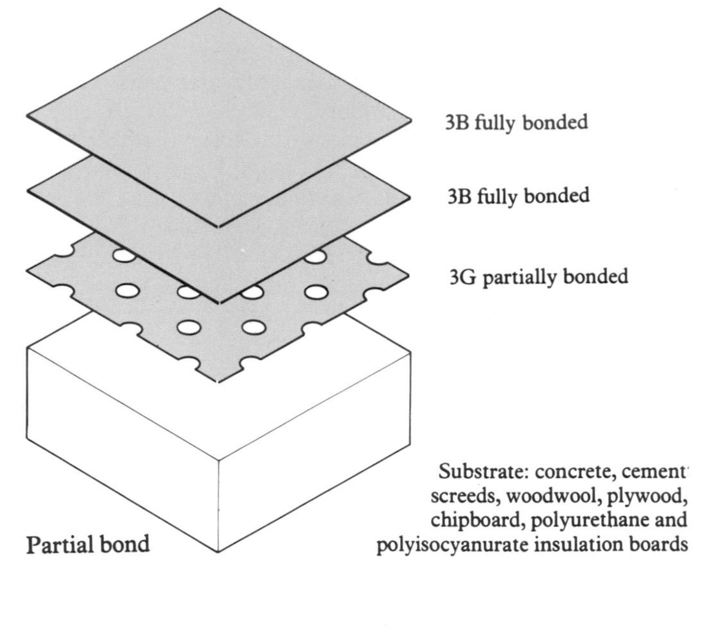

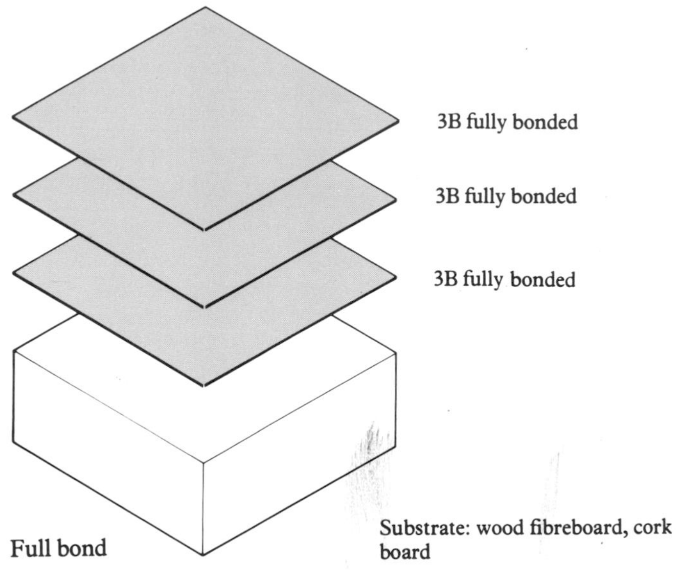

UPGRADED SPECIFICATION

The 3-layer BS 747 glass base specifications described above are the minimum acceptable quality built-up roofing systems.The next step forward is to substitute a layer of polyester base roofing or similar high performance roofing for the cap sheet. This is a useful reinforcement of the system and it can be taken as the minimum specification on those insulations which do not provide suitable support for an all BS 747 type 3B glass base specification, that is glass fibre, mineral wool and perlite board.

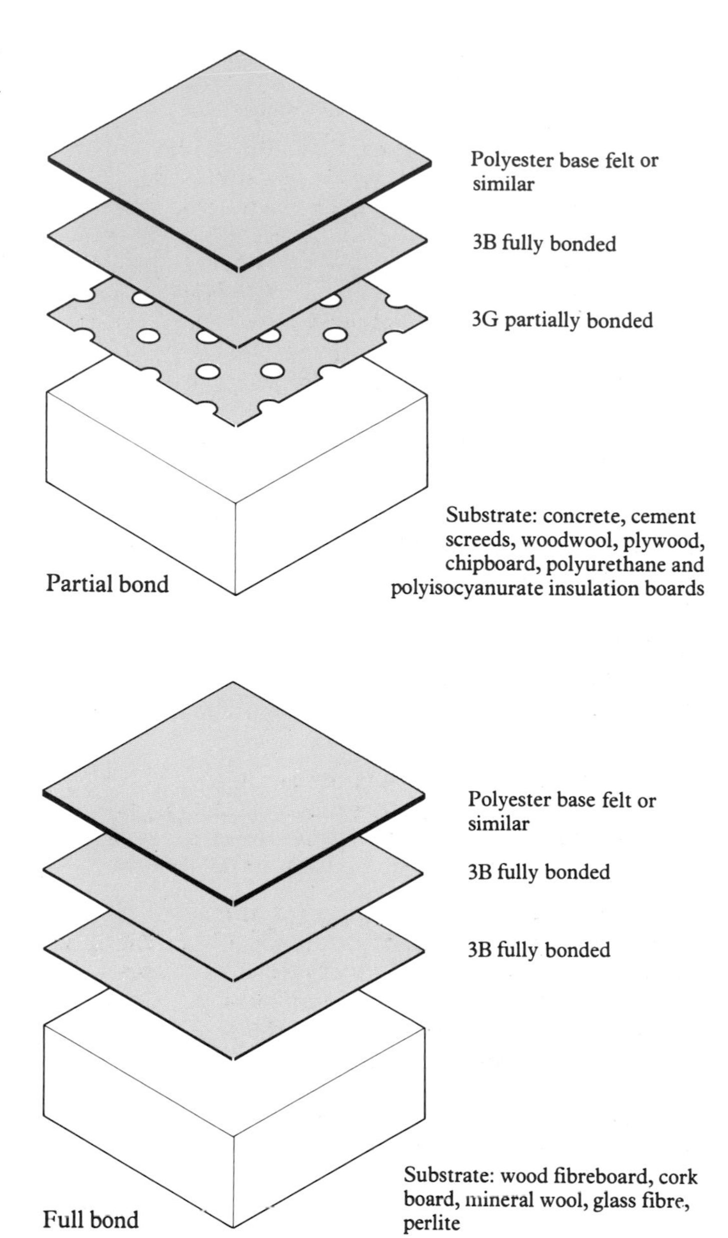

Polyester base roofing systems

Polyester base systems have established themselves as true high performance specifications. Fully bonded specifications consist of two layers of polyester base roofing. Part bonded specifications always include in addition a first layer of BS 747 type 3G perforated gritted roofing, making three layers in total.

Bitumen polymer/pitch polymer membranes

There are a number of systems where the high performance material is used as a single layer on a base of BS 747 felts. These include pitch polymer or bitumen polymer sheeting. The fully bonded specification will normally comprise one layer of glass base type 3B roofing with one layer of the high performance material. For a partial bond, the addition of a first layer of type 3G perforated gritted roofing gives the part bond and again results in a three-layer overall specification. A single layer of high performance material on a single layer of type 3G is not normally recommended.

Polyester base roofing systems

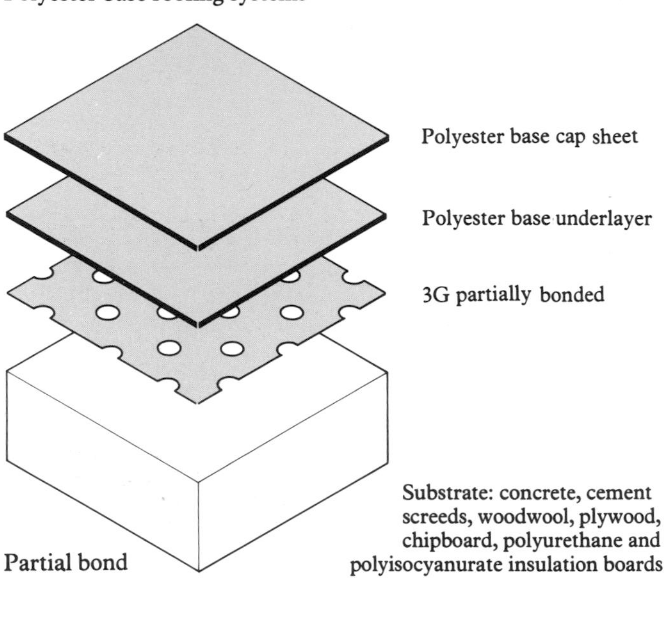

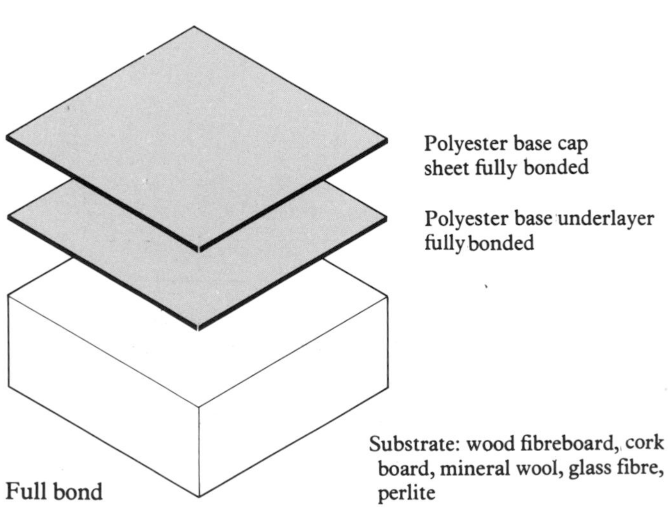

Bitumen polymer/pitch polymer membranes

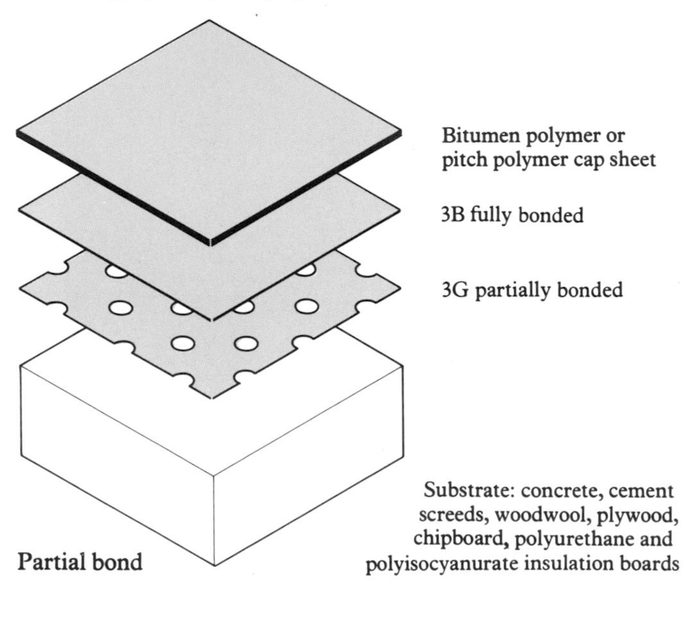

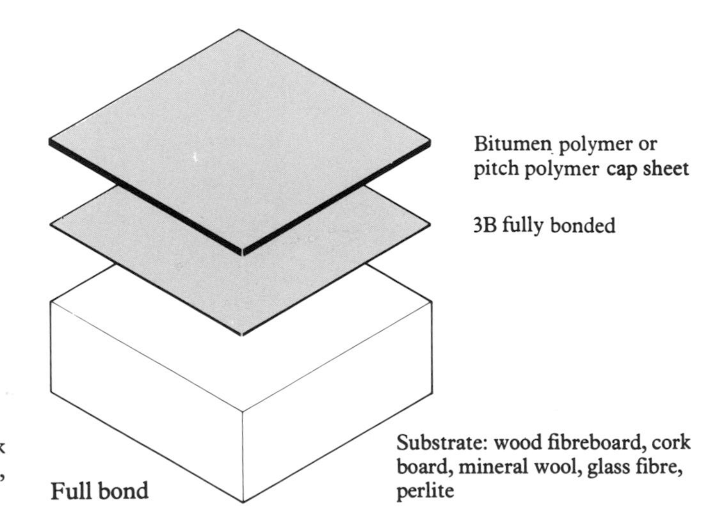

Torch-on roofing systems
Torch-on materials are normally applied as a cap sheet to fully or partially bonded underlays depending on the nature of the substrate.

It will be usual to use a first layer of high performance or BS 747 type 3B roofing bonded in bitumen for a full bond to the substrate. If the substrate is one which normally requires a partial bond, this will require an additional layer of BS 747 type 3G.

Two layer systems of torch-on materials are also available. For part bonding, the under surface of the first layer may include polystyrene beads which provide the separation. The torch is applied to patches at about 450mm centres in both directions to melt the bitumen and polystyrene and form the partial bond. These materials are generally more substantial than BS 747 type 3G perforated gritted and being unperforated are suitable for a two layer part bond system direct to deck or for re-roofing on top of existing membranes.

Torch-on roofing

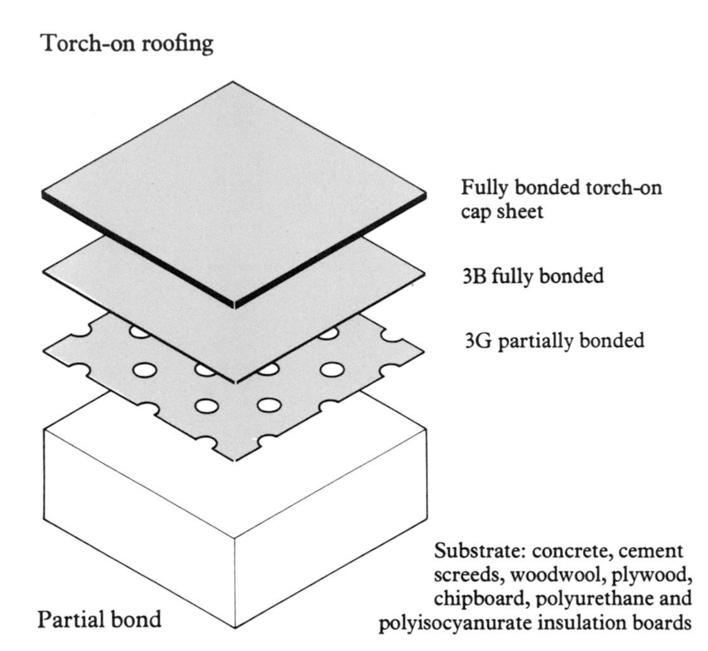

Partial bond

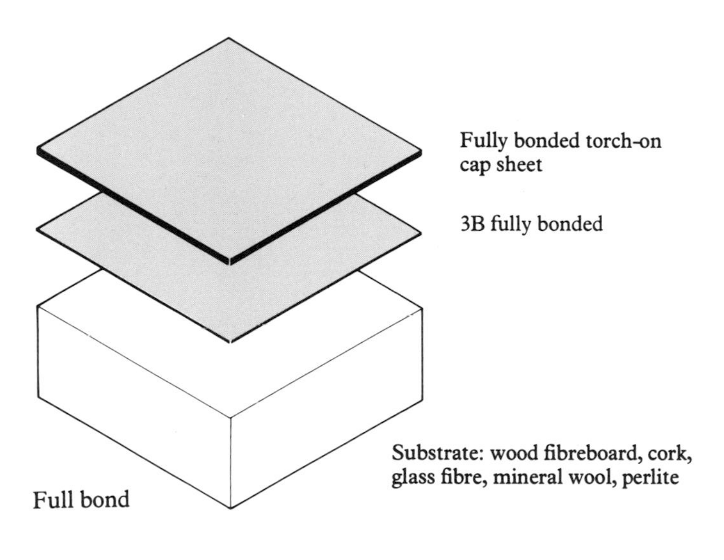

Full bond

Torch-on roofing

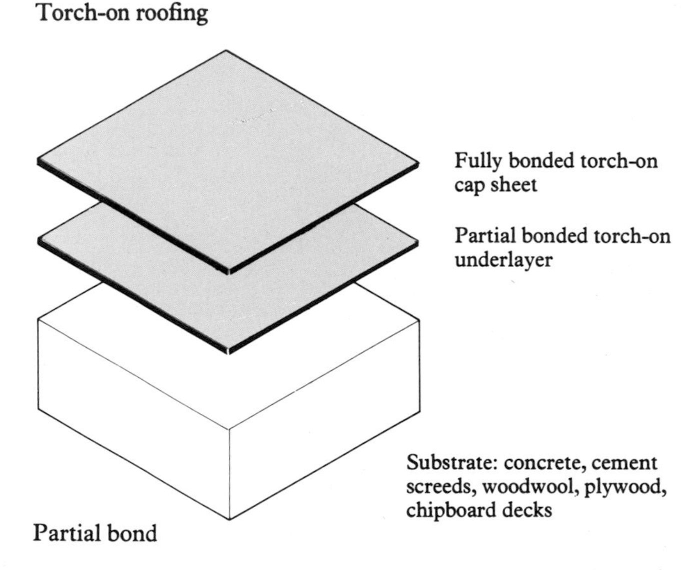

Partial bond

TYPICAL SPECIFICATIONS

In the following pages, typical built-up roofing specifications are illustrated for the most commonly encountered forms of flat roof construction.

Specifications are grouped according to the structural deck material in the following sequence:

Concrete decks (in-situ or precast)
Plywood and chipboard decks
Timber boarded decks
Woodwool decks
Metal decking

The methods of component assembly are shown for warm roof constructions (waterproofing in conjunction with over-deck insulation), cold roof constructions (waterproofing direct to deck) and for the protected membrane roof system where appropriate.

The principles described and illustrated for the sequence and method of waterproofing attachment are fundamental to good flat roofing practice. The advice of a reputable roofing contractor or industry body should be sought if a specification is proposed that changes these principles.

BUILT-UP ROOFING

Standard specification: BS 747 Class 3 felts

TOP LAYER:	Type 3B glass fibre base felt, bonded in bitumen.
2nd LAYER:	Type 3B glass fibre base felt, bonded in bitumen.
1st LAYER:	Type 3B glass fibre base felt, bonded in bitumen.

Upgraded specification

TOP LAYER:	Polyester base roofing or similar, bonded in bitumen.
2nd LAYER:	Type 3B glass fibre base felt, bonded in bitumen.
1st LAYER:	Type 3B glass fibre base felt, bonded in bitumen.

High performance specifications

TOP LAYER:	Polyester base cap sheet*, bonded in bitumen.
1st LAYER:	Polyester base underlayer*, bonded in bitumen.

* Polyester base roofing systems usually incorporate a heavyweight cap sheet and a lighter weight underlayer

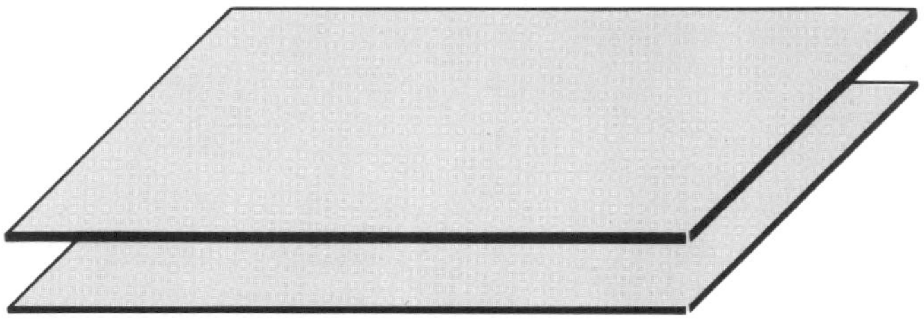

TOP LAYER:	Bitumen polymer or pitch polymer roofing bonded in bitumen (pour and roll method), OR an APP modified bitumen roofing (torch applied).
1st LAYER:	Type 3B glass fibre base felt, bitumen bonded.

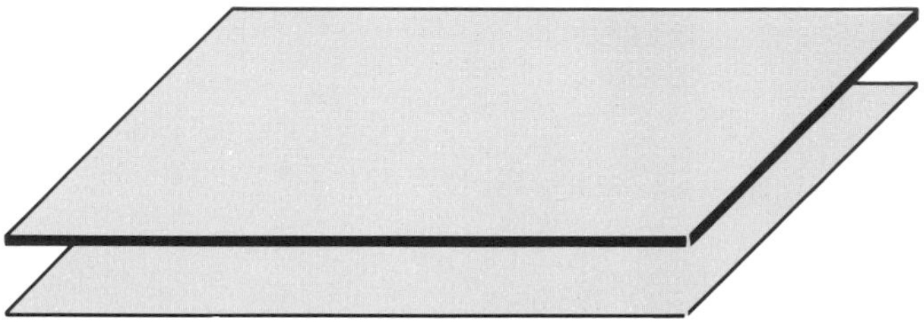

1 SPECIFICATIONS 10° SLOPE OR OVER

Standard:	Type 2E mineral surfaced cap sheet Type 2B fully bonded.
Upgraded:	High performance mineral surfaced cap sheet Type 2B or 3B fully bonded.
High performance:	Mineral surfaced polyester base cap sheet Polyester base underlayer fully bonded.

2 DECK

2.1 In-situ cast slab (illustrated)
A suitable surface of slab or screed for roofing is provided by a wood float finish. Temporary drainage holes should be formed in the structural slab and the deck should be adequately drained and dry before roofing. Apply a bitumen based primer to bind damp or dusty surfaces.

2.2 Precast concrete units
Joints between units may be taped or grouted according to manufacturers' instructions and the deck primed as necessary.

Depending on the accuracy of manufacture and laying, precast units may require a screed to even out deck irregularities. A screed may also be needed to create falls.

3 UNDERLAY

A fully bonded layer of BS 747 type 2B asbestos base or type 3B glass fibre base felt is always recommended to act as a damp proof underlay to the insulation while initial excess moisture dries out from an in-situ cast slab and screed or from a screed applied to precast units.

4 VAPOUR CHECK AND VAPOUR BARRIER

The requirements for a vapour check or vapour barrier with concrete deck constructions are indicated in Tables 1.6 and 1.7, Section 1.3, Vapour Barrier Design Guide.

4.1 Vapour check
The recommended specification for a vapour check on all concrete decks is a single layer of BS 747 type 2B asbestos base or type 3B glass fibre base felt bonded in bitumen.

4.2 Vapour barrier
For a vapour barrier, a two layer system is recommended as follows:
In-situ cast slab
First layer: BS 747 type 2B asbestos base or type 3B glass fibre base felt bonded in bitumen.
Second layer: High performance roofing bonded in bitumen.

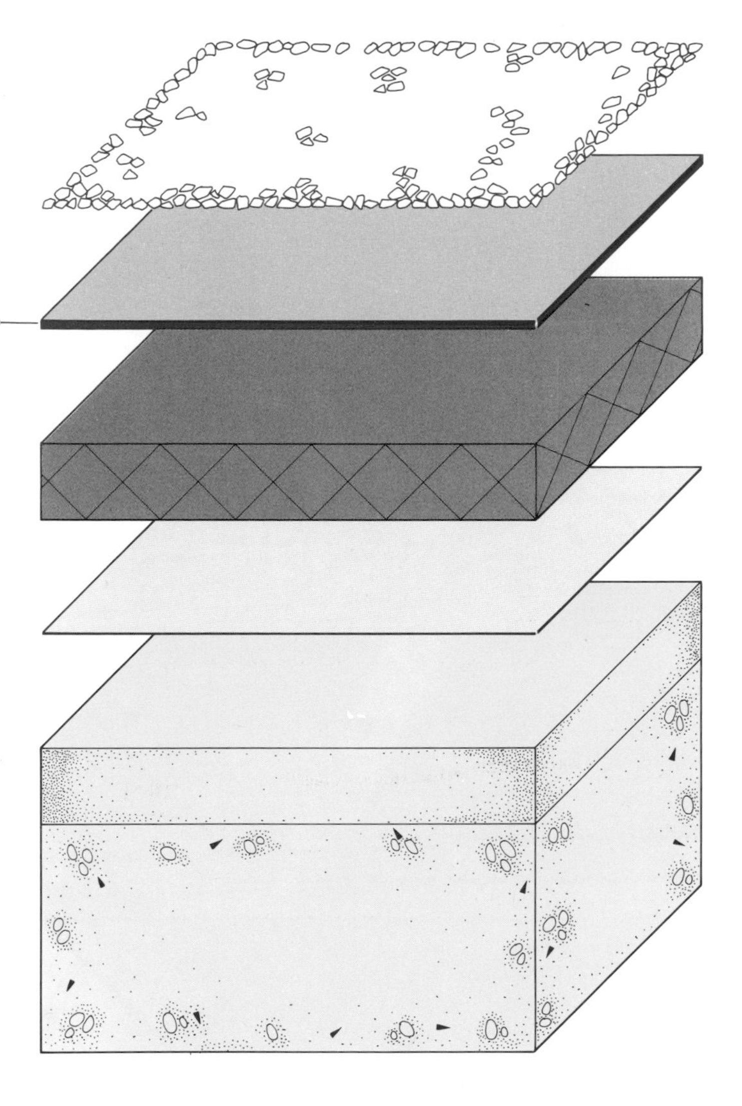

Precast concrete units
First layer: BS 747 type 3G glass fibre base perforated felt partially bonded.
Second layer: High performance roofing bonded in bitumen.

5 THERMAL INSULATION

The following thicknesses (mm) of insulation are required to achieve specified U-values for the total roof construction, calculated from the thermal values listed in Table 4 Appendix A.

5.1 Deck: 150mm In-situ cast slab and screed (illustrated)

Insulation	U-value (W/m² °C) 0.7	0.6	0.5	0.4	0.3
Wood fibreboard	56	68	85	110	151
Cork board	47	57	71	92	127
Glass fibre roofboard	38	46	58	75	103
Mineral wool slab	38	46	58	75	103
Cellular glass slab	51	61	76	99	136
Perlite board	56	68	85	110	151
Expanded polystyrene*	29	38	49	66	94

* U-value allows for additional 13mm wood fibreboard overlay

5.2 Deck: 100mm Precast lightweight concrete units

Insulation	U-value (W/m² °C) 0.7	0.6	0.5	0.4	0.3
Wood fibreboard	30	42	59	84	125
Cork board	25	35	49	70	105
Glass fibre roofboard	21	29	40	57	85
Mineral wool slab	21	29	40	57	85
Cellular glass slab	27	38	53	75	113
Perlite board	30	42	59	84	125
Expanded polystyrene*	12	20	31	48	77

* U-value allows for additional 13mm wood fibreboard overlay

Standard specification: BS 747 Class 3 felts

TOP LAYER: Type 3B glass fibre base felt, bonded in bitumen.

2nd LAYER: Type 3B glass fibre base felt, bonded in bitumen.

1st LAYER: Type 3G glass fibre base perforated felt, partially bonded.

Upgraded specification

TOP LAYER: Polyester base roofing or similar, bonded in bitumen.

2nd LAYER: Type 3B glass fibre base felt, bonded in bitumen.

1st LAYER: Type 3G glass fibre base perforated felt, partially bonded.

High performance specifications

TOP LAYER: Polyester base cap sheet*, bonded in bitumen.

2nd LAYER: Polyester base underlayer*, bonded in bitumen.

1st LAYER: Type 3G glass fibre base perforated felt, partially bonded.

* Polyester base roofing systems usually incorporate a heavyweight cap sheet and a lighter weight underlayer

TOP LAYER: Bitumen polymer or pitch polymer roofing bonded in bitumen (pour and roll method), OR an APP modified bitumen roofing (torch applied).

2nd LAYER: Type 3B glass fibre base felt, bonded in bitumen.

1st LAYER: Type 3G glass fibre base felt, partially bonded in bitumen.

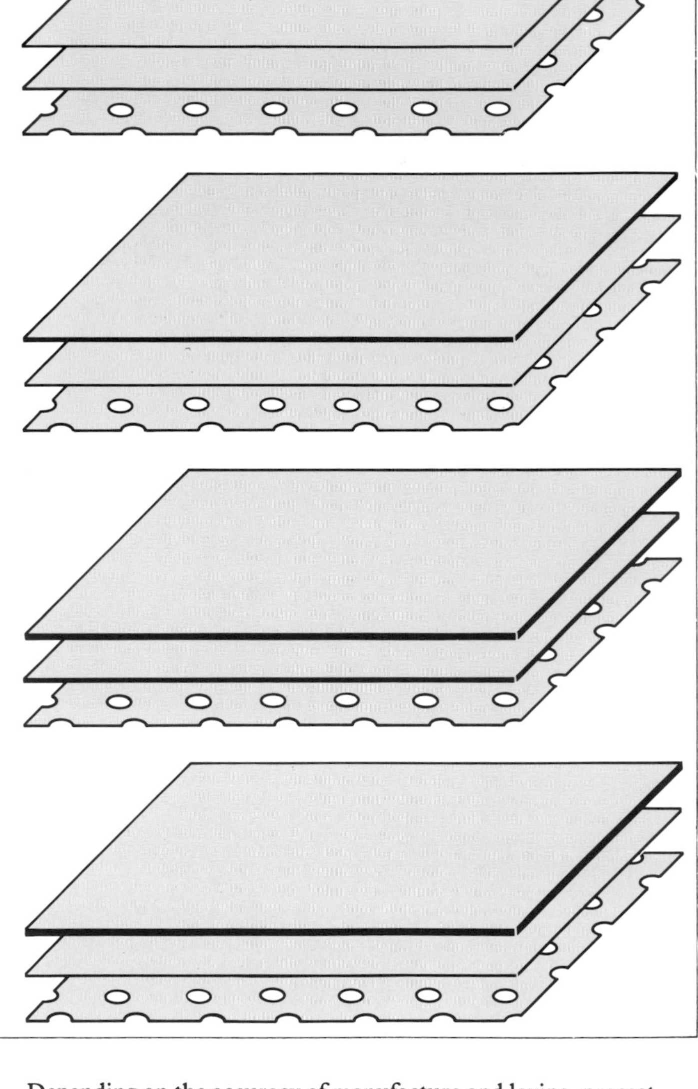

1 SPECIFICATIONS 10° SLOPE OR OVER

Overlay with wood fibreboard or cork to allow a full bond.

Standard: Type 2E mineral surfaced cap sheet
Type 2B fully bonded.

Upgraded: High performance mineral surfaced cap sheet
Type 2B or 3B fully bonded.

High performance: Mineral surfaced polyester base cap sheet
Polyester base underlayer fully bonded.

2 DECK

2.1 In-situ cast slab (illustrated)
A suitable surface of slab or screed for roofing is provided by a wood float finish. Temporary drainage holes should be formed in the structural slab and the deck should be adequately drained and dry before roofing. Apply a bitumen based primer to bind damp or dusty surfaces.

2.2 Precast concrete units
Joints between units may be taped or grouted according to manufacturers' instructions and the deck primed as necessary.

Depending on the accuracy of manufacture and laying, precast units may require a screed to even out deck irregularities. A screed may also be needed to create falls.

3 UNDERLAY

A fully bonded layer of BS 747 type 2B asbestos base or type 3B glass fibre base felt is always recommended to act as a damp proof underlay to the insulation while initial excess moisture dries out from an in-situ cast slab and screed or from a screed applied to precast units.

4 VAPOUR CHECK AND VAPOUR BARRIER

The requirements for a vapour check or vapour barrier with concrete deck constructions are indicated in Tables 1.6 and 1.7, Section 1.3, Vapour Barrier Design Guide.

4.1 Vapour check
The recommended specification for a vapour check on all concrete decks is a single layer of BS 747 type 2B asbestos base or type 3B glass fibre base felt bonded in bitumen.

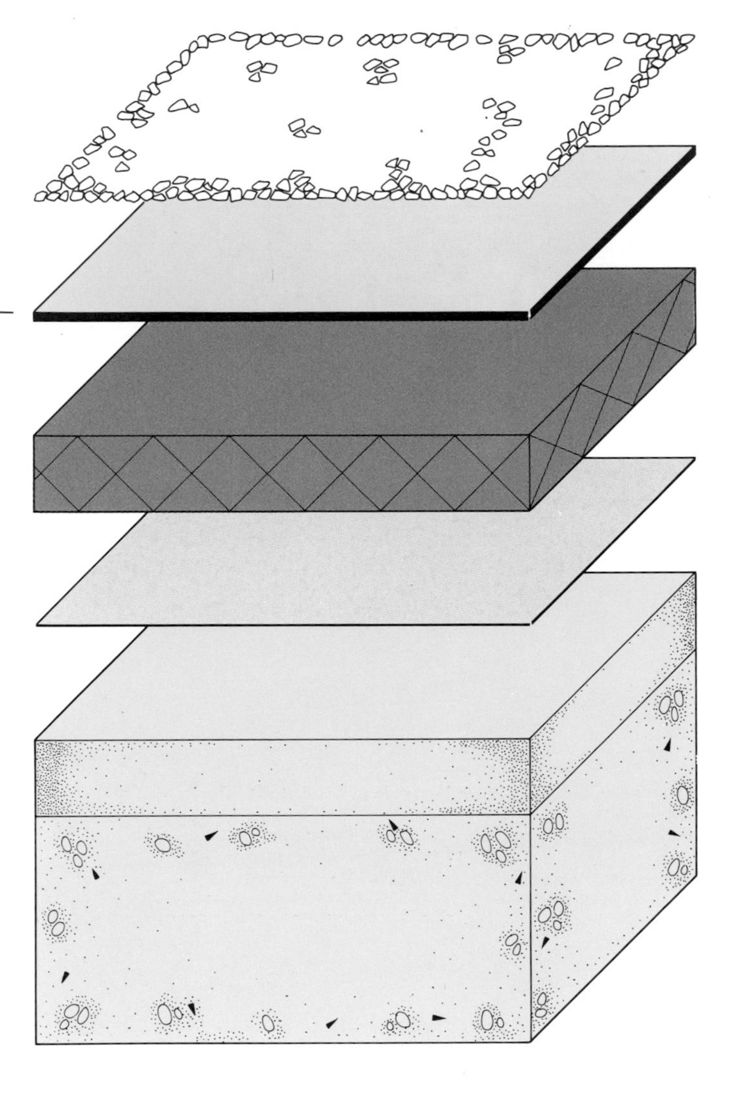

INSULATION
Polyurethane and Polyisocyanurate roofboards (glass tissue faced)

4.2 Vapour barrier
For a vapour barrier, a two layer system is recommended as follows:

In-situ cast slab
First layer: BS 747 type 2B asbestos base or type 3B glass fibre base felt bonded in bitumen.
Second layer: High performance roofing bonded in bitumen.

Precast concrete units
First layer: BS 747 type 3G glass fibre base perforated felt partially bonded.
Second layer: High performance roofing bonded in bitumen.

5 THERMAL INSULATION

The following thicknesses (mm) of insulation are required to achieve specified U-values for the total roof construction, calculated from the thermal values listed in Table 4 Appendix A.

5.1 Deck: 150mm In-situ cast slab and screed (illustrated)

Insulation	U-value (W/m^2 °C) 0.7	0.6	0.5	0.4	0.3
Polyurethane board	25	30	38	49	67
Polyisocyanurate board	25	30	38	49	67

5.2 Deck: 100mm Precast lightweight concrete units

Insulation	U-value (W/m^2 °C) 0.7	0.6	0.5	0.4	0.3
Polyurethane board	14	19	26	37	55
Polyisocyanurate board	14	19	26	37	55

Upgraded specification

TOP LAYER: Polyester base roofing or similar, bonded in bitumen.

2nd LAYER: Type 3B glass fibre base felt, bonded in bitumen.

1st LAYER: Type 3G glass fibre base perforated felt, partially bonded.

High performance specifications

TOP LAYER: Polyester base cap sheet*, bonded in bitumen.

2nd LAYER: Polyester base underlayer*, bonded in bitumen.

1st LAYER: Type 3G glass fibre base perforated felt, partially bonded.

* Polyester base roofing systems usually incorporate a heavyweight cap sheet and a lighter weight underlayer

TOP LAYER: Bitumen polymer or pitch polymer roofing bonded in bitumen (pour and roll method), OR an APP modified bitumen roofing (torch applied).

2nd LAYER: Type 3B glass fibre base felt, bonded in bitumen.

1st LAYER: Type 3G glass fibre base felt, partially bonded in bitumen.

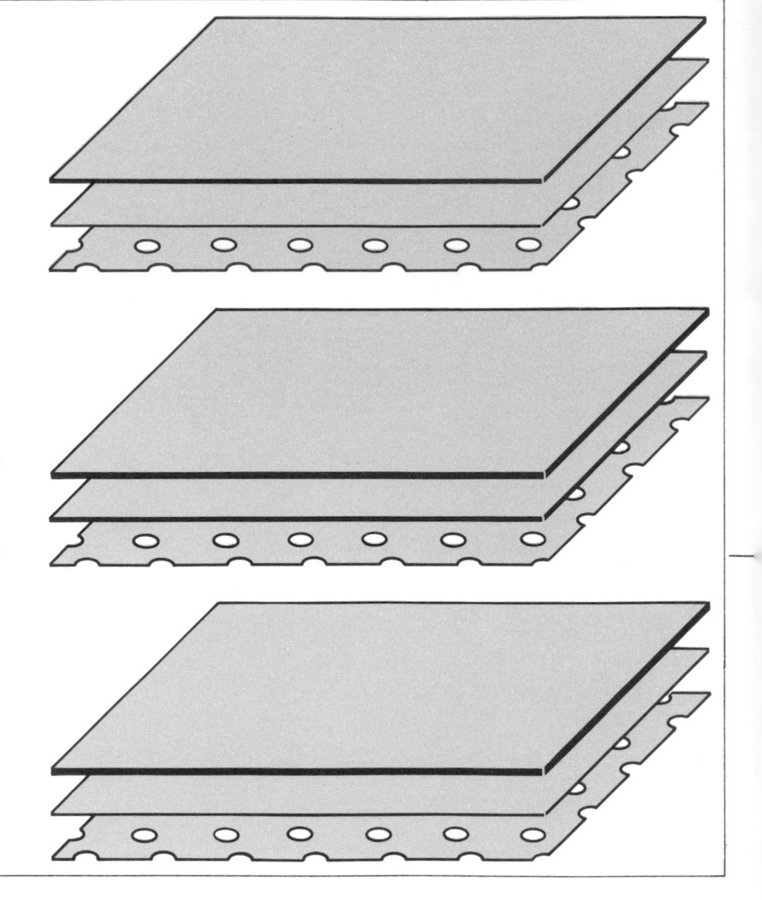

1 SLOPING SPECIFICATIONS

Protected membrane roofs are not suitable for sloping specifications. Normal maximum falls will be in the order of 3°.

2 DECK

2.1 In-situ cast slab (illustrated)
A suitable surface of slab or screed for roofing is provided by a wood float finish. Temporary drainage holes should be formed in the structural slab and the deck should be adequately drained and dry before roofing. Apply a bitumen based primer to bind damp or dusty surfaces.

2.2 Precast concrete units
Joints between units may be taped or grouted according to manufacturers' instructions and the deck primed as necessary.

Depending on the accuracy of manufacture and laying, precast units may require a screed to even out deck irregularities. A screed may also be needed to create falls.

3 UNDERLAY

A loose-laid underlay may be applied to even out surface irregularities or, if the waterproofing has a rough or sharp surface, to prevent abrasion to the underside of the board. The insulation manufacturer will advise on a suitable underlay material.

4 THERMAL INSULATION

When calculating board thickness to achieve a specific U-value, it is necessary to allow for loss of efficiency once the board is installed due to the effect of rainwater draining under the insulation.

Manufacturers may take this into account by including a 20% increase in board thickness. If no such allowance has been made, the loss of efficiency should be taken into account by adjusting the calculated U-value according to the thermal resistance below the level of the insulation.

Standard thicknesses of insulation board provide the following U-values (W/m^2°C) for the total construction, calculated from the thermal values listed in Table 4 Appendix A, with an adjustment for the effect of rainwater drainage below the insulation as indicated in Table 1 Appendix A.

Deck	Thickness of insulation (mm)				
	40	50	60	75	80
150mm in-situ cast slab and screed	0.68	0.58	0.50	0.43	0.41
100mm precast lightweight concrete units	0.51	0.45	0.40	0.35	0.33

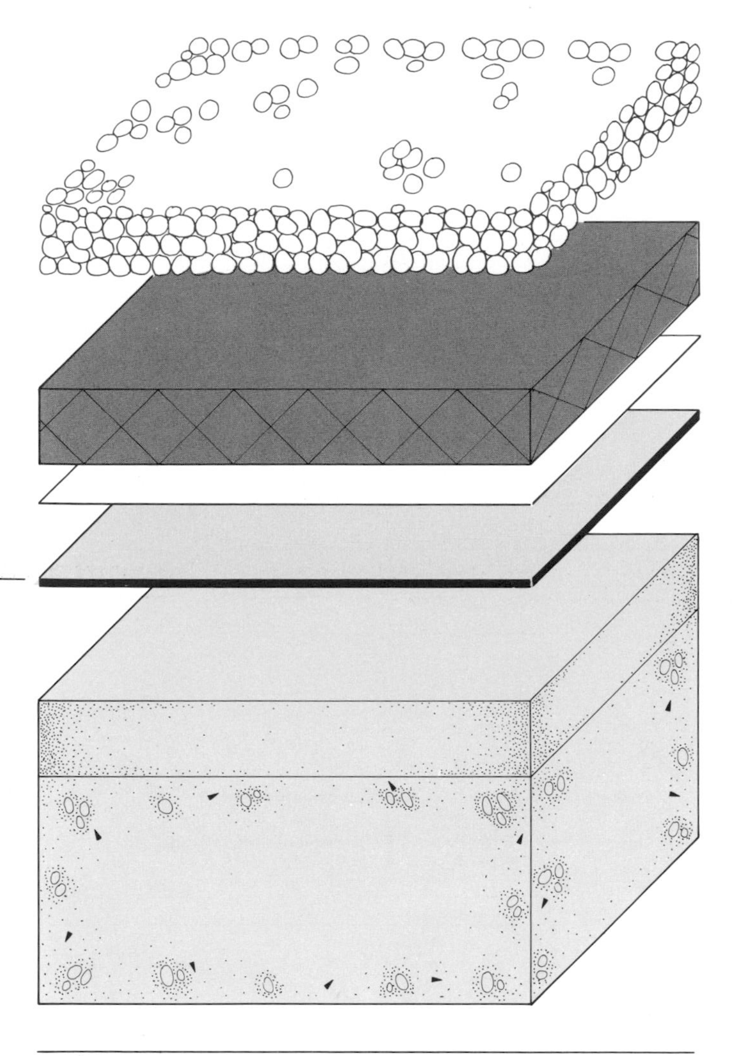

Minimum 50mm layer of gravel (20-30mm nominal diameter) or 50mm concrete paving (Note 5)

Extruded polystyrene insulation board (minimum 50mm) laid loose (Note 4)

INSULATION
Extruded polystyrene

Underlay sheet if required (Note 3)

Built-up roofing

Sand and cement screed to falls

In-situ cast dense concrete slab or precast lightweight concrete deck units (Note 2)

5 SURFACE FINISH

If significant quantities of fine gravel will be present or if paving slabs are to be bedded in sand, it will be necessary to add a filter layer above the insulation to prevent fine material working through the joints and accumulating on the underside of the boards. The manufacturer of the insulation will advise a suitable material.

A 50mm depth of gravel or concrete paving will prevent flotation of the insulation, provided an efficient drainage has been incorporated in the design. If drainage is inadequate, a heavier loading coat is advisable to prevent flotation and a depth of gravel or paving equal to the thickness of the insulation will be required.

BUILT-UP ROOFING

Standard specification: BS 747 Class 3 felts

TOP LAYER:	Type 3B glass fibre base felt, bonded in bitumen.
2nd LAYER:	Type 3B glass fibre base felt, bonded in bitumen.
1st LAYER:	Type 3G glass fibre base perforated felt, partially bonded.

Upgraded specification

TOP LAYER:	Polyester base roofing or similar, bonded in bitumen.
2nd LAYER:	Type 3B glass fibre base felt, bonded in bitumen.
1st LAYER:	Type 3G glass fibre base perforated felt, partially bonded.

High performance specifications

TOP LAYER:	Polyester base cap sheet*, bonded in bitumen.
2nd LAYER:	Polyester base underlayer*, bonded in bitumen.
1st LAYER:	Type 3G glass fibre base perforated felt, partially bonded.

* Polyester base roofing systems usually incorporate a heavyweight cap sheet and a lighter weight underlayer

TOP LAYER:	Bitumen polymer or pitch polymer roofing bonded in bitumen (pour and roll method), OR an APP modified bitumen roofing (torch applied).
2nd LAYER:	Type 3B glass fibre base felt, bonded in bitumen.
1st LAYER:	Type 3G glass fibre base felt, partially bonded in bitumen.

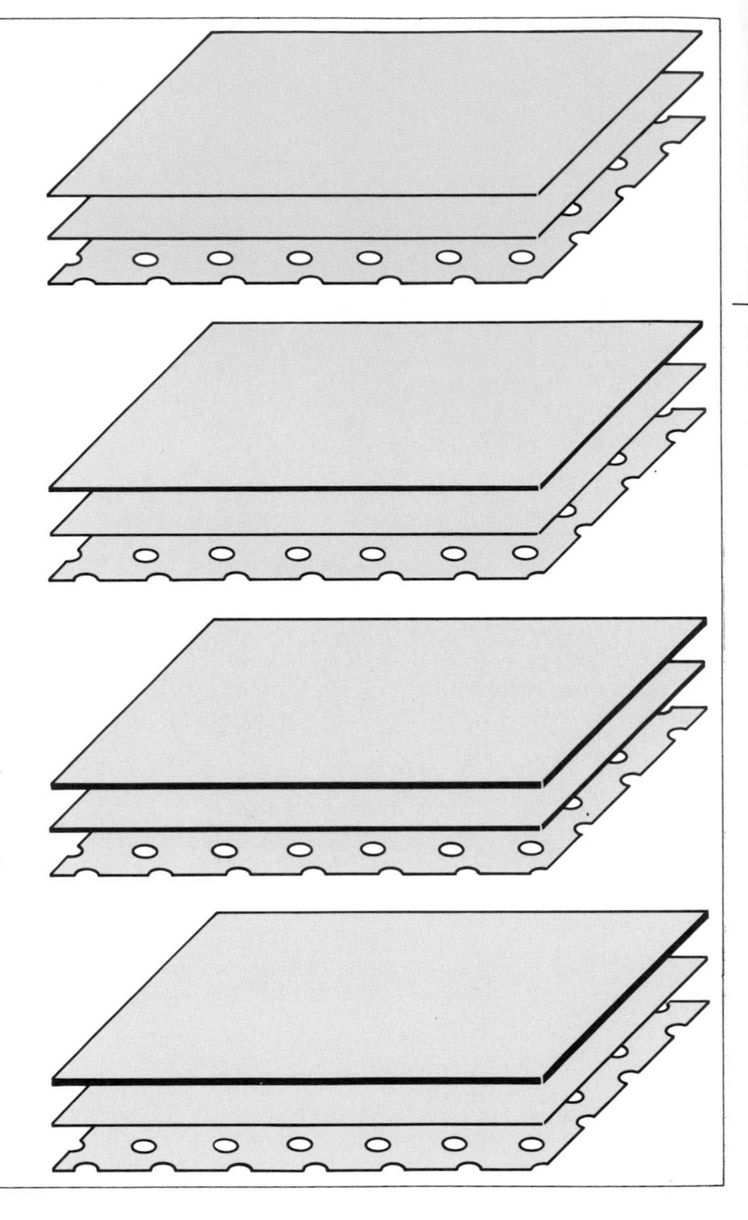

1 SPECIFICATIONS 10° SLOPE OR OVER

Overlay with wood fibreboard or cork to allow a full bond.

Standard:	Type 2E mineral surfaced cap sheet Type 2B fully bonded.
Upgraded:	High performance mineral surfaced cap sheet Type 2B or 3B fully bonded.
High performance:	Mineral surfaced polyester base cap sheet Polyester base underlayer fully bonded.

2 DECK

2.1 In-situ cast slab (illustrated)
A suitable surface of slab or screed for roofing is provided by a wood float finish. Temporary drainage holes should be formed in the structural slab and the deck should be adequately drained and dry before roofing. Apply a bitumen based primer to bind damp or dusty surfaces.

2.2 Precast concrete units
Joints between units may be taped or grouted according to manufacturers' instructions and the deck primed as necessary.

Depending on the accuracy of manufacture and laying, precast units may require a screed to even out deck irregularities. A screed may also be needed to create falls.

3 THERMAL DESIGN

The contribution of the construction to a required U-value will depend on the type and thickness of deck and screed.

Manufacturers of precast lightweight concrete units will indicate the thickness of deck to achieve specific U-values without additional above-deck insulation.

The effects of adding a ceiling or insulation at ceiling level are discussed in Section 1.2 Thermal Design.

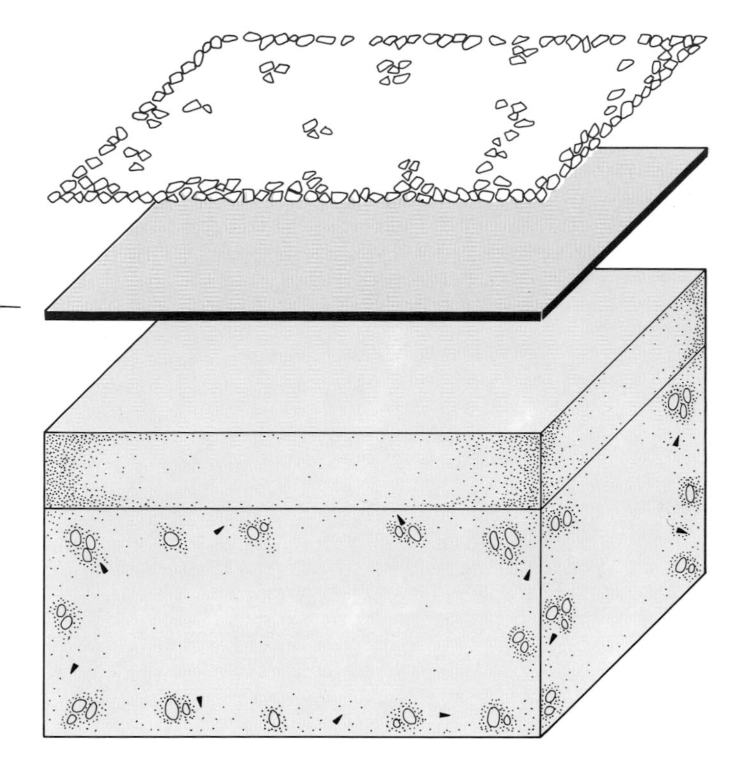

Layer of 10mm stone chippings
in bitumen based adhesive compound

Built-up roofing

Sand and cement screed to falls

In-situ cast dense concrete slab or precast lightweight concrete deck units (Note 2)

BUILT-UP ROOFING

Standard specification: BS 747 Class 3 felts

TOP LAYER:	Type 3B glass fibre base felt, bonded in bitumen.
2nd LAYER:	Type 3B glass fibre base felt, bonded in bitumen.
1st LAYER:	Type 3B glass fibre base felt, bonded in bitumen.

Upgraded specification

TOP LAYER:	Polyester base roofing or similar, bonded in bitumen.
2nd LAYER:	Type 3B glass fibre base felt, bonded in bitumen.
1st LAYER:	Type 3B glass fibre base felt, bonded in bitumen.

High performance specifications

TOP LAYER:	Polyester base cap sheet*, bonded in bitumen.
1st LAYER:	Polyester base underlayer*, bonded in bitumen.

* Polyester base roofing systems usually incorporate a heavyweight cap sheet and a lighter weight underlayer

TOP LAYER:	Bitumen polymer or pitch polymer roofing bonded in bitumen (pour and roll method), OR an APP modified bitumen roofing (torch applied).
1st LAYER:	Type 3B glass fibre base felt, bitumen bonded.

1 SPECIFICATIONS 10° SLOPE OR OVER

Standard:	Type 2E mineral surfaced cap sheet Type 2B fully bonded.
Upgraded:	High performance mineral surfaced cap sheet Type 2B or 3B fully bonded.
High performance:	Mineral surfaced polyester base cap sheet Polyester base underlayer fully bonded.

2 DECK

2.1 Plywood
Plywood should be of exterior grade WBP to BS 1455. Panels should be well nailed to timber joists and noggings. Any deck joints not closed off by the support system should be taped.

2.2 Chipboard
Chipboard decking should be prefelted and of type II/III or type III to BS 5669. Panels should be well nailed to timber joists and noggings and supported at all edges. A full taping of the deck joints will be necessary to complete a temporary waterproof covering.

3 UNDERLAY

A fully bonded layer of BS 747 type 2B asbestos base or type 3B glass fibre base felt may be applied to provide temporary waterproofing cover to plywood decking before the main roofing specification is applied.

4 VAPOUR CHECK AND VAPOUR BARRIER

The requirements for a vapour check or vapour barrier with plywood and chipboard deck constructions are indicated in Table 1.8, Section 1.3, Vapour Barrier Design Guide.

4.1 Vapour check
The recommended specification for a vapour check is a single layer of BS 747 type 2B asbestos base or type 3B glass fibre base felt bonded in bitumen.

4.2 Vapour barrier
Chipboard decking will not normally be suitable for use in high humidity conditions.

For a vapour barrier over plywood decking a two layer system is recommended.
First layer: BS 747 type 3G glass fibre base perforated felt partially bonded.
Second layer: High performance roofing bonded in bitumen.

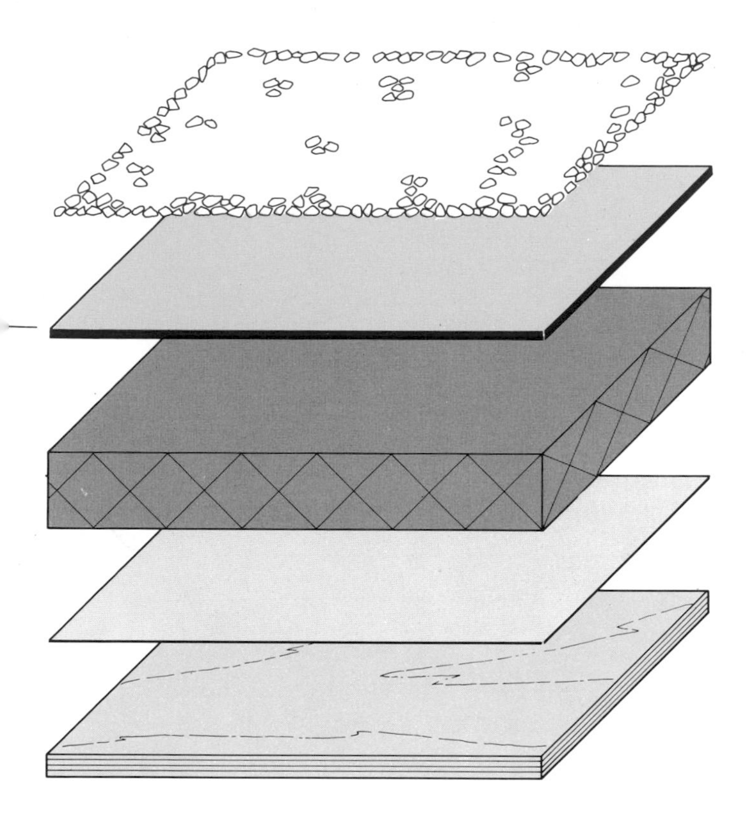

INSULATION
Wood fibreboard
Cork board
Glass fibre roofboard
Mineral wool slab
Cellular glass slab
Perlite board
Expanded polystyrene (HD grade, prefelted)

5 THERMAL INSULATION

The following average thicknesses (mm) of insulation are required to achieve specified U-values for the total roof construction, calculated from standard thermal values listed in Table 4 Appendix A.

Deck: 19mm plywood or chipboard

Insulation	U-value (W/m² °C)				
	0.7	0.6	0.5	0.4	0.3
Wood fibreboard	55	67	83	108	150
Cork board	46	56	70	91	126
Glass fibre roofboard	37	45	57	74	102
Mineral wool slab	37	45	57	74	102
Cellular glass slab	49	60	75	97	135
Perlite board	55	67	83	108	150
Expanded polystyrene*	29	37	48	65	93

* U-value allows for additional 13mm wood fibreboard overlay

Standard specification: BS 747 Class 3 felts

TOP LAYER: Type 3B glass fibre base felt, bonded in bitumen.

2nd LAYER: Type 3B glass fibre base felt, bonded in bitumen.

1st LAYER: Type 3G glass fibre base perforated felt, partially bonded.

Upgraded specification

TOP LAYER: Polyester base roofing or similar, bonded in bitumen.

2nd LAYER: Type 3B glass fibre base felt, bonded in bitumen.

1st LAYER: Type 3G glass fibre base perforated felt, partially bonded.

High performance specifications

TOP LAYER: Polyester base cap sheet*, bonded in bitumen.

2nd LAYER: Polyester base underlayer*, bonded in bitumen.

1st LAYER: Type 3G glass fibre base perforated felt, partially bonded.

* Polyester base roofing systems usually incorporate a heavyweight cap sheet and a lighter weight underlayer

TOP LAYER: Bitumen polymer or pitch polymer roofing bonded in bitumen (pour and roll method), OR an APP modified bitumen roofing (torch applied).

2nd LAYER: Type 3B glass fibre base felt, bonded in bitumen.

1st LAYER: Type 3G glass fibre base felt, partially bonded in bitumen.

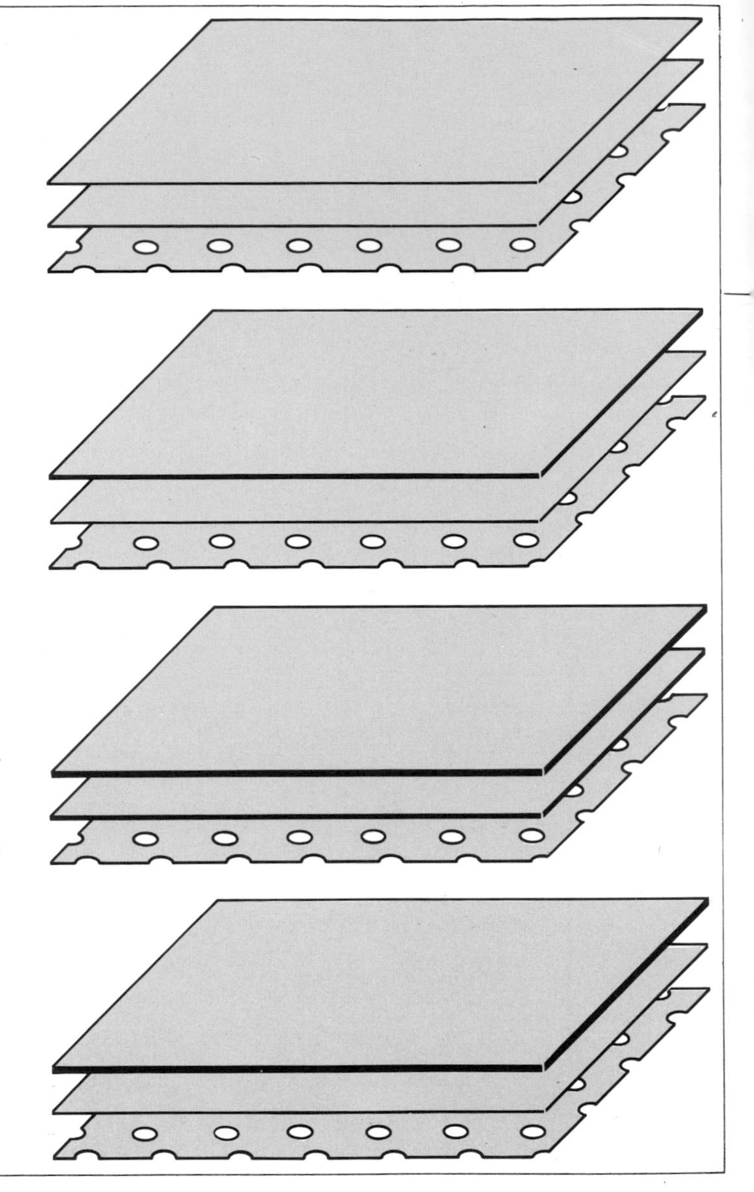

1 SPECIFICATIONS 10° SLOPE OR OVER

Overlay with wood fibreboard or cork to allow a full bond.

Standard: Type 2E mineral surfaced cap sheet
Type 2B fully bonded.

Upgraded: High performance mineral surfaced cap sheet
Type 2B or 3B fully bonded.

High performance: Mineral surfaced polyester base cap sheet
Polyester base underlayer fully bonded.

2 DECK

2.1 Plywood
Plywood should be of exterior grade WBP to BS 1455. Panels should be well nailed to timber joists and noggings. Any deck joints not closed off by the support system should be taped.

2.2 Chipboard
Chipboard decking should be prefelted and of type II/III or type III to BS 5669. Panels should be well nailed to timber joists and noggings and supported at all edges. A full taping of the deck joints will be necessary to complete a temporary waterproof covering.

3 UNDERLAY

A fully bonded layer of BS 747 type 2B asbestos base or type 3B glass fibre base felt may be applied to provide temporary waterproofing cover to plywood decking before the main roofing specification is applied.

4 VAPOUR CHECK AND VAPOUR BARRIER

The requirements for vapour check or vapour barrier with plywood and chipboard deck constructions are indicated in Table 1.8, Section 1.3, Vapour Barrier Design Guide.

4.1 Vapour check
The recommended specification for a vapour check is a single layer of BS 747 type 2B asbestos base or type 3B glass fibre base felt bonded in bitumen.

4.2 Vapour barrier
Chipboard decking will not normally be suitable for use in high humidity conditions.

For a vapour barrier over plywood decking a two layer system is recommended.

PLYWOOD, CHIPBOARD DECKS

WARM ROOF

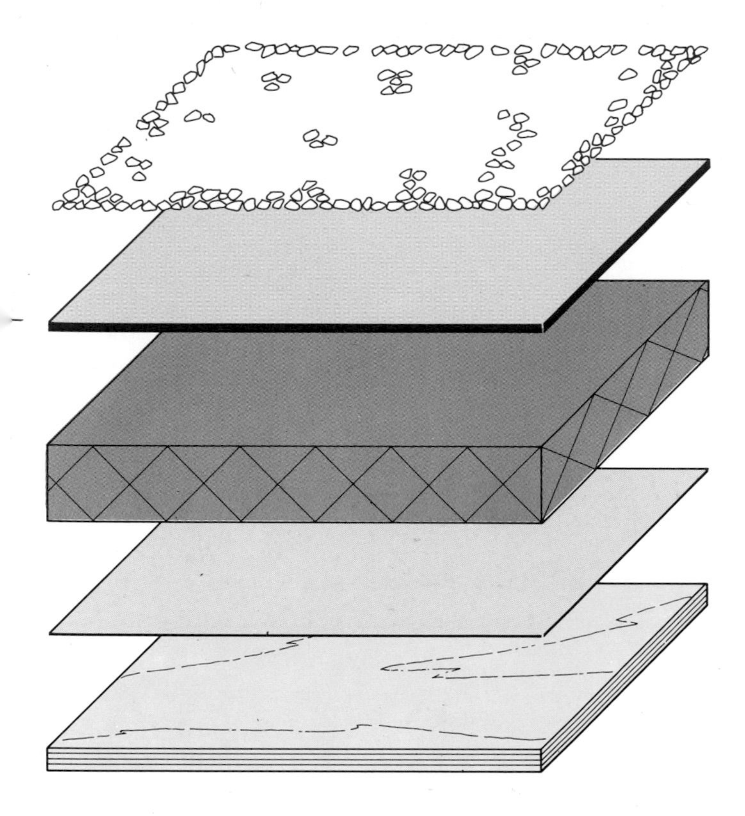

Layer of 10mm stone chippings
in bitumen based adhesive compound

Built-up roofing

Insulation board bonded in hot
bitumen (Note 5)

INSULATION
Polyurethane and
Polyisocyanurate roofboards
(glass tissue faced)

Underlay, vapour check or
vapour barrier as required
(Notes 3 and 4)

Plywood or chipboard decking laid to falls (Note 2)

First layer: BS 747 type 3G glass fibre base perforated felt
partially bonded.
Second layer: High performance roofing bonded in bitumen.

5 THERMAL INSULATION

The following thicknesses (mm) of insulation are required to achieve specified U-values for the total roof construction, calculated from the thermal values listed in Table 4 Appendix A.

Deck: 19mm Plywood or chipboard

Insulation	U-value (W/m² °C)				
	0.7	0.6	0.5	0.4	0.3
Polyurethane board	24	30	37	48	66
Polyisocyanurate board	24	30	37	48	66

BUILT-UP ROOFING

Standard specification: BS 747 Class 3 felts

TOP LAYER:	Type 3B glass fibre base felt, bonded in bitumen.
2nd LAYER:	Type 3B glass fibre base felt, bonded in bitumen.
1st LAYER:	Type 3G glass fibre base perforated felt, partially bonded.

Upgraded specification

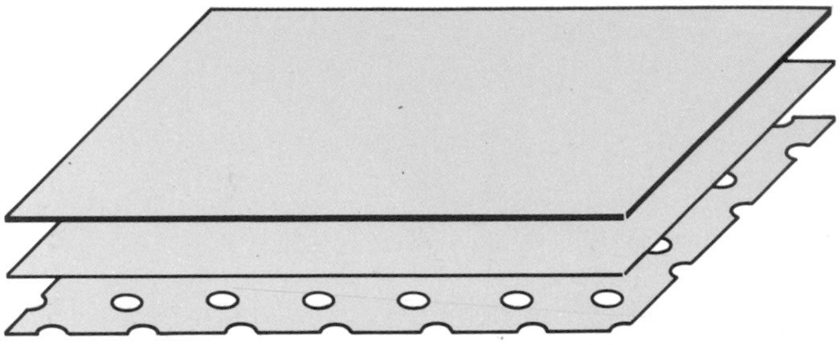

TOP LAYER:	Polyester base roofing or similar, bonded in bitumen.
2nd LAYER:	Type 3B glass fibre base felt, bonded in bitumen.
1st LAYER:	Type 3G glass fibre base perforated felt, partially bonded.

High performance specifications

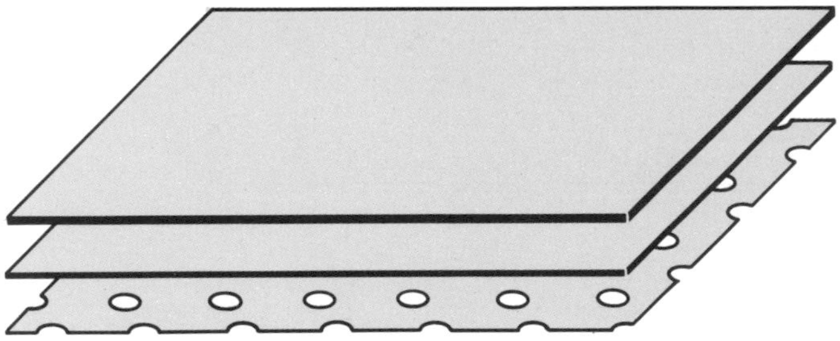

TOP LAYER:	Polyester base cap sheet*, bonded in bitumen.
2nd LAYER:	Polyester base underlayer*, bonded in bitumen.
1st LAYER:	Type 3G glass fibre base perforated felt, partially bonded.

* Polyester base roofing systems usually incorporate a heavyweight cap sheet and a lighter weight underlayer

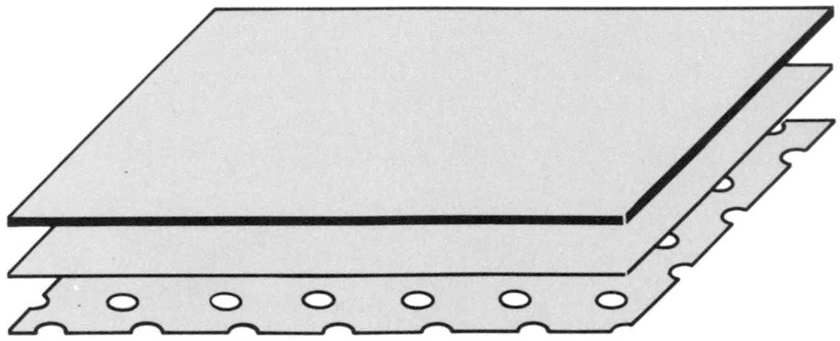

TOP LAYER:	Bitumen polymer or pitch polymer roofing bonded in bitumen (pour and roll method), OR an APP modified bitumen roofing (torch applied).
2nd LAYER:	Type 3B glass fibre base felt, bonded in bitumen.
1st LAYER:	Type 3G glass fibre base felt, partially bonded in bitumen.

1 SPECIFICATIONS 10° SLOPE OR OVER

Overlay with wood fibreboard or cork to allow a full bond.

Standard:	Type 2E mineral surfaced cap sheet Type 2B fully bonded.
Upgraded:	High performance mineral surfaced cap sheet Type 2B or 3B fully bonded.
High performance:	Mineral surfaced polyester base cap sheet Polyester base underlayer fully bonded.

2 DECK

2.1 Plywood
Plywood should be of exterior grade WBP to BS 1455. Panels should be well nailed to timber joists and noggings. Any deck joints not closed off by the support system should be taped.

2.2 Chipboard
Chipboard decking should be prefelted and of type II/III or type III to BS 5669. Panels should be well nailed to timber joists and noggings and supported at all edges. A full taping of the deck joints will complete a temporary waterproofing but this should not be counted as part of the final waterproofing specification.

Chipboard decking will not normally be suitable for use in high humidity conditions.

3 THERMAL DESIGN

To reduce the risk of condensation the cavity space formed between the deck and suspended ceiling should be ventilated to the outside air. Ventilation openings equivalent to 4% of the plan roof area are suggested.

If cavity ventilation is likely to be restricted it is advisable to change the design to a warm roof construction.

PLYWOOD, CHIPBOARD DECKS

COLD ROOF

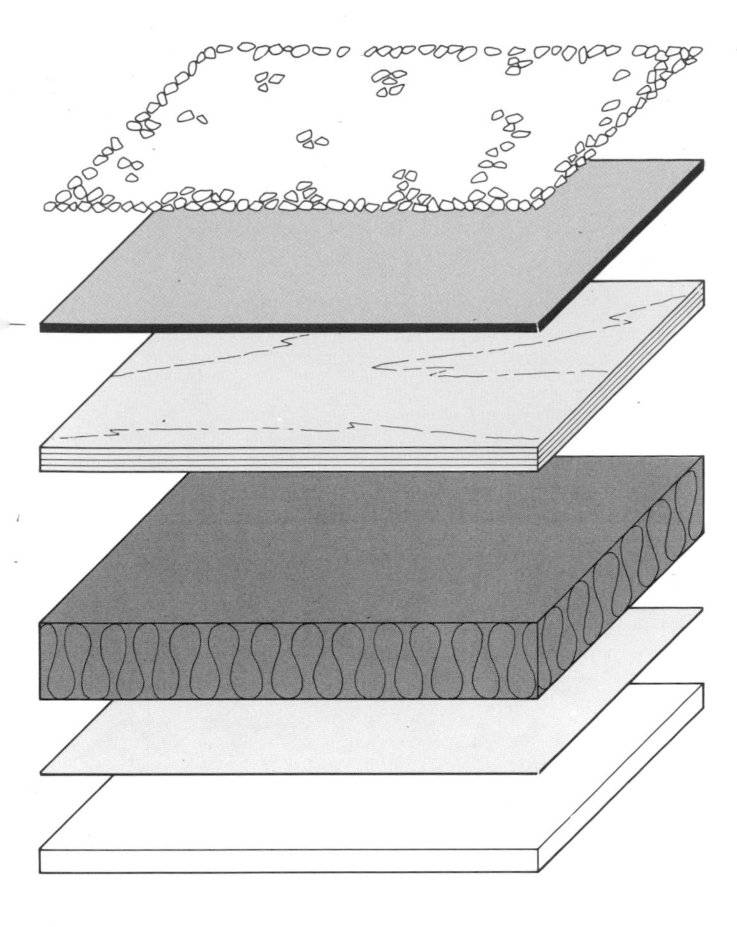

Layer of 10mm stone chippings
in bitumen based adhesive compound

Built-up roofing

Plywood or chipboard decking laid to falls (Note 2)

Cavity ventilated to the outside air (Note 3)

Insulation

Vapour check where required

Suspended ceiling

BUILT-UP ROOFING

Standard specification: BS 747 Class 3 felts

TOP LAYER: Type 3B glass fibre base felt, bonded in bitumen.

2nd LAYER: Type 3B glass fibre base felt, bonded in bitumen.

1st LAYER: Type 3B glass fibre base felt, bonded in bitumen.

Upgraded specification

TOP LAYER: Polyester base roofing or similar, bonded in bitumen.

2nd LAYER: Type 3B glass fibre base felt, bonded in bitumen.

1st LAYER: Type 3B glass fibre base felt, bonded in bitumen.

High performance specifications

TOP LAYER: Polyester base cap sheet*, bonded in bitumen.

1st LAYER: Polyester base underlayer*, bonded in bitumen.

* Polyester base roofing systems usually incorporate a heavyweight cap sheet and a lighter weight underlayer

TOP LAYER: Bitumen polymer or pitch polymer roofing bonded in bitumen (pour and roll method), OR an APP modified bitumen roofing (torch applied).

1st LAYER: Type 3B glass fibre base felt, bitumen bonded.

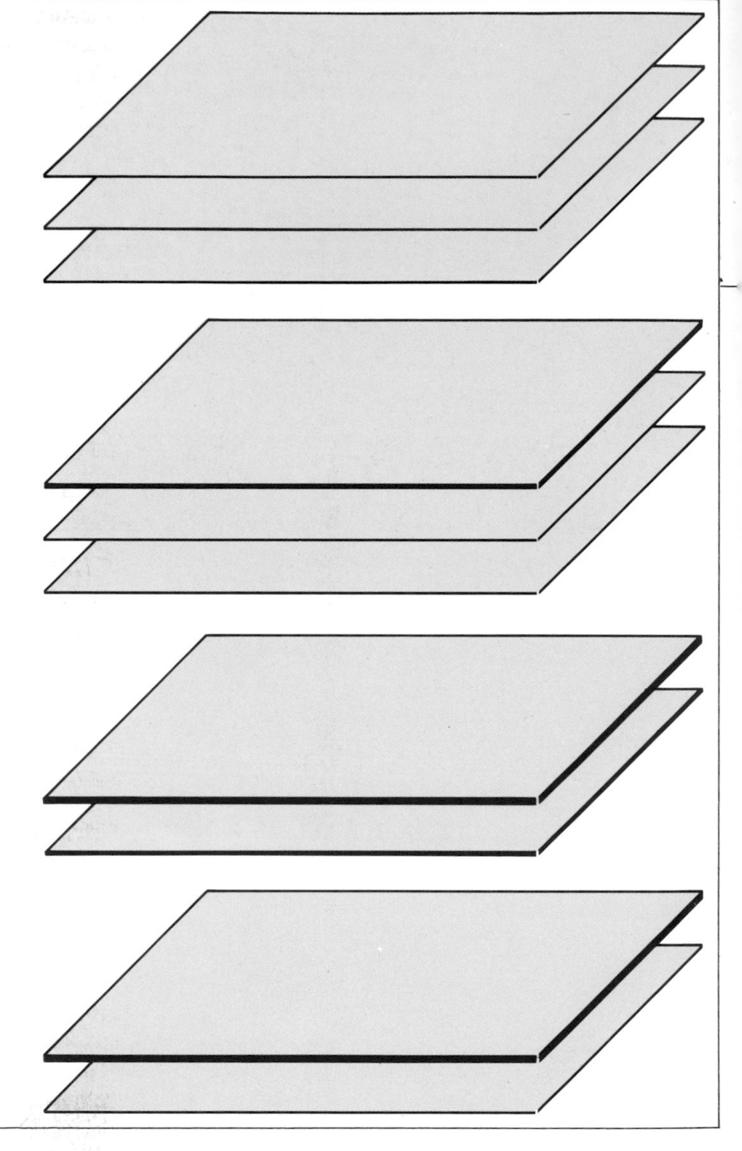

1 SPECIFICATIONS 10° SLOPE OR OVER

Standard: Type 2E mineral surfaced cap sheet
Type 2B fully bonded.

Upgraded: High performance mineral surfaced cap sheet
Type 2B or 3B fully bonded.

High performance: Mineral surfaced polyester base cap sheet
Polyester base underlayer fully bonded.

2 DECK

Timber boards should be well seasoned and not less than 19mm nominal thickness. Tongued and grooved boarding is essential, with boards closely clamped and securely nailed. Plain edge boards should not be used.

3 UNDERLAY

A nailed layer of BS 747 type 2B asbestos base felt is required to provide a suitable surface for bitumen bonding the insulation boards.

4 VAPOUR CHECK AND VAPOUR BARRIER

The requirements for a vapour check or vapour barrier with timber boarded constructions are indicated in Table 1.9, Section 1.3, Vapour Barrier Design Guide.

4.1 Vapour check
The nailed underlay of BS 747 type 2B asbestos base felt will also function as a vapour check layer.

4.2 Vapour barrier
For a vapour barrier, a two layer system is recommended.
First layer: High performance roofing nailed.
Second layer: BS 747 type 2B asbestos base or type 3B glass fibre base felt bonded in bitumen.

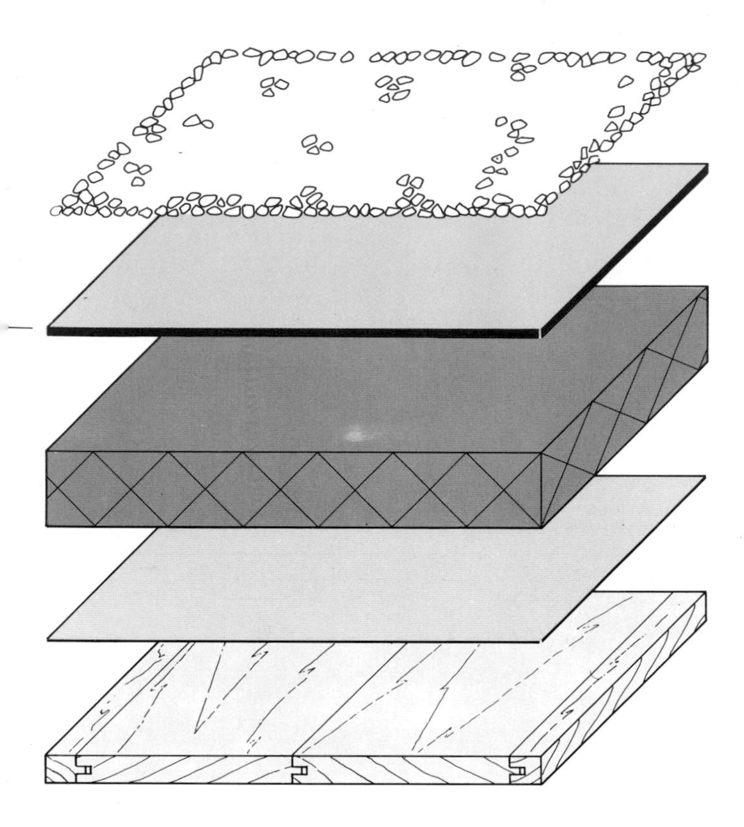

INSULATION
Wood fibreboard
Cork board
Glass fibre roofboard
Mineral wool slab
Cellular glass slab
Perlite board
Expanded polystyrene (HD grade, prefelted)

5 THERMAL INSULATION

The following thicknesses (mm) of insulation are required to achieve specified U-values for the total roof construction, calculated from the thermal values listed in Table 4 Appendix A.

Deck: 19mm Tongued and grooved boarding

Insulation	U-value (W/m^2 °C)				
	0.7	0.6	0.5	0.4	0.3
Wood fibreboard	55	67	83	108	150
Cork board	46	56	70	91	126
Glass fibre roofboard	37	45	57	74	102
Mineral wool slab	37	45	57	74	102
Cellular glass slab	49	60	75	97	135
Perlite board	55	67	83	108	150
Expanded polystyrene*	29	37	48	65	93

* U-value allows for additional 13mm wood fibreboard overlay

BUILT-UP ROOFING

Standard specification: BS 747 Class 3 felts

TOP LAYER: Type 3B glass fibre base felt, bonded in bitumen.

2nd LAYER: Type 3B glass fibre base felt, bonded in bitumen.

1st LAYER: Type 3G glass fibre base perforated felt, partially bonded.

Upgraded specification

TOP LAYER: Polyester base roofing or similar, bonded in bitumen.

2nd LAYER: Type 3B glass fibre base felt, bonded in bitumen.

1st LAYER: Type 3G glass fibre base perforated felt, partially bonded.

High performance specifications

TOP LAYER: Polyester base cap sheet*, bonded in bitumen.

2nd LAYER: Polyester base underlayer*, bonded in bitumen.

1st LAYER: Type 3G glass fibre base perforated felt, partially bonded.

* Polyester base roofing systems usually incorporate a heavyweight cap sheet and a lighter weight underlayer

TOP LAYER: Bitumen polymer or pitch polymer roofing bonded in bitumen (pour and roll method), OR an APP modified bitumen roofing (torch applied).

2nd LAYER: Type 3B glass fibre base felt, bonded in bitumen.

1st LAYER: Type 3G glass fibre base felt, partially bonded in bitumen.

1 SPECIFICATIONS 10° SLOPE OR OVER

Overlay with wood fibreboard or cork to allow a full bond.

Standard: Type 2E mineral surfaced cap sheet
Type 2B fully bonded.

Upgraded: High performance mineral surfaced cap sheet
Type 2B or 3B fully bonded.

High performance: Mineral surfaced polyester base cap sheet
Polyester base underlayer fully bonded.

2 DECK

Timber boards should be well seasoned and not less than 19mm nominal thickness. Tongued and grooved boarding is essential, with boards closely clamped and securely nailed. Plain edge boards should not be used.

3 UNDERLAY

A nailed layer of BS 747 type 2B asbestos base felt is required to provide a suitable surface for bitumen bonding the insulation boards.

4 VAPOUR CHECK AND VAPOUR BARRIER

The requirements for a vapour check or vapour barrier with timber boarded constructions are indicated in Table 1.9, Section 1.3, Vapour Barrier Design Guide.

4.1 Vapour check
The nailed underlay of BS 747 type 2B asbestos base felt will also function as a vapour check layer.

4.2 Vapour barrier
For a vapour barrier, a two layer system is recommended.
First layer: High performance roofing nailed.
Second layer: BS 747 type 2B asbestos base or type 3B glass fibre base felt bonded in bitumen.

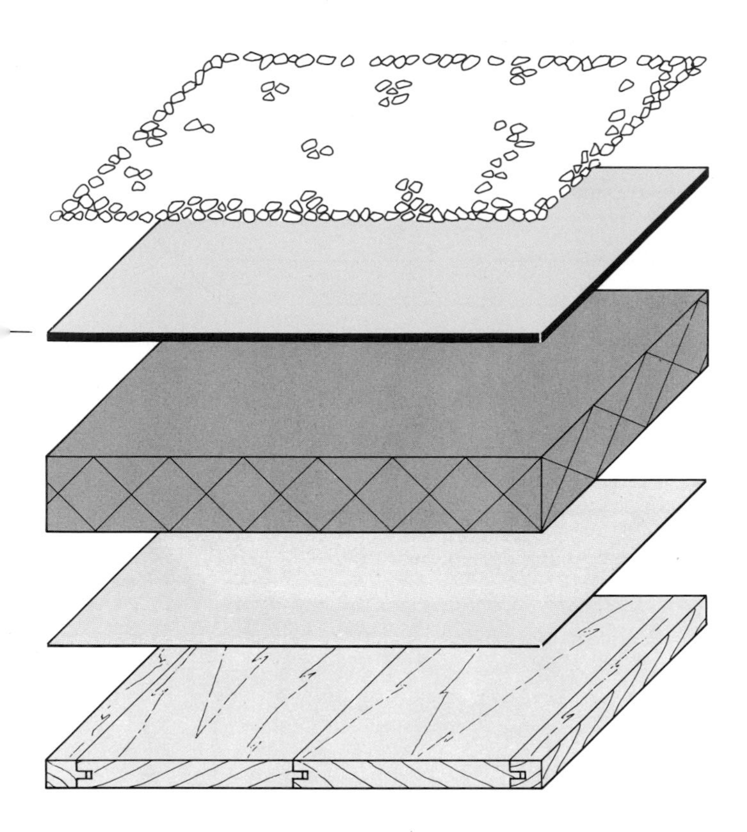

INSULATION
Polyurethane and
Polyisocyanurate roofboards
(glass tissue faced)

5 THERMAL INSULATION

The following thicknesses (mm) of insulation are required to achieve specified U-values for the total roof construction, calculated from the thermal values listed in Table 4 Appendix A.

Deck: 19mm Tongued and grooved boarding

Insulation	U-value (W/m^2 °C)				
	0.7	0.6	0.5	0.4	0.3
Polyurethane board	24	30	37	48	66
Polyisocyanurate board	24	30	37	48	66

BUILT-UP ROOFING

Standard specification: BS 747 Class 3 felts

TOP LAYER:	Type 3B glass fibre base felt, bonded in bitumen.
2nd LAYER:	Type 3B glass fibre base felt, bonded in bitumen.
1st LAYER:	Type 2B asbestos base felt, nailed.

Upgraded specification

TOP LAYER:	Polyester base roofing or similar, bonded in bitumen.
2nd LAYER:	Type 3B glass fibre base felt, bonded in bitumen.
1st LAYER:	Type 2B asbestos base felt, nailed.

High performance specifications

TOP LAYER:	Polyester base cap sheet*, bonded in bitumen.
2nd LAYER:	Polyester base underlayer*, bonded in bitumen.
1st LAYER:	Type 2B asbestos base felt, nailed.

*Polyester base roofing systems usually incorporate a heavyweight cap sheet and a lighter weight underlayer.

TOP LAYER:	Bitumen polymer or pitch polymer roofing bonded in bitumen (pour and roll method), OR an APP modified bitumen roofing (torch applied).
2nd LAYER:	Type 3B glass fibre base felt, bonded in bitumen.
1st LAYER:	Type 2B asbestos base felt, nailed.

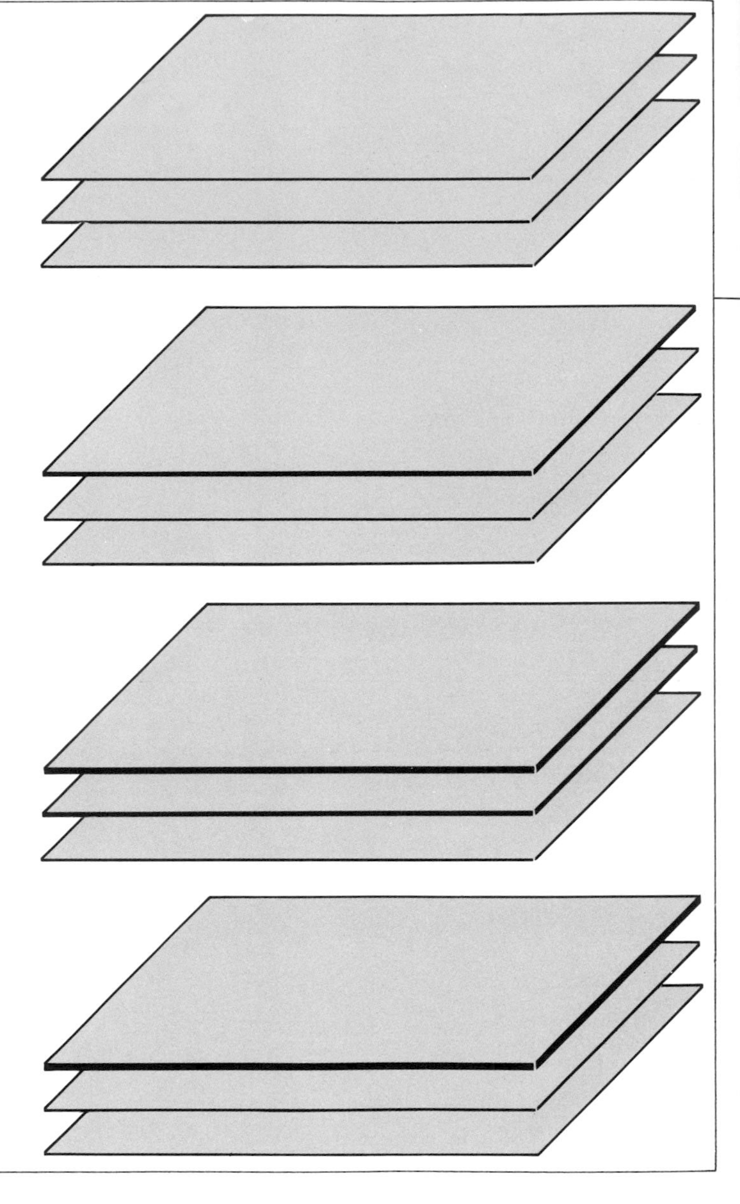

1 SPECIFICATIONS 10° SLOPE OR OVER

Standard:	Type 2E mineral surfaced cap sheet Type 2B fully bonded Type 2B nailed.
Upgraded:	High performance mineral surfaced cap sheet Type 2B or 3B fully bonded Type 2B nailed.
High performance:	Mineral surfaced polyester base cap sheet Polyester base underlayer fully bonded Type 2B nailed.

2 DECK

Timber boards should be well seasoned and not less than 19mm nominal thickness. Tongued and grooved boarding is essential, with boards closely clamped and securely nailed. Plain edge boards should not be used.

3 THERMAL DESIGN

To reduce the risk of condensation the cavity space formed between the deck and suspended ceiling should be ventilated to the outside air. Ventilation openings equivalent to 4% of the plan roof area are suggested.

If cavity ventilation is likely to be restricted it is advisable to change the design to a warm roof construction.

TIMBER BOARDED DECK

COLD ROOF

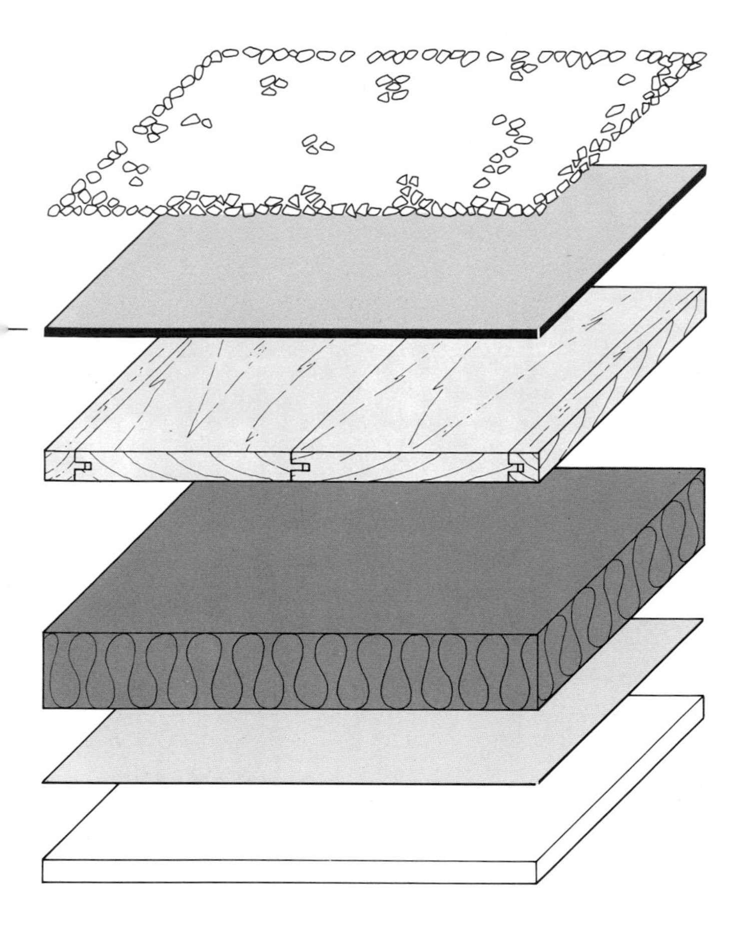

Layer of 10mm stone chippings
in bitumen based adhesive compound

Built-up roofing

Timber boarded deck laid to falls (Note 2)

Cavity ventilated to the outside air (Note 3)

Insulation

Vapour check where required

Suspended ceiling

Standard specification: BS 747 Class 3 felts

TOP LAYER: Type 3B glass fibre base felt, bonded in bitumen.

2nd LAYER: Type 3B glass fibre base felt, bonded in bitumen.

1st LAYER: Type 3B glass fibre base felt, bonded in bitumen.

Upgraded specification

TOP LAYER: Polyester base roofing or similar, bonded in bitumen.

2nd LAYER: Type 3B glass fibre base felt, bonded in bitumen.

1st LAYER: Type 3B glass fibre base felt, bonded in bitumen.

High performance specifications

TOP LAYER: Polyester base cap sheet*, bonded in bitumen.

1st LAYER: Polyester base underlayer*, bonded in bitumen.

* Polyester base roofing systems usually incorporate a heavyweight cap sheet and a lighter weight underlayer

TOP LAYER: Bitumen polymer or pitch polymer roofing bonded in bitumen (pour and roll method), OR an APP modified bitumen roofing (torch applied).

1st LAYER: Type 3B glass fibre base felt, bitumen bonded.

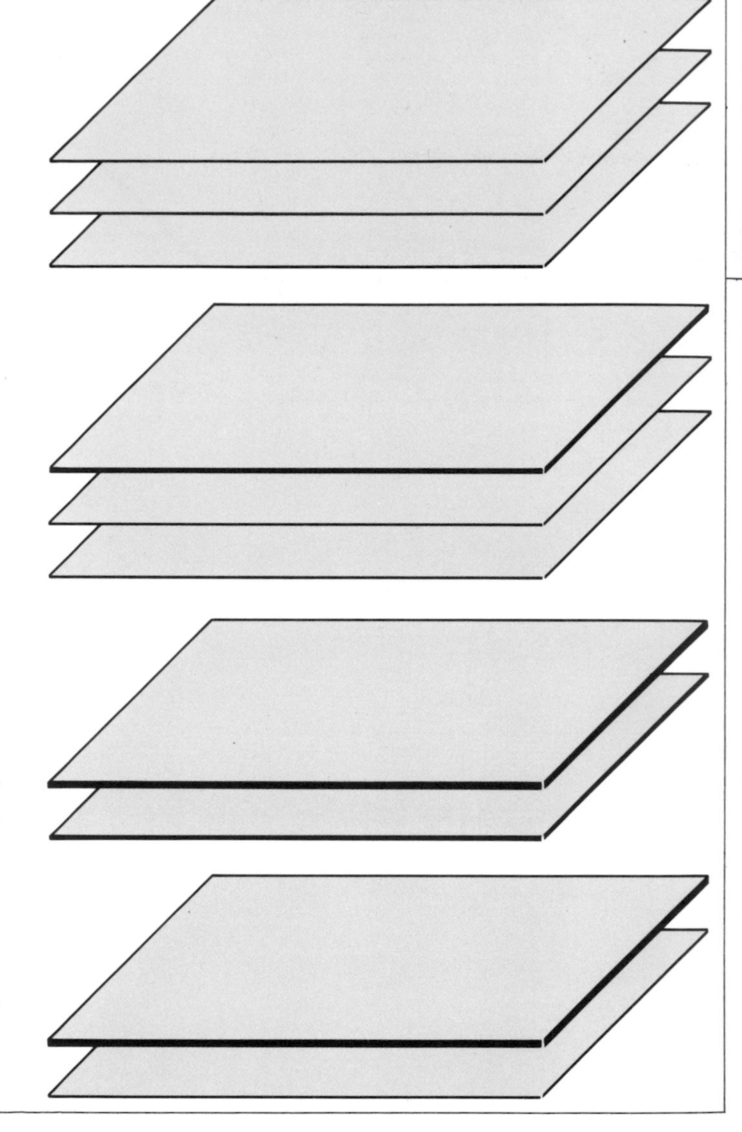

1 SPECIFICATIONS 10° SLOPE OR OVER

Standard:	Type 2E mineral surfaced cap sheet Type 2B fully bonded.
Upgraded:	High performance mineral surfaced cap sheet Type 2B or 3B fully bonded.
High performance:	Mineral surfaced polyester base cap sheet Polyester base underlayer fully bonded.

2 DECK

Woodwool slabs must be type SB to BS 1105 and preferably have a prefelted or prescreeded finish. The joints of all slabs should be taped according to manufacturers' instructions. In the case of prefelted slabs, taping will be necessary to complete the temporary waterproof covering. Where a plain finish slab is used, a sand and cement screed or slurry should be applied to provide a suitable surface for the roofing specification.

3 UNDERLAY

A fully bonded layer of BS 747 type 2B asbestos base or type 3B glass fibre base felt may be applied to provide waterproofing cover to prescreeded and plain finish slabs before the main roofing specification is applied.

4 VAPOUR CHECK AND VAPOUR BARRIER

The requirements for a vapour check or vapour barrier with woodwool constructions are indicated in Table 1.10, Section 1.3, Vapour Barrier Design Guide.

4.1 Vapour check
The recommended specification for a vapour check is a single layer of BS 747 type 2B asbestos base or type 3B glass fibre base felt bonded in bitumen. The facing to prefelted woodwool slabs will also function as a vapour check if taped.

4.2 Vapour barrier
For a vapour barrier, a two layer system is recommended.
First layer: BS 747 type 3G glass fibre base perforated felt partially bonded.
Second layer: High performance roofing bonded in bitumen.

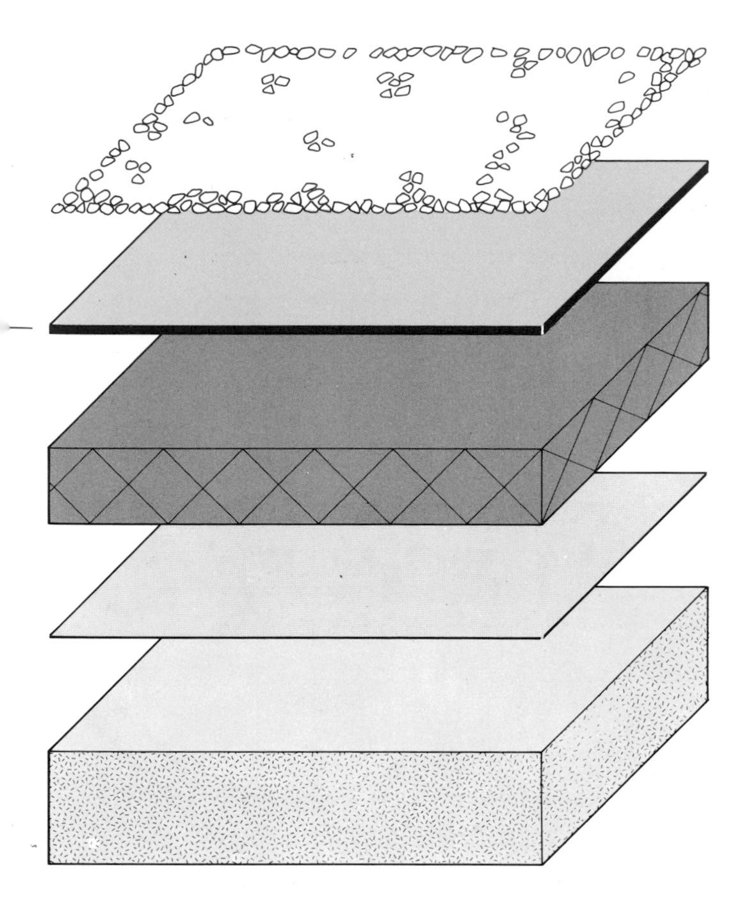

INSULATION
Wood fibreboard
Cork board
Glass fibre roofboard
Mineral wool slab
Cellular glass slab
Perlite board
Expanded polystyrene (HD grade, prefelted)

5 THERMAL INSULATION

The following thicknesses (mm) of insulation are required to achieve specified U-values for the total roof construction calculated from the thermal values listed in Table 4 Appendix A.

Deck: 50mm Woodwool slab

Insulation	U-value (W/m^2 °C)				
	0.7	0.6	0.5	0.4	0.3
Wood fibreboard	35	46	63	88	130
Cork board	29	39	53	74	109
Glass fibre roofboard	24	32	43	60	88
Mineral wool slab	24	32	43	60	88
Cellular glass slab	31	42	57	79	117
Perlite board	35	46	63	88	130
Expanded polystyrene*	15	23	34	51	80

* U-value allows for additional 13mm wood fibreboard overlay

BUILT-UP ROOFING

Standard specification: BS 747 Class 3 felts

TOP LAYER:	Type 3B glass fibre base felt, bonded in bitumen.
2nd LAYER:	Type 3B glass fibre base felt, bonded in bitumen.
1st LAYER:	Type 3G glass fibre base perforated felt, partially bonded.

Upgraded specification

TOP LAYER:	Polyester base roofing or similar, bonded in bitumen.
2nd LAYER:	Type 3B glass fibre base felt, bonded in bitumen.
1st LAYER:	Type 3G glass fibre base perforated felt, partially bonded.

High performance specifications

TOP LAYER:	Polyester base cap sheet*, bonded in bitumen.
2nd LAYER:	Polyester base underlayer*, bonded in bitumen.
1st LAYER:	Type 3G glass fibre base perforated felt, partially bonded.

* Polyester base roofing systems usually incorporate a heavyweight cap sheet and a lighter weight underlayer

TOP LAYER:	Bitumen polymer or pitch polymer roofing bonded in bitumen (pour and roll method), OR an APP modified bitumen roofing (torch applied).
2nd LAYER:	Type 3B glass fibre base felt, bonded in bitumen.
1st LAYER:	Type 3G glass fibre base felt, partially bonded in bitumen.

1 SPECIFICATIONS 10° SLOPE OR OVER

Overlay with wood fibreboard or cork to allow a full bond.

Standard:	Type 2E mineral surfaced cap sheet Type 2B fully bonded.
Upgraded:	High performance mineral surfaced cap sheet Type 2B or 3B fully bonded.
High performance:	Mineral surfaced polyester base cap sheet Polyester base underlayer fully bonded.

2 DECK

Woodwool slabs must be type SB to BS 1105 and preferably have a prefelted or prescreeded finish. The joints of all slabs should be taped according to manufacturers' instructions. In the case of prefelted slabs, taping will be necessary to complete the temporary waterproof covering. Where a plain finish slab is used, a sand and cement screed or slurry should be applied to provide a suitable surface for the roofing specification.

3 UNDERLAY

A fully bonded layer of BS 747 type 2B asbestos base or type 3B glass fibre base felt may be applied to provide waterproofing cover to prescreeded and plain finish slabs before the main roofing specification is applied.

4 VAPOUR CHECK AND VAPOUR BARRIER

The requirements for a vapour check or vapour barrier with woodwool constructions are indicated in Table 1.10, Section 1.3, Vapour Barrier Design Guide.

4.1 Vapour check
The recommended specification for a vapour check is a single layer of BS 747 type 2B asbestos base or type 3B glass fibre base felt bonded in bitumen. The facing to prefelted woodwool slabs will also function as a vapour check if taped.

4.2 Vapour barrier
For a vapour barrier, a two layer system is recommended.
First layer: BS 747 type 3G glass fibre base perforated felt partially bonded.
Second layer: High performance roofing bonded in bitumen.

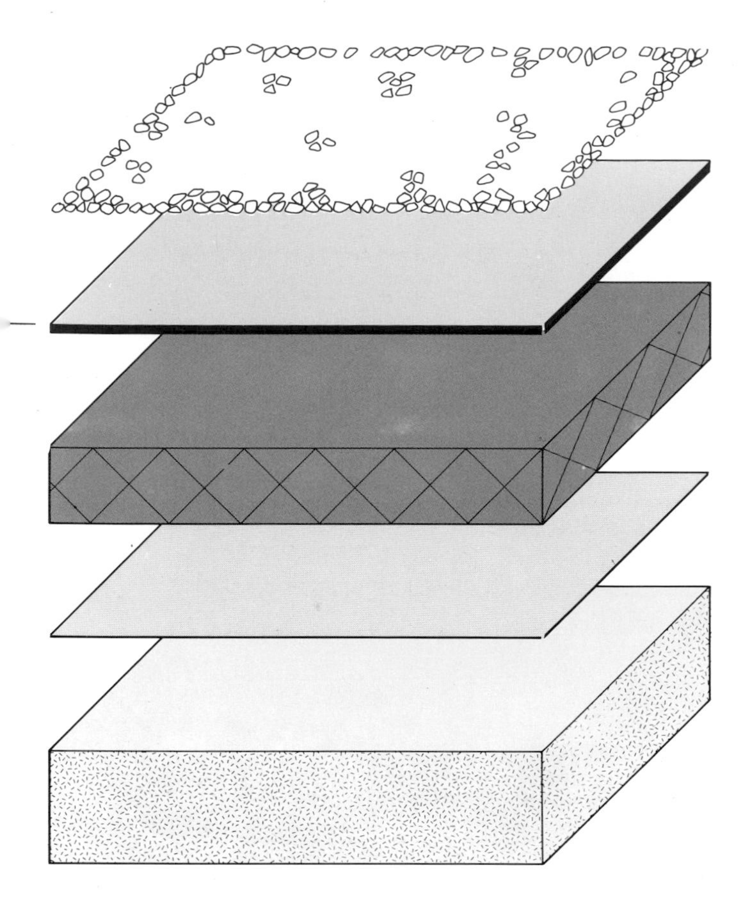

INSULATION
Polyurethane and Polyisocyanurate roofboards (glass tissue faced)

5 THERMAL INSULATION

The following thicknesses (mm) of insulation are required to achieve specified U-values for the total roof construction calculated from the thermal values listed in Table 4 Appendix A.

Deck: 50mm Woodwool slab

Insulation	U-value (W/m² °C) 0.7	0.6	0.5	0.4	0.3
Polyurethane board	15	21	28	39	57
Polyisocyanurate board	15	21	28	39	57

BUILT-UP ROOFING

Standard specification: BS 747 Class 3 felts

TOP LAYER: Type 3B glass fibre base felt, bonded in bitumen.

2nd LAYER: Type 3B glass fibre base felt, bonded in bitumen.

1st LAYER: Type 3G glass fibre base perforated felt, partially bonded.

Upgraded specification

TOP LAYER: Polyester base roofing or similar, bonded in bitumen.

2nd LAYER: Type 3B glass fibre base felt, bonded in bitumen.

1st LAYER: Type 3G glass fibre base perforated felt, partially bonded.

High performance specifications

TOP LAYER: Polyester base cap sheet*, bonded in bitumen.

2nd LAYER: Polyester base underlayer*, bonded in bitumen.

1st LAYER: Type 3G glass fibre base perforated felt, partially bonded.

* Polyester base roofing systems usually incorporate a heavyweight cap sheet and a lighter weight underlayer

TOP LAYER: Bitumen polymer or pitch polymer roofing bonded in bitumen (pour and roll method), OR an APP modified bitumen roofing (torch applied).

2nd LAYER: Type 3B glass fibre base felt, bonded in bitumen.

1st LAYER: Type 3G glass fibre base felt, partially bonded in bitumen.

1 SPECIFICATIONS 10° SLOPE OR OVER

Overlay with wood fibreboard or cork to allow a full bond.

Standard: Type 2E mineral surfaced cap sheet
Type 2B fully bonded.

Upgraded: High performance mineral surfaced cap sheet
Type 2B or 3B fully bonded.

High performance: Mineral surfaced polyester base cap sheet
Polyester base underlayer fully bonded.

2 DECK

Woodwool slabs must be type SB to BS 1105 and preferably with a prefelted or prescreeded finish. The joints of all slabs should be taped according to manufacturers' instructions. In the case of prefelted slabs, taping will be necessary to complete temporary waterproof covering. On no account however should the prefelted finish be counted as part of the final waterproofing specification. Where a plain finish slab is used, a sand and cement screed or slurry should be applied to provide a suitable surface for the roofing specification.

3 THERMAL DESIGN

Standard thicknesses of woodwool slab provide the following U-values for the total construction, calculated from the thermal values listed in Table 4 Appendix A.

Thickness of woodwool (mm)	50	75	100	125
U-value (W/m^2 °C)	1.34	0.99	0.78	0.65

The effects of adding a ceiling or insulation at ceiling level are discussed in Section 1.2 Thermal Design.

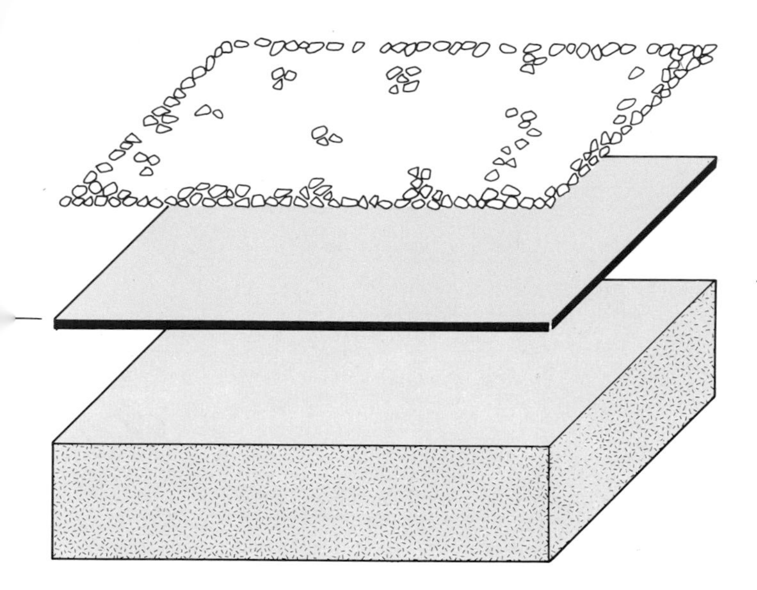

Layer of 10mm stone chippings
in bitumen based adhesive compound

Built-up roofing

Woodwool deck slabs laid to falls (Note 2)

BUILT-UP ROOFING

Standard specification: BS 747 Class 3 felts

TOP LAYER:	Type 3B glass fibre base felt, bonded in bitumen.
2nd LAYER:	Type 3B glass fibre base felt, bonded in bitumen.
1st LAYER:	Type 3B glass fibre base felt, bonded in bitumen.

Upgraded specification

TOP LAYER:	Polyester base roofing or similar, bonded in bitumen.
2nd LAYER:	Type 3B glass fibre base felt, bonded in bitumen.
1st LAYER:	Type 3B glass fibre base felt, bonded in bitumen.

High performance specifications

TOP LAYER:	Polyester base cap sheet*, bonded in bitumen.
1st LAYER:	Polyester base underlayer*, bonded in bitumen.

* Polyester base roofing systems usually incorporate a heavyweight cap sheet and a lighter weight underlayer

TOP LAYER:	Bitumen polymer or pitch polymer roofing bonded in bitumen (pour and roll method), OR an APP modified bitumen roofing (torch applied).
1st LAYER:	Type 3B glass fibre base felt, bitumen bonded.

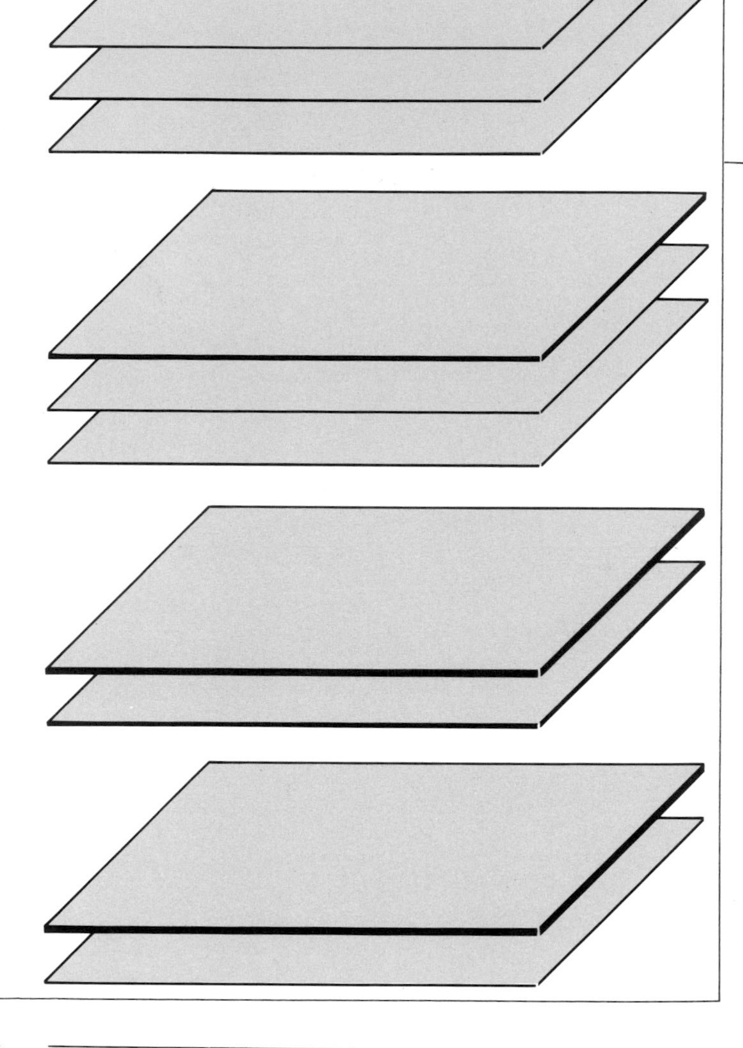

1 SPECIFICATIONS 10° SLOPE OR OVER

Standard:	Type 2E mineral surfaced cap sheet Type 2B fully bonded.
Upgraded:	High performance mineral surfaced cap sheet Type 2B or 3B fully bonded.
High performance:	Mineral surfaced polyester base cap sheet Polyester base underlayer fully bonded.

2 DECK

Metal decking is usually fixed by a specialist contractor as part of a single responsibility roofing operation comprising deck, thermal insulation and waterproofing.

For high humidity conditions, aluminium decks are recommended. If steel decks are used under these conditions, a protective coating to both sides of the deck should be specified.

3 UNDERLAY

Depending on the thickness and type of insulation board and the deck profile, an underlay may be required to support the board over the deck troughs. The specialist contractor will advise on the need for a suitable underlay and will recommend a suitable material.

An underlay is always required beneath expanded polystyrene to improve the bitumen bond of insulation to deck, and may also be required beneath insulations of low laminar strength to act as the critical layer for wind attachment, see Section 1.7, Wind Attachment Design Guide.

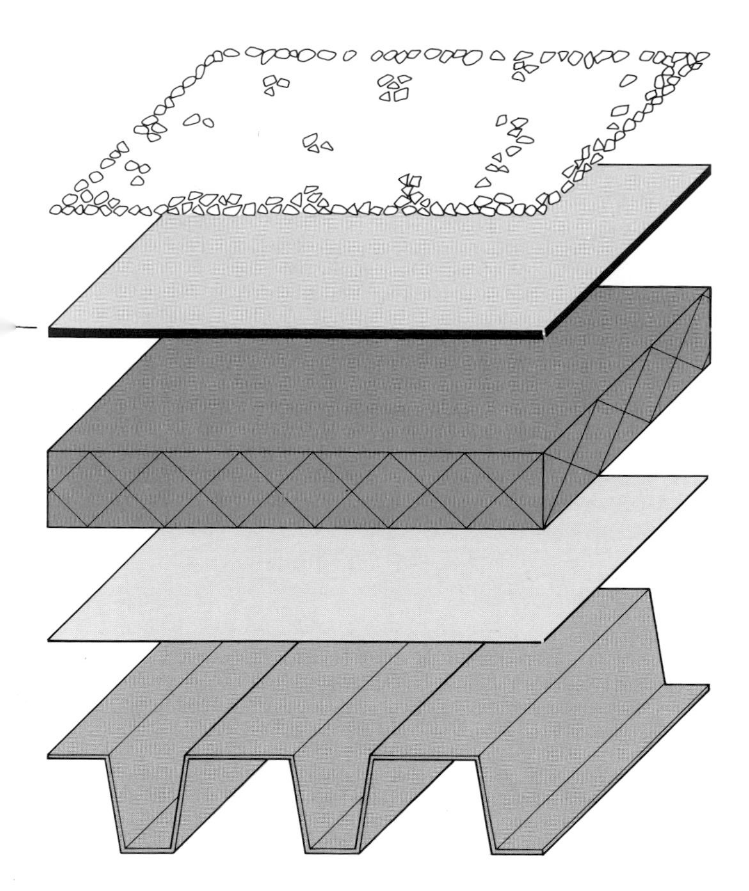

Layer of 10mm stone chippings in bitumen based adhesive compound

Built-up roofing

Insulation board bonded in hot bitumen (Note 5)

Underlay, vapour check or vapour barrier as required (Notes 3 and 4)

Metal trough decking (Note 2)

INSULATION
Wood fibreboard
Cork board
Glass fibre roofboard
Mineral wool slab
Cellular glass slab
Perlite board
Expanded polystyrene (HD grade, prefelted)

4 VAPOUR CHECK AND VAPOUR BARRIER

The requirements for a vapour check or vapour barrier with metal deck constructions are indicated in Table 1.11, Section 1.3, Vapour Barrier Design Guide.

4.1 Vapour check
The recommended specification for a vapour check is a single layer high performance roofing or a hessian or glass fibre reinforced felt bonded in bitumen to the top flats of the deck.

4.2 Vapour barrier
For a vapour barrier a two layer system is recommended.
First layer: High performance roofing bonded in bitumen to the top flats of the deck.
Second layer: BS 747 type 2B asbestos base or type 3B glass fibre base felt bonded in bitumen.

5 THERMAL INSULATION

The following thicknesses (mm) of insulation are required to achieve specified U-values for the total roof construction, calculated from the thermal values listed in Table 4 Appendix A.

Deck: steel or aluminium decking

Insulation	U-value (W/m^2 °C)				
	0.7	0.6	0.5	0.4	0.3
Wood fibreboard	59	71	88	113	154
Cork board	50	60	74	95	130
Glass fibre roofboard	40	48	60	77	105
Mineral wool slab	40	48	60	77	105
Cellular glass slab	53	64	79	102	139
Perlite board	59	71	88	113	154
Expanded polystyrene*	32	40	51	68	96

* U-value allows for additional 13mm wood fibreboard overlay

Standard specification: BS 747 Class 3 felts

TOP LAYER: Type 3B glass fibre base felt, bonded in bitumen.

2nd LAYER: Type 3B glass fibre base felt, bonded in bitumen.

1st LAYER: Type 3G glass fibre base perforated felt, partially bonded.

Upgraded specification

TOP LAYER: Polyester base roofing or similar, bonded in bitumen.

2nd LAYER: Type 3B glass fibre base felt, bonded in bitumen.

1st LAYER: Type 3G glass fibre base perforated felt, partially bonded.

High performance specifications

TOP LAYER: Polyester base cap sheet*, bonded in bitumen.

2nd LAYER: Polyester base underlayer*, bonded in bitumen.

1st LAYER: Type 3G glass fibre base perforated felt, partially bonded.

* Polyester base roofing systems usually incorporate a heavyweight cap sheet and a lighter weight underlayer

TOP LAYER: Bitumen polymer or pitch polymer roofing bonded in bitumen (pour and roll method), OR an APP modified bitumen roofing (torch applied).

2nd LAYER: Type 3B glass fibre base felt, bonded in bitumen.

1st LAYER: Type 3G glass fibre base felt, partially bonded in bitumen.

1 SPECIFICATIONS 10° SLOPE OR OVER

Overlay with wood fibreboard or cork to allow a full bond.

Standard: Type 2E mineral surfaced cap sheet
Type 2B fully bonded.

Upgraded: High performance mineral surfaced cap sheet
Type 2B or 3B fully bonded.

High performance: Mineral surfaced polyester base cap sheet
Polyester base underlayer fully bonded.

2 DECK

Metal decking is usually fixed by a specialist contractor as part of a single responsibility roofing operation comprising deck, thermal insulation and waterproofing.

For high humidity conditions, aluminium decks are recommended. If steel decks are used under these conditions, a protective coating to both sides of the deck should be specified.

3 UNDERLAY

Depending on the thickness of insulation board and the deck profile, an underlay may be required to support the board over the deck troughs. The specialist contractor will advise on the need for a suitable underlay and will recommend a suitable material.

4 VAPOUR CHECK AND VAPOUR BARRIER

The requirements for a vapour check or vapour barrier with metal deck constructions are indicated in Table 1.11, Section 1.3, Vapour Barrier Design Guide.

4.1 Vapour check
The recommended specification for a vapour check is a single layer high performance roofing or a hessian or glass fibre reinforced felt bonded in bitumen to the top flats of the deck.

4.2 Vapour barrier
For a vapour barrier a two layer system is recommended.
First layer: High performance roofing bonded in bitumen to the top flats of the deck.

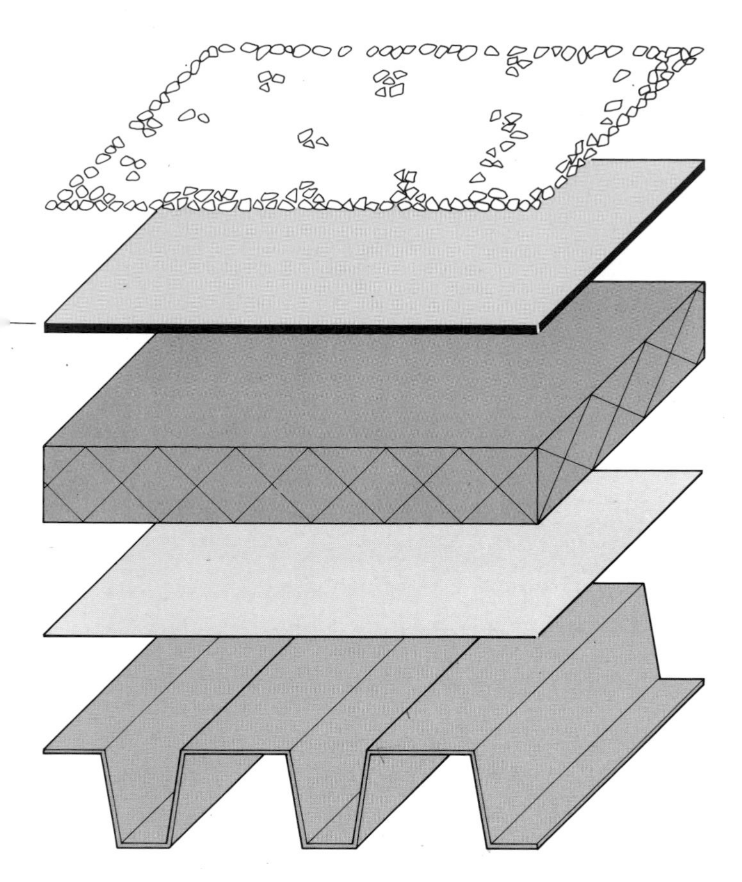

INSULATION
Polyurethane and
Polyisocyanurate roofboards
(glass tissued faced)

Second layer: BS 747 type 2B asbestos base or type 3B glass fibre base felt bonded in bitumen.

5 THERMAL INSULATION

The following thicknesses (mm) of insulation are required to achieve specified U-values for the total roof construction, calculated from the thermal values listed in Table 4 Appendix A.

Deck: Steel or aluminium decking

Insulation	U-value (W/m² °C) 0.7	0.6	0.5	0.4	0.3
Polyurethane board	26	32	39	50	68
Polyisocyanurate board	26	32	39	50	68

SECTION 4 MASTIC ASPHALT

4.1 MATERIALS

INTRODUCTION

Asphalt is available in two main basic forms, rolled asphalt or mastic asphalt. Rolled asphalt is used for road surfacing and paving and as the name implies, it is compacted by rolling. Rolled asphalt resists the passage of water but it cannot be described as an efficient waterproofing and is not used for roofing work.

Mastic asphalt is composed of suitably graded aggregates bound together with bitumens to make a dense material with no voids. It cannot be compacted, and is spread rather than rolled. It is fully waterproof and in terms of weathering is virtually indestructible.

Mastic asphalt is used for roof waterproofing, bridge-deck waterproofing and for wearing courses above the waterproofing on decks subject to traffic. It is also used for basement tanking, flooring, paving and for some heavy duty road surfacing.

The original mastic asphalts were all based on compositions containing natural rock asphalt. This is a naturally occurring limestone rock impregnated with bitumen and is found in geological formations mainly in France, Switzerland, Italy and Germany. In the natural form it cannot be melted down but can be crushed to form a fine aggregate which is blended with bitumen to form mastic asphalt.

Lake asphalt is a naturally occurring bitumen including 36% by weight of finely divided clay. The principal source of supply for the UK is from the asphalt lake in Trinidad.

Lake asphalt contains a higher proportion of bitumen than natural rock asphalt and occurs as a soft material which, although firm enough to walk on at normal temperatures, becomes liquid when heated.

Lake asphalt was first used with natural rock asphalt but it is now combined with bitumen and limestone to make some of the current specifications of asphalt.

ASPHALT MANUFACTURE

The manufacture of mastic asphalt is a batch production process. The raw materials include fine and coarse aggregates and asphaltic cement, which is the bitumen or mixture of bitumen and lake asphalt which binds the aggregate. These materials are pre-heated and fed in controlled amounts into a heated mixer capable of processing up to 10 tonnes of material. The mixture is continually agitated at a controlled temperature for 3-4 hours and, following satisfactory laboratory testing of samples, is discharged into steel moulds to form convenient sized blocks for transport and re-melting on site.

It is also possible to deliver asphalt in bulk as a hot charge, in areas which are close to the manufacturing plant.

Mastic asphalt is hard and brittle at freezing temperatures and soft, plastic and pliable in hot, sunny weather. The hardness can be varied over a wide range by changing the formulation of bitumens and aggregates during manufacture. The hardness can also be increased by the addition of coarse aggregate on site. Asphalt is hardened as a result of re-melting on site, and the higher the temperature and the longer the duration of heating, the harder the asphalt becomes.

Asphalt is normally delivered in block form and an allowance is made in manufacture for the hardening which will take place during re-melting.

When delivered as a hot charge, the asphalt is manufactured to the hardness required in service and there is no significant change in hardness during the time required for transport and application.

The standard quality control criterion for mastic asphalt is the hardness number. This denotes the depth in tenths of a millimetre to which a flat-ended steel rod 6.35mm diameter will indent the asphalt under a load of 9.8MN/m^2 applied for 60 seconds at a specified temperature.

Asphalt is cast into blocks

TYPES OF MASTIC ASPHALT

Types of mastic asphalt are normally defined by British Standards which set down the specification for manufacture including the composition, proportions and performance of the constituents, together with a stated hardness range for the finished material. Roofing and paving grades are available as follows:

Type	Description	Asphaltic cement compositition
BS 988B	Roofing grade	100% Bitumen
BS 988T		75% Bitumen 25% Lake asphalt
BS 1447	Paving grade	
Column 1		100% Bitumen
Column 2		100% Lake asphalt
Column 3		50% Bitumen 50% Lake asphalt
BS 1162	Natural rock roofing grade	
Column 1		100% Lake asphalt
Column 2		50% Bitumen 50% Lake asphalt
BS 1446	Natural rock paving grade	
Column 1		100% Bitumen
Column 2		100% Lake asphalt
Column 3		50% Bitumen 50% Lake asphalt

Mastic asphalt BS 988
Often referred to as roofing grade, BS 988 asphalt consists primarily of refinery produced bitumen, compounded with limestone powder and limestone aggregate. It is used for all waterproofing layers other than for basement tanking.

Up to 1973 BS 988 roofing grade was made in several types but the British Standard was rationalised in that year to include only two types - BS 988B and BS 988T. The asphaltic cement in BS 988B is 100% bitumen but for BS 988T it includes 75% bitumen and 25% lake asphalt.

BS 988T is generally regarded as the better quality. It has a silky texture and consequently improved laying properties. It is also thought by some designers to provide better thermal stability to the finished asphalt. Superfine particles of clay are very thoroughly dispersed in the natural lake asphalt material which has the effect of providing a thixotropic characteristic to the bitumen binder. It also confers unusual characteristics on the finished asphalt that cause the bitumen rich skin on the exposed surface to weather away more rapidly, and expose the light coloured aggregate. This surface skin comes to the top of all mastic asphalt grades during the spreading operation. The highly glossy film which would result if it were left untreated is broken down by rubbing sand into the surface while the asphalt is still hot. BS 988B asphalt based on 100% refined bitumen tends to maintain a darker colour as the surface skin requires a longer period of weathering before it is removed.

The permissible range of hardness at manufacture and hardness at laying are stipulated by British Standards. The hardness of roofing grade BS 988 is to be not less than 40 and not more than 80 at 25°C at manufacture, and not less than 30 at 25°C at time of laying. This allows a hardening of at least 10 during re-melting, but most manufacturers will keep the hardness at manufacture above 50 to allow an increased tolerance on site.

Hardness at time of laying can lie between 30 and about 70 at 25°C and this range has been carefully chosen to represent the optimum for installed roofing grade. Natural ageing will further harden the asphalt in service and the British Standard requirements are chosen with this aspect in mind. Roofing grade will normally age to a hardness number of about 20 after a few years and to about 10 after many years service.

It is clearly important that the asphalt is not over-heated or heated for too long during re-melting. This is normally easy to control on site but difficulties arise when unexpected weather changes prevent the application of asphalt which has already been prepared. It is sometimes necessary to keep the mastic hot for an extra day or even more. Under these circumstances it may be necessary to add a small quantity of bitumen to the mixer under controlled conditions, in order to soften the asphalt and restore it to a workable consistency.

The tolerances allowed for aggregate content in manufacture are fairly wide and a normal roofing grade will contain up to 20% of coarse aggregate. CP 144: Part 4 allows the incorporation of 5% to 10% additional coarse aggregate on site to harden the formulation to provide for a measure of increased wear and resistance to slumping and indentation. Although no value is specified by the Code, it can be expected that the decrease in hardness due to grit reinforcement will be in the order of 10 to 20.

Mastic asphalt to BS 1447
BS 1447 asphalt is known as paving grade. This type of asphalt is used for roads and footways and also as a heavy duty wearing course on top of roofing grade material to provide a suitable surface for roofs which are accessible to the public or subject to wheeled traffic.

Three paving grades are described in Table 1 of BS 1447 according to the composition of the asphaltic cement. Column 1 material is made from an asphaltic cement which is 100% bitumen, column 2 is from 100% fluxed lake asphalt and column 3 is from 50% bitumen and 50% lake asphalt. Column 3 material is most often specified as it spreads well and has skid resistant properties. Column 2 material has few advantages for the extra cost and is seldom used. Column 1 material can be hard to spread, is less skid resistant and would not normally be considered best value for money.

Paving grades contain a harder bitumen, more aggregate and a greater size of aggregate than roofing grade, to make them harder and more suitable as a wearing course. Their increased hardness makes paving grades too susceptible to thermal contraction in cold weather, or cracking from thermal shock, to be reliable as a waterproofing. The waterproofing function must always be provided by including a roofing grade layer under the paving grade.

No hardness requirements are specified in the British Standard for paving grade asphalts and the proportion of coarse aggregate can vary from 15% to 50% by weight. Thus there is scope for widely differing formulations, calling for specialist skills in the design and selection of the paving. A high proportion of aggregate will harden the wearing course and reduce indentation but the risk of cracking at the joints through contraction in cold weather will increase. Designers may therefore decide that it would be better to accept some degree of indentation rather than increase the risk of joint opening by adding more aggregate. The choice is personal and even experienced asphalters will disagree on how much aggregate to add. The designer must balance the risk at his own discretion, and if in doubt it is probably better to add a lower proportion of aggregate on the basis that a measure of indentation may prove less of a problem than joint opening.

Mastic asphalt to BS 1162 and BS 1446
BS 1162 and BS 1446 specifications relate to natural rock asphalt, roofing and paving grades respectively. They are the natural rock equivalents of BS 988 and BS 1447 and are used for specialised high quality applications.

THE SEPARATING LAYER

Section 1.5, Movement and Membranes, describes the need for a separating layer on all flat roof areas before the application of the mastic asphalt. Mention is also made of the need to present a relatively smooth clean substrate to receive the separating layer.

The purpose of the separating layer is to isolate the asphalt from joint movement in the substrate but still provide a significant friction to help restrain the asphalt against contraction in cold weather. It must also allow a free lateral passage for hot air and moisture vapour during the application of the hot asphalt and act as a long term vapour pressure release layer.

Separation is normally provided by a black sheathing felt conforming with BS 747 Type 4A (1), which is laid entirely loose with lap joints of 50mm. Sheathing felt has ideal characteristics as a separating layer for normal roof specifications but it allows a small measure of compression under traffic. A glass tissue separating layer is therefore applied to provide firm support under paving specifications and again this is laid entirely loose with lap joints of 50mm.

Care must be taken to ensure that local points of restraint are not formed by indentations or roughness of surface of the deck or by stepped joints between insulation panels. It is also necessary to ensure that the separating layer does not adhere to the substrate. Pre-felted decks, prefelted insulations, any bituminised surfaces or spillages of bitumen would lead to adhesion of the separating layer as well as allowing the bitumen to migrate into and pollute the asphalt. The non-bituminous facings of certain insulations can also lead to adhesion particularly if coated with a heat sensitive plastic film.

On surfaces which are likely to lead to adhesion, one or more layers of non-bituminous building paper beneath the sheathing felt will prevent adhesion and also guard against contamination of the asphalt by underlying bitumen.

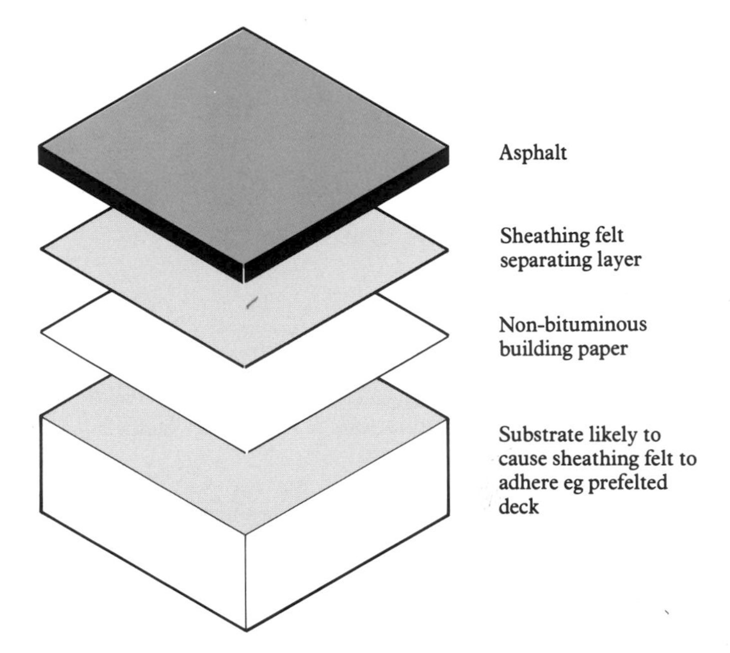

Black sheathing felt is normally used as a separating layer beneath asphalt roofing.

Heat sensitive insulations, such as expanded polystyrene, require an overlay of wood fibreboard, cork, perlite or similar heat-resistant board to act as a separating layer. The overlay boards should be tightly butted to prevent heat strike through open joints. Sheathing felt is then applied in the normal way before the asphalt.

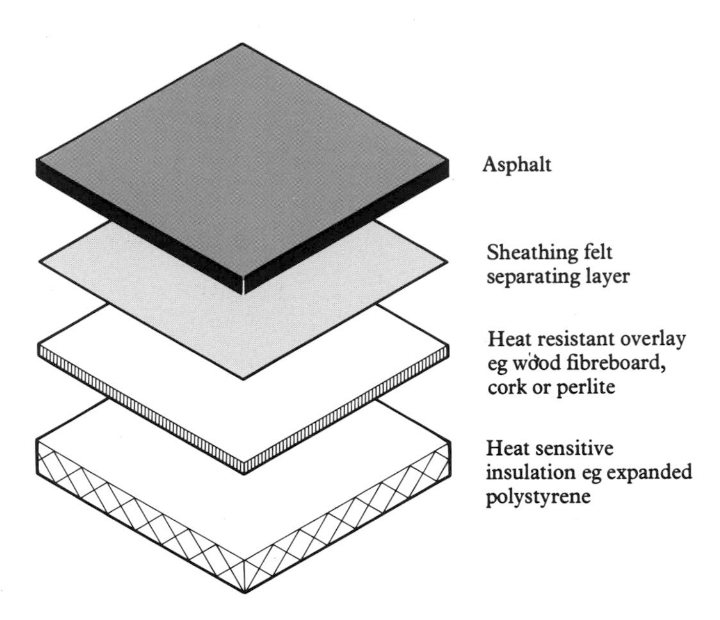

Whilst roofing grade asphalt will tolerate a small amount of compression in the insulation, paving grade will not. Asphalt also becomes soft in service when applied on an insulation and will not support traffic without indentation. Paving grade should not therefore be applied to insulation other than certain lightweight screeds which are poor insulators, have high compressive strength and a relatively high thermal mass. Some experimental work is proceeding with insulated paving specifications but designers will be well advised to keep to traditional procedures and add the insulation on the underside of the structural deck in the form of sprayed or proprietary linings.

4.2 ASPHALTING TECHNIQUES

Asphalting is a craft industry with fine traditions and asphalters learn their trade from long apprenticeships and years of experience under skilled supervision. The application of asphalt requires considerable judgement and hard work. Codes of Practice are published giving general advice but it has never proved possible to write down all the skills of the trade and it has to be expected that site conditions will lead asphalters to a wide variety of different judgements and techniques.

ASPHALT PREPARATION

The re-melting of asphalt blocks on site is carried out in mechanically agitated mixers which are large heated boilers with moving paddles to stir the mixture and ensure the thorough dispersion of the aggregate without settlement. For small work, cauldrons may be used and these are agitated manually.

The asphalt blocks are broken into lumps and added to the mixer to be heated until molten. The agitator is then started and heating continues until the application temperature of about 230°C is reached. The material is drawn off as required and it will be normal to add new broken blocks to the mix to keep the mixer full. In this way the mixerman can heat and deliver approximately 2½ times the capacity of the mixer to the roof during a working day.

Paving grade cannot be supplied in block form to the final specification as it would be too hard to successfully re-melt the blocks on site. The coarse aggregate is therefore added on site during agitation.

The asphalt is drawn off from the mixers or cauldrons into buckets or barrows for transport to the roof. The inside of the bucket or barrow will be dusted with cement or limestone to break adhesion and make it easy to remove the asphalt.

Temperatures are judged in relation to the weather, the time it takes to transport the hot asphalt to the working position and the type of work. For detail and vertical work the asphalt will be used at a lower temperature so that it will hold in position with minimum flow.

Detail work is normally carried out after horizontal work. However, in the early morning when the temperature of the asphalt is too low for application to flat areas, the detail work may be carried out in advance.

There is no technical advantage in following a particular sequence of work and the judgement is at the discretion of the asphalter.

If the mixer is insulated, it can be left partially charged with hot asphalt at the end of the day and this will remain hot overnight.

ASPHALT SPREADING

On level surfaces asphalt is applied at its highest working temperature and is guided and worked into position with a float. Asphalt is not only very hot when applied but holds its heat for a long time. In the hot state it will have a natural tendency to flow, and on flat areas will settle to a dead level top surface. The asphalter can use the flow characteristics to take out or ride over undulations.

The thickness of the layers of the flat areas are judged with only minimal help from battening, which acts both as formwork and a depth gauge.

Asphalt is applied in a thick layer, and retains a high temperature for a long period after application. The result is that entrapped moisture and air at the surface of the deck immediately expand and force their way under or through the asphalt as it is applied to appear as a blow or blister on the asphalt surface. The most common specification for roofing grade is two-coat work to a 20mm

The mixerman draws off the hot asphalt into buckets or barrows for transport to the roof

Asphalt is spread by hand using a wooden float

final thickness. The first 10mm coat will take up undulations and irregularities in the surface. It will be applied on a separating layer of sheathing felt which will isolate the asphalt from substrate movement and reduce the incidence of blowing by allowing the lateral escape of trapped air and moisture. Blowing is most likely to occur on detail work to brick or concrete when a bond is required and it is not possible to incorporate a layer of sheathing felt.

The separating layer will normally control blowing on flat areas to acceptable proportions providing the surface is not unduly wet. Any affected areas are pierced to release the pressures which have been formed and it may be necessary to work backwards and forwards over the asphalt for some time to create a smooth voidless finish.

The asphalter can only control a width of asphalt which he can reach from a kneeling position. He therefore works in bays of 2.5 or 3 metres width. Battens will normally be used to mark out bays for the first and second coats. No more first coat will be laid than can be covered the same day with the second coat.

The formation of the bay joints in all asphalt work is critical, as a weakness in the formation of the asphalt will allow the joint to be pulled open by the contraction of the material in cold weather. The bay joints between coats are staggered so that the first coat of asphalt acts as a strapping for the bay joints of the second coat. The need to form the strapping arrangement and the need to control blowing are the main reasons for the application of mastic asphalt in two coats.

The Code of Practice recommends that the junction between the two adjoining bays of a coat of mastic asphalt shall not be less than 150mm from a corresponding junction in a preceding coat. This is good practice particularly if the second layer can be applied the same day as the first.

Care has to be taken with the removal of the batten to ensure that all timber is removed. If thin slivers of timber break off and remain embedded in the edge of the asphalt it is possible that a crack may develop in the asphalt later. Fortunately, such cracks are normally localised and although unsightly do not lead to a breakdown of the waterproofing.

As the first layer proceeds from bay to bay, it is necessary to join the new hot asphalt to the asphalt which has previously been laid and is cold. The hot asphalt in a new bay is taken over the adjacent cold section for approximately 50mm and left in position for a few minutes to heat and soften the cold asphalt at the junction. The asphalter then works over the joint removing most of this 50mm band and rubbing the remainder well in to ensure complete fusion between old and new work.

The staggering of bay joints is also necessary when a layer of paving grade is applied on two coats of roofing grade to provide for heavy duty access or wheeled traffic.

Some years ago it was the practice to isolate the paving grade with a layer of glass tissue. Under these circumstances the joints of the paving grade were not strapped by the under-lying asphalt and failure at the joints occured too frequently. Joint opening was unsightly and in some cases would allow water to pass through and seep between the paving and roofing layers. This led to increased joint failure and finally to a breakdown of the paving at the joints. The accepted practice today is to bond the paving grade to the roofing grade with no separating layer. This allows the roofing grade to form a strapping for the paving and prevents a flow of water between the two grades.

After the application of the final coat of asphalt, sand is scattered over the top surface and rubbed in with a wooden float to break up the bitumen rich layer which will have come to the surface.

A hot poultice of asphalt is applied to warm and soften the edge of the previously laid bay to ensure complete fusion.

Sand rubbing the final coat

4.3 SURFACE PROTECTION

A sand rubbed surface will finish the application of all horizontal areas. If the asphalt is applied direct to the deck with no insulation, there is no need for further treatment to protect the asphalt.

Where mastic asphalt is used on a warm roof construction with an efficient insulation beneath the asphalt the membrane will operate within an increased temperature range. It will be hotter in the sun due to the low heat loss from the underside and colder in the winter due to the absence of a stabilising thermal mass beneath. Although there are a large number of asphalt roofs that perform satisfactorily without surface protection, it is now generally accepted that a protective surfacing should be applied to all flat areas of asphalt over an efficient insulation.

Asphalt hardens during the first few years of service but in the newly applied and unprotected state it is relatively soft at high temperatures and feels spongy underfoot on a hot sunny day. It is in the first few years of service that a reflective surfacing is most required.

This may be achieved by the addition of stone chippings, promenade tiles or paving, or a paint coating.

STONE CHIPPINGS

Where asphalt is laid on top of an efficient thermal insulation a layer of 10mm or 14mm stone chippings provides a good general purpose surfacing. New light coloured chippings can give a similar efficiency to that achieved using white reflective paint. Even though light coloured chippings normally darken with dirt after a few years, they will act as a permanent protection for the full life of the asphalt. The efficiency of reflective coatings is compared in the charts in Section 3.3 page 75.

A disadvantage of using stone chippings is in the location and repair of leakage. When the chippings are well bonded to the asphalt their removal can be difficult and it is better to provide only a minimum bond of the chippings to ensure that they remain in place. This can be achieved using a cold applied bitumen solution or mastic.

REFLECTIVE COATINGS

The practice of applying reflective paints has become a common treatment for asphalt skirtings and flat areas of asphalt applied to an insulation. Such a coating usefully reduces the temperature of the asphalt in service for as long as the coating lasts. A year or two of useful service is all that is necessary to allow the asphalt to settle, for stresses to be relieved, and to start the natural weathering process which changes the asphalt from black to a light grey. The ultimate failure of the reflective coating is not detrimental to the aged asphalt and it need not be replaced unless it is required to contribute to the temperature control of the building in the long term.

Aluminium coatings can be efficient when new but this will depend on the composition of the material. Ageing can quickly turn the aluminium into a medium grey finish which will offer little temperature reduction. The manufacturer should be consulted about the surface emissivity of both the new and aged materials and the frequency of recoating required.

PROMENADE SURFACING

Where regular traffic is expected, for example to and from a plant room, a walkway can be defined by tiles or pavings. The smooth level surface of mastic asphalt forms a good base for the application of concrete, asbestos or glass reinforced concrete tiles which are applied in hot bitumen.

For complete paved areas concrete, asbestos or glass reinforced concrete tiles may be bedded in bitumen and as the tiles are laid with a gap of a few millimetres at each joint, intermediate expansion joints are not normally required.

Concrete tiles or paving slabs can also be bedded in sand and cement in which case an isolating layer of building paper is necessary to allow for differential movement between the asphalt and the surfacing. Allowance for expansion should also be made with tiles or slabs set back 25mm from the roof perimeter and around major details and with intermediate joints at 3m centres.

An alternative method of supporting the slabs is by plastic or rubber corner supports. These proprietary support systems have the advantage of allowing drainage of rainwater beneath the paving, isolating the paving from the asphalt and making inspection and repair of the waterproofing easier. In buildings where vandalism is likely it may however be wise to adopt a fully bedded paving.

In-situ sand and cement paving is laid by specialist companies. This is normally cut in a tile pattern before the material hardens, and must be separated from the asphalt with building paper to allow for differential movements.

For all tiled or paved roofs, a minimum fall of 1:80 is required to ensure adequate drainage. For tiles and slabs bedded in bitumen a maximum fall of 1:40 is advisable to reduce the possibility of slippage in hot weather.

ROOF GARDENS AND TERRACING

Three-coat mastic asphalt is the most suitable specification for waterproofing to roof gardens and terracing where the location and repair of leaks will be difficult and expensive.

The asphalt must be applied under planned, favourable conditions and completed in full before the construction of the gardens or terraces. Before it is covered, the asphalt must be inspected carefully for visual faults and test cuts must be made if there is any doubt about the formation of the work. The specification must be designed to be free of the risk of blistering before it is covered and a full water test should be carried out on the membrane before it is accepted as suitable for construction of the garden or terrace. The test should last for approximately 48 hours and include water spray on surfaces which are to be covered but which cannot be included in the static water test.

All construction on top of the asphalt should be designed to be removable in sections to minimise disturbance or expense as far as this is possible. Whilst the roofing contractor may accept a contract which makes him responsible for certain repair work, he is unlikely to enter a contract accepting the unlimited costs of removal and replacement of superimposed materials for the purposes of inspection and repair.

The waterproofing under the insulation of a protected membrane roof is not easily accessible and whilst the cost of removal of the insulation and surfacing for inspection and repair of the membrane is not so great as in the case of a roof garden, this aspect should be borne in mind by the designer and the specification and detailing maintained at a high standard accordingly.

Asphalt is normally the most suitable waterproofing for roof gardens

4.4 TYPICAL SPECIFICATIONS

Mastic asphalt roofing specifications will include applications of two or three layers of asphalt and different types may be brought together. For example roofing and paving grades may be used together to provide an efficient waterproofing combined with a suitable wearing surface for heavy traffic.

ROOFING GRADE

For normal roofing purposes BS 988 roofing grade is applied in two coats of approximately 10mm to give a 20mm nominal thickness. Extra security is achieved by three coats of approximately 10mm to give a 30mm nominal thickness and this is recommended for work which is inaccessible for inspection and repair.

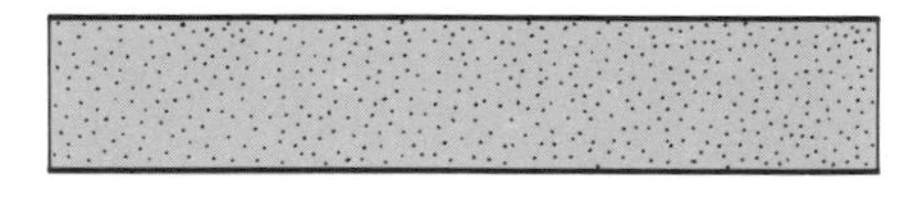

20mm two-coat asphalt on sheathing felt separating membrane

Roofing grade specifications will accept maintenance pedestrian traffic but will not be suitable for stiletto heels or static loads such as chairs or tables unless fully in the shade.

The addition of 5%-10% grit provides a harder and more resistant material for terraces and balconies. Specifications including grit reinforced roofing will normally be to 25mm finished thickness. The first coat will be a nominal 10mm thickness of normal roofing grade but may also include added grit. The second coat will be 15mm grit reinforced roofing to give added strength and wear resistance. Some asphalters believe that roofing grade with added grit is always preferable to the standard roofing grade as it holds up better to form a good thickness during application and is thought to have greater stability and firmness, particularly when the asphalt is applied to an insulated surface.

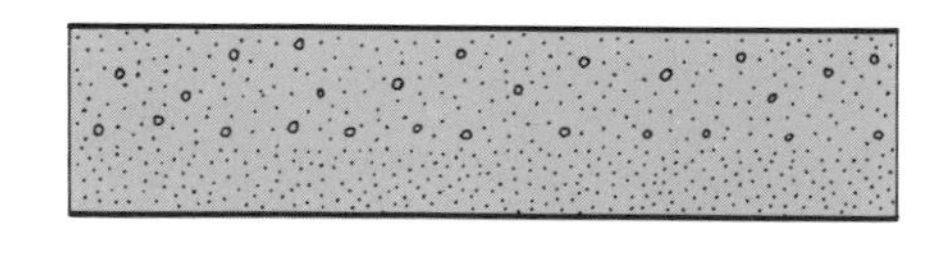

2nd coat roofing grade to 15mm with 5%-10% additional grit
1st coat roofing grade to 10mm on sheathing felt separating membrane

In order to further increase the wear and traffic resistance of asphalt roofing it is necessary to apply a suitable surface protection such as tiles or paving or turn to a specification with a paving grade applied on top of two-coat 20mm roofing.

PAVING GRADE

25mm is regarded as the minimum thickness of paving grade and this will provide a suitable surface for public access decks and podia. The coarse aggregate content of the asphalt will be in the region of 25%-30% for this thickness of paving grade.

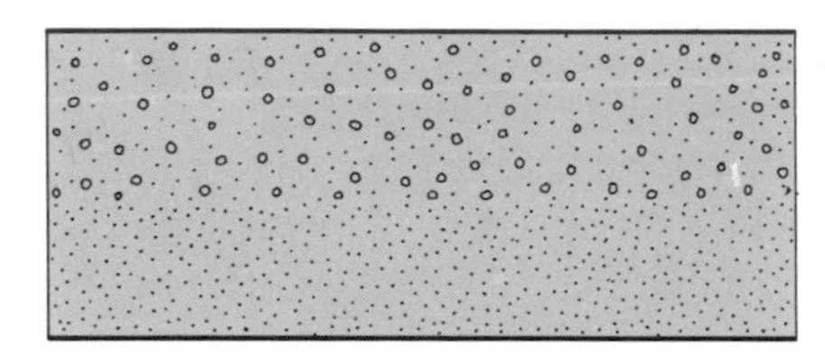

25mm paving grade asphalt

20mm two coat roofing grade laid on a glass fibre tissue separating layer

It is generally accepted that 30mm is the desirable thickness of paving grade for roofs designed to accept lightweight vehicle traffic and provide rooftop car parking, and for this specification the coarse aggregate content will be in the order of 30% to 40%.

Occasionally, a specifier will choose a 35mm thickness of paving grade for which the aggregate content would be in the order of 35% to 45%.

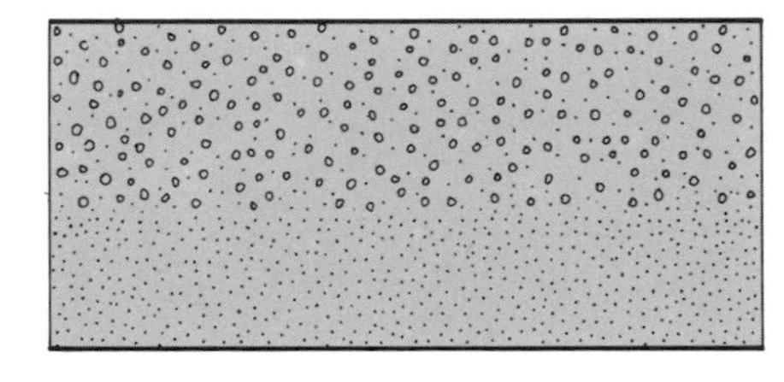

30mm paving grade asphalt

20mm two-coat roofing grade laid on a glass fibre tissue separating layer

A heavy duty paving for elevated roadways, loading bays and similar heavily trafficked areas will be a 40mm thickness of paving grade with an aggregate content in the range of 40% to 50%.

There is little to be said in favour of a thickness of paving in excess of 40mm, but local areas subjected to extreme point loads can be strengthened with proprietary metal reinforcements.

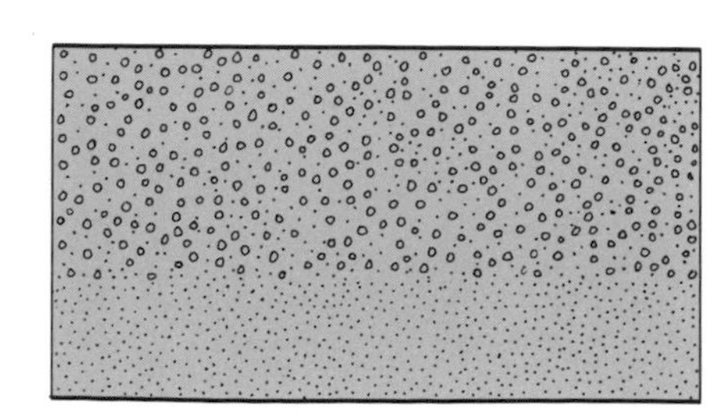

40mm paving grade asphalt

20mm two coat roofing grade laid on a glass fibre tissue separating layer

Further resistance to indentation can be achieved by the application of pre-coated chippings, scattered over the surface and rolled well in. This will also give improved traction for wheeled traffic. Increased traction and skid resistance can also be obtained by the application of proprietary surface dressings, normally based on epoxy resins and fine grit, or other skid resistant aggregate. Ramps to car parks can be so treated if required.

Surface warming may be achieved by the incorporation of electrical warming elements. The temperatures will be low and will have little effect on the hardness of the asphalt but specialist design will be necessary taking into account the form of the heating element, the slope and many other factors.

FLEXIBLE ASPHALT

A proprietary form of reinforced flexible asphalt is available which includes woven glass reinforcement between layers of an asphalt specification comprising approximately 50% bitumen and 50% fine aggregate. The high bitumen content and glass reinforcement add strength and flexibility to the asphalt membrane.

All surfaces are prepared by the application of BS 747 glass based roofing felt bonded or part bonded according to the substrate. Three layers of flexible asphalt and two layers of reinforcement are applied in successive layers to form the standard specification which has a thickness of approximately 10mm. Four layers of asphalt and three layers of reinforcement are applied for heavy duty specifications or for membranes which are to be buried beneath heavy surfaces. In either case the specification is regarded as a high performance membrane.

A protective surfacing of stone chippings or paving is always applied and the specification is suitable for slopes up to 10°. Skirtings are protected by mineral surfaced or aluminium foil faced high performance flashings.

TYPICAL SPECIFICATIONS

In the following pages, typical mastic asphalt specifications are illustrated for the most commonly encountered forms of flat roof construction.

Specifications are grouped according to the structural deck material in the following sequence:

Concrete decks (in-situ or precast)
Plywood and chipboard decks
Timber boarded deck
Woodwool deck
Metal deck

The methods of component assembly are shown for warm roof constructions (waterproofing in conjunction with over-deck insulation), cold roof constructions (waterproofing direct to deck) and for the protected membrane roof system where appropriate.

1 DECK

1.1 In-situ cast slab (illustrated)
A suitable surface of slab or screed for roofing is provided by a wood float finish. Temporary drainage holes should be formed in the structural slab and the deck should be adequately drained and dry before roofing. Apply a bitumen based primer to bind damp or dusty surfaces.

1.2 Precast concrete units
Joints between units may be taped or grouted according to manufacturers' instructions and the deck primed as necessary.

Depending on the accuracy of manufacture and laying, precast units may require a screed layer to even out deck irregularities. A screed may also be needed to create falls.

2 UNDERLAY

A fully bonded layer of BS 747 type 2B asbestos base or type 3B glass fibre base felt is always recommended to act as a damp proof underlay to the insulation while initial excess moisture dries out from an in-situ cast slab and screed or from a screed applied to precast units.

3 VAPOUR CHECK AND VAPOUR BARRIER

The requirements for a vapour check or vapour barrier with concrete deck constructions are indicated in Tables 1.6 and 1.7, Section 1.3, Vapour Barrier Design Guide.

4.1 Vapour check
The recommended specification for a vapour check on all concrete decks is a single layer of BS 747, type 2B asbestos base or type 3B glass fibre base felt bonded in bitumen.

4.2 Vapour barrier
For a vapour barrier, a two layer system is recommended as follows:
In-situ cast slab
First layer: BS 747 type 2B asbestos base or type 3B glass fibre base felt bonded in bitumen.
Second layer: High performance roofing bonded in bitumen.

Precast concrete units
First layer: BS 747 type 3G glass fibre base perforated felt, partially bonded.
Second layer: High performance roofing bonded in bitumen.

4 THERMAL INSULATION

Mastic asphalt requires continuous firm support. Most grades of mineral wool and glass fibre must be overlaid with a firm heat resistant board such as wood fibreboard, perlite board or cork. Heat sensitive boards such as expanded polystyrene or polyurethane foam must be overlaid in a similar fashion.

The following thicknesses of insulation are required to achieve specified U-values for the total roof construction, calculated from the thermal values listed in Table 4 Appendix A.

4.1 Deck: 150mm In-situ cast slab and screed (illustrated)

Insulation	U-value (W/m² °C) 0.7	0.6	0.5	0.4	0.3
Wood fibreboard	56	68	85	110	151
Cork board	47	57	71	92	127
Glass fibre roofboard	38	46	58	75	103
Mineral wool slab	38	46	58	75	103
Cellular glass slab†	51	61	76	99	136
Perlite board	56	68	85	110	151
Expanded polystyrene*	29	38	49	66	94
Polyisocyanurate roofboard	25	30	38	49	67

* U-value allows for additional 13mm wood fibreboard overlay

4.2 Deck: 100mm Precast lightweight concrete units

Insulation	U-value (W/m² °C) 0.7	0.6	0.5	0.4	0.3
Wood fibreboard	30	42	59	84	125
Cork board	25	35	49	70	105
Glass fibre roofboard	21	29	40	57	85
Mineral wool slab	21	29	40	57	85
Cellular glass slab†	27	38	53	75	113
Perlite board	30	42	59	84	125
Expanded polystyrene*	12	20	31	48	77
Polyisocyanurate roofboard	14	19	26	37	55

† Cellular glass slabs are mopped with hot bitumen after laying and two layers of non-bituminised paper are therefore required between the sheathing felt separating layer and the insulation to prevent the felt adhering to the slabs.

*U-value allows for additional 13mm wood fibreboard overlay

CONCRETE DECKS

WARM ROOF

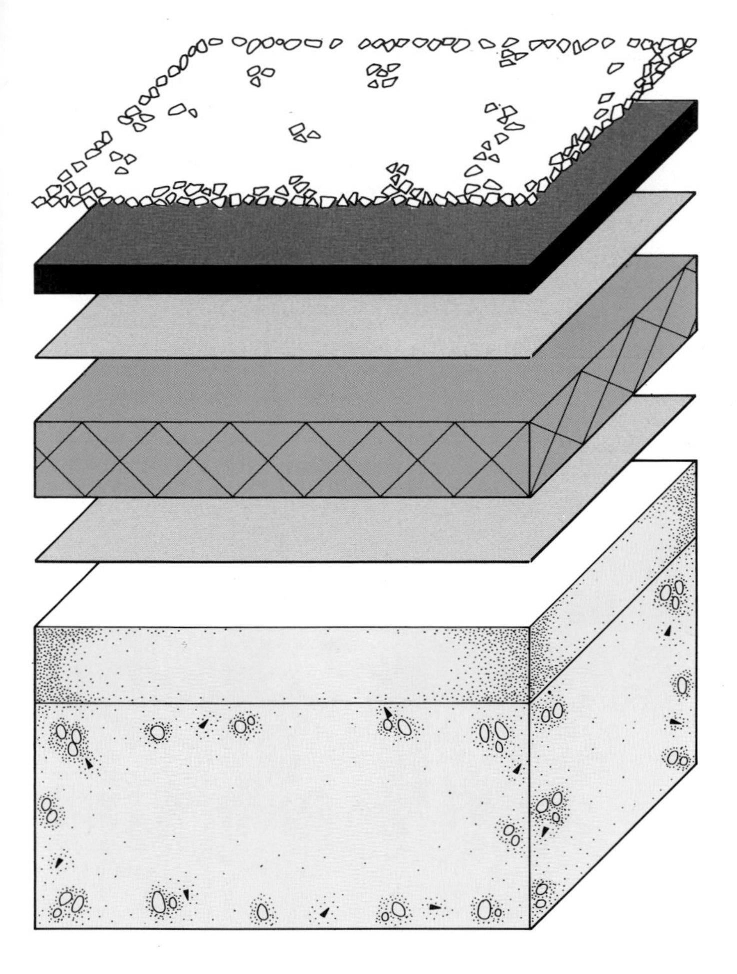

Layer of 10mm stone chippings
in bitumen based adhesive compound

20mm two-coat mastic asphalt
BS 988T recommended

Separating layer of loose-laid sheathing felt

Insulation board bonded
in hot bitumen (Note 4)

Underlay, vapour check or
vapour barrier as required
(Notes 2 and 3)

Sand and cement
screed to falls

INSULATION
Wood fibreboard
Cork board
Glass fibre roofboard
Mineral wool slab
Cellular glass slab
Perlite board
Expanded polystyrene (HD grade, prefelted)
Polyisocyanurate roofboard

In-situ cast dense concrete slab
or precast lightweight concrete deck units (Note 1)

1 DECK

1.1 In-situ cast slab (illustrated)
A suitable surface of slab or screed for roofing is provided by a wood float finish. Temporary drainage holes should be formed in the structural slab and the deck should be adequately drained and dry before roofing.

1.2 Precast concrete units
Joints between units may be taped or grouted according to manufacturers' instructions.

Depending on the accuracy of manufacture and laying, precast units may require a screed to even out deck irregularities. A screed may also be needed to create falls.

2 THERMAL DESIGN

The contribution of the construction to a required U-value will depend on the type and thickness of deck and screed used. Manufacturers of precast lightweight concrete units will indicate the thicknesses of deck to achieve specific U-values without additional above-deck insulation.

The effects of adding a ceiling or insulation at ceiling level are discussed in Section 1.2 Thermal Design.

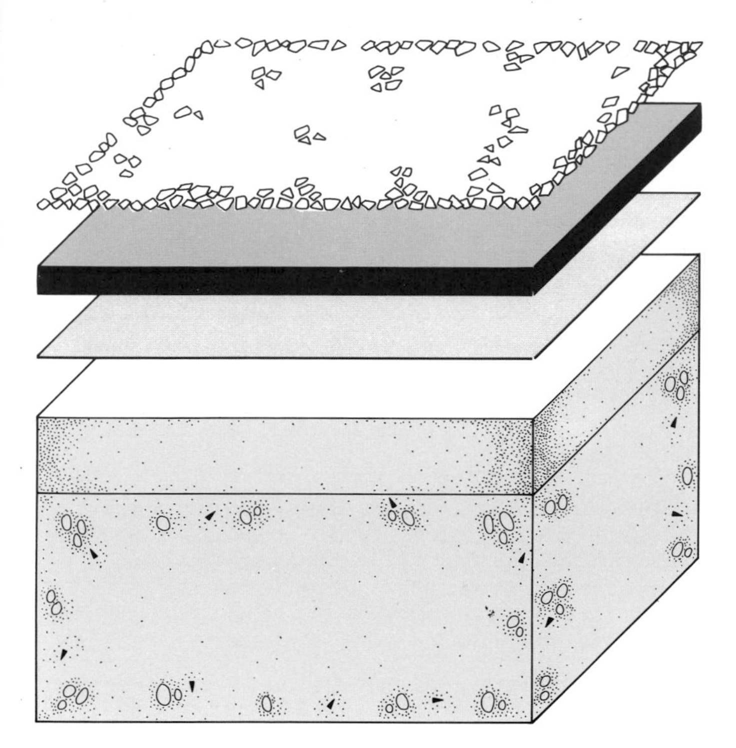

Layer of 10mm stone chippings
in bitumen based adhesive compound

20mm two-coat mastic asphalt
BS 988T recommended

Separating layer of loose-laid sheathing felt

Sand and cement screed to falls

In-situ cast dense concrete slab or precast lightweight concrete deck units (Note 1)

1 DECK

1.1 In-situ cast slab (illustrated)
A suitable surface of slab or screed for roofing is provided by a wood float finish. Temporary drainage holes should be formed in the structural slab and the deck should be adequately drained and dry before roofing.

1.2 Precast concrete units
Joints between units may be taped or grouted according to manufacturers' instructions.
Depending on the accuracy of manufacture and laying, precast units may require a screed layer to even out deck irregularities. A screed may also be needed to create falls.

2 THERMAL INSULATION

When calculating board thickness to achieve a specific U-value, it is necessary to allow for loss of efficiency once the board is installed due to the effect of rainwater draining under the insulation.

Manufacturers may take this into account by including a 20% increase in board thickness. If no such allowance has been made, the loss of efficiency should be taken into account by adjusting the calculated U-value according to the thermal resistance below the level of the insulation, see Appendix A.

Standard thicknesses of insulation board provide the following U-values (W/m$^{2\circ}$C) for the total construction, calculated from the thermal values listed in Table 4 Appendix A, with an adjustment for the effect of rainwater drainage below the insulation as indicated in Table 1 Appendix A.

Deck	Thickness of insulation (mm)				
	40	50	60	75	80
150mm in-situ cast slab and screed	0.68	0.58	0.50	0.43	0.41
100mm precast lightweight concrete units	0.51	0.45	0.40	0.35	0.33

3 SURFACE FINISH

If significant quantities of fine gravel will be present or if paving slabs are to be bedded in sand, it will be necessary to add a filter layer above the insulation to prevent fine material working through the joints and accumulating on the underside of the boards. The manufacturer of the insulation will advise a suitable material.

A 50mm depth of gravel or concrete paving will prevent flotation of the insulation, provided an efficient drainage has been incorporated in the design. If drainage is inadequate, a depth of gravel or paving equal to the thickness of the insulation will be required.

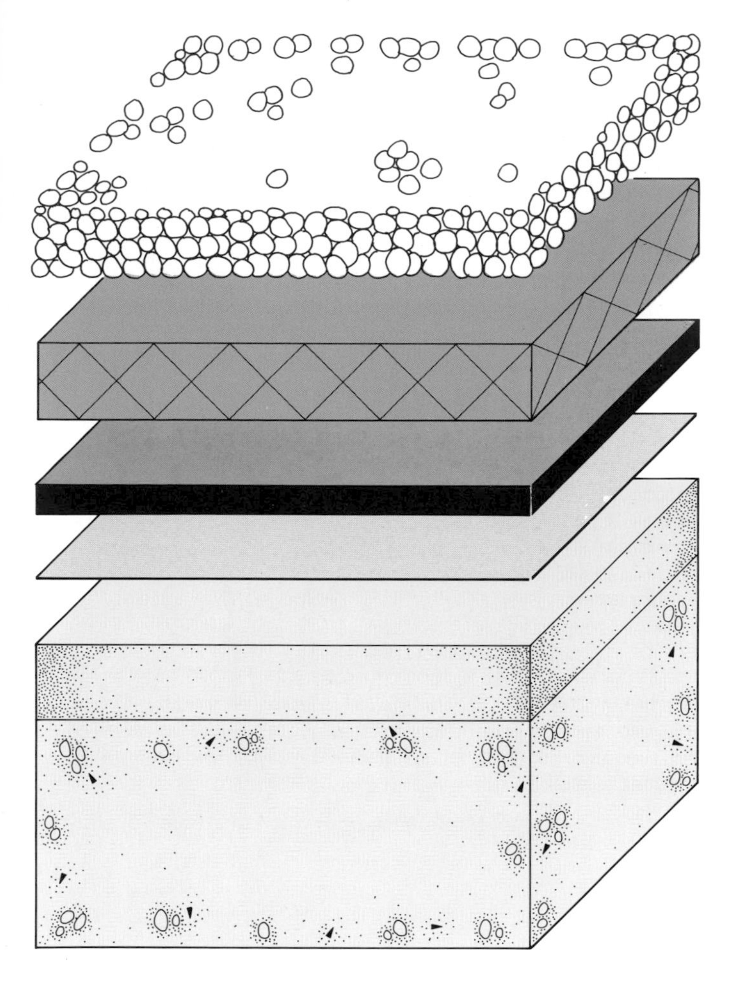

Minimum 50mm layer of gravel (20-30mm nominal diameter) or 50mm paving (Note 3)

Extruded polystyrene insulation board (minimum 50mm) laid loose (Note 2)

INSULATION
Extruded polystyrene

20mm two-coat mastic asphalt
BS 988T recommended

Separating layer of loose-laid sheathing felt

Sand and cement screed to falls

In-situ cast dense concrete slab or precast lightweight concrete deck units (Note 1)

1 DECK

1.1 Plywood
Plywood should be of exterior grade WBP to BS 1455. Panels should be well nailed to timber joists and noggings. Any deck joints not closed off by the support system should be taped.

1.2 Chipboard
Chipboard decking should be prefelted and of type II/III or type III to BS 5669. Panels should be well nailed to timber joists and noggings and supported at all edges. A full taping of the deck joints will be necessary to complete a temporary waterproof covering.

2 UNDERLAY

A fully bonded layer of BS 747 type 2B asbestos base or type 3B glass fibre base felt may be applied to provide waterproofing cover to plywood decking before the main roofing specification is applied.

3 VAPOUR CHECK AND VAPOUR BARRIER

The requirements for a vapour check or vapour barrier with plywood and chipboard deck constructions are indicated in Table 1.8, Section 1.3 Vapour Barrier Design Guide.

3.1 Vapour check
The recommended specification for a vapour check is a single layer of BS 747 type 2B asbestos base or type 3B glass fibre base felt bonded in bitumen.

3.2 Vapour barrier
Chipboard decking will not normally be suitable for use in high humidity conditions.

For a vapour barrier over plywood decking a two layer system is recommended.
First layer: BS 747 type 3G glass fibre base perforated felt partially bonded.
Second layer: High performance roofing bonded in bitumen.

4 THERMAL INSULATION

Mastic asphalt requires continuous firm support. Most grades of mineral wool and glass fibre must be overlaid with a firm heat resistant board such as wood fibreboard, perlite board or cork. Heat sensitive boards such as expanded polystyrene or polyurethane foam must be overlaid in a similar fashion.

The following thicknesses (mm) of insulation are required to achieve specified U-values for the total roof construction, calculated from the thermal values listed in Table 4 Appendix A.

Deck: 19mm plywood or chipboard

Insulation	U-value (W/m^2°C)				
	0.7	0.6	0.5	0.4	0.3
Wood fibreboard	55	67	83	108	150
Cork board	46	56	70	91	126
Glass fibre roofboard	37	45	57	74	102
Mineral wool slab	37	45	57	74	102
Cellular glass slab†	49	60	75	97	135
Perlite board	55	67	83	108	150
Expanded polystyrene*	29	37	48	65	93
Polyisocyanurate roofboard	24	30	37	48	66

†Cellular glass slabs are mopped with hot bitumen after laying and two layers of non-bituminised paper are therefore required between the sheathing felt separating layer and the insulation to prevent the felt adhering to the slabs.

* U-value allows for additional 13mm wood fibreboard overlay

PLYWOOD, CHIPBOARD DECKS

WARM ROOF

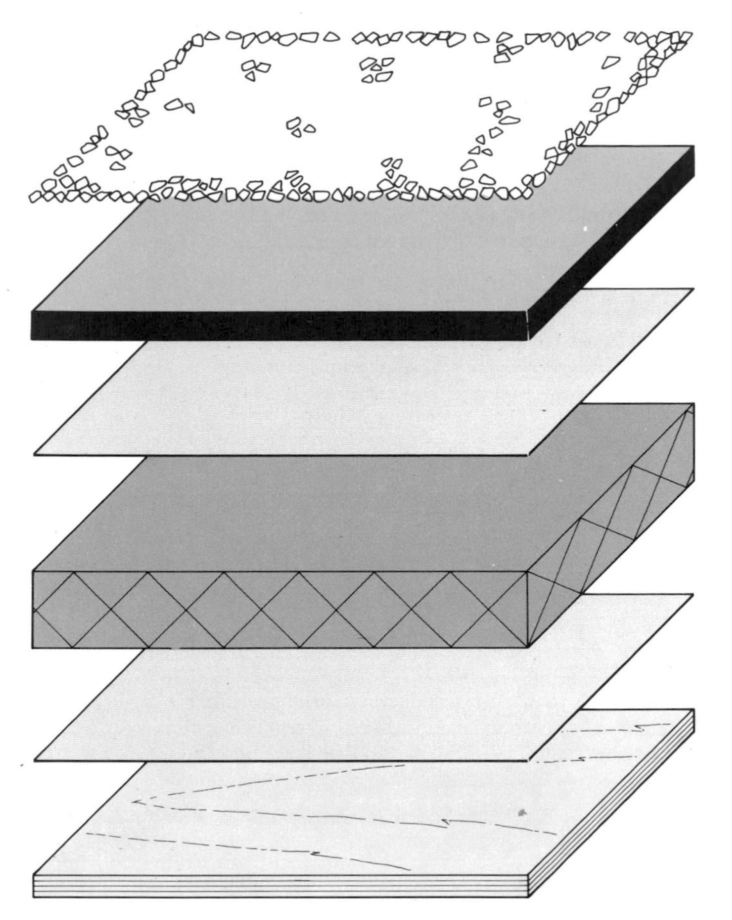

Layer of 10mm stone chippings in bitumen based adhesive compound

20mm two-coat mastic asphalt BS 988T recommended

Separating layer of loose-laid sheathing felt

Insulation board bonded in hot bitumen (Note 4)

Underlay, vapour check or vapour barrier as required (Notes 2 and 3)

Plywood or chipboard decking laid to falls (Note 1)

INSULATION
Wood fibreboard
Cork board
Glass fibre roofboard
Mineral wool slab
Cellular glass slab
Perlite board
Expanded polystyrene (HD grade, prefelted)
Polyisocyanurate roofboard

1 DECK

Timber boards should be well seasoned and not less than 19mm nominal thickness. Tongued and grooved boarding is essential, with boards closely clamped and securely nailed. Plain edge boards should not be used.

2 UNDERLAY

A nailed layer of BS 747 type 2B asbestos base felt is required to provide a suitable surface for bitumen bonding the insulation boards.

3 VAPOUR CHECK AND VAPOUR BARRIER

The requirements for a vapour check or vapour barrier with timber boarded constructions are indicated in Table 1.9, Section 1.3 Vapour Barrier Design Guide.

3.1 Vapour check
The nailed underlay of BS 747 type 2B asbestos base felt will also function as a vapour check layer.

3.2 Vapour barrier
For a vapour barrier, a two layer system is recommended.
First layer: High performance roofing nailed.
Second layer: BS 747 type 2B asbestos base or type 3B glass fibre base felt bonded in bitumen.

4 THERMAL INSULATION

Mastic asphalt requires continuous firm support. Most grades of mineral wool and glass fibre must be overlaid with a firm heat resistant board such as wood fibreboard, perlite board or cork. Heat sensitive boards such as expanded polystyrene or polyurethane foam must be overlaid in a similar fashion.

The following thicknesses (mm) of insulation are required to achieve specified U-values for the total roof construction, calculated from the thermal values listed in Table 4 Appendix A.

Deck: 19mm Tongued and grooved boarding

Insulation	U-value (W/m² °C)				
	0.7	0.6	0.5	0.4	0.3
Wood fibreboard	55	67	83	108	150
Cork board	46	56	70	91	126
Glass fibre roofboard	37	45	57	74	102
Mineral wool slab	37	45	57	74	102
Cellular glass slab†	49	60	75	97	135
Perlite board	55	67	83	108	150
Expanded polystyrene*	29	37	48	65	93
Polyisocyanurate roofboard	24	30	37	48	66

† Cellular glass slabs are mopped with hot bitumen after laying and two layers of non-bituminised paper are therefore required between the sheathing felt separating layer and the insulation to prevent the felt adhering to the slabs.

* U-value allows for additional 13mm wood fibreboard overlay

TIMBER BOARDED DECK

WARM ROOF

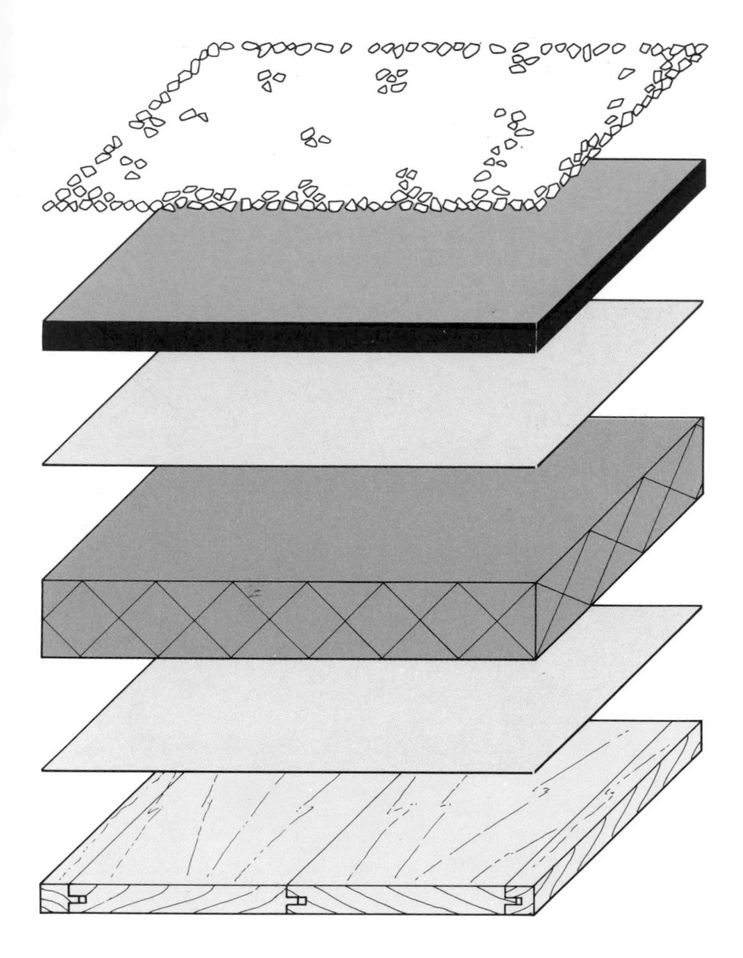

Layer of 10mm stone chippings
in bitumen based adhesive compound

20mm two-coat mastic asphalt
BS 988T recommended

Separating layer of loose-laid sheathing felt

Insulation board bonded in hot bitumen (Note 4)

INSULATION
Wood fibreboard
Cork board
Glass fibre roofboard
Mineral wool slab
Cellular glass slab
Perlite board
Expanded polystyrene (HD grade, prefelted)
Polyisocyanurate roofboard

Underlay, vapour check or vapour barrier as required (Notes 2 and 3)

Timber boarded deck laid to falls (Note 1)

MASTIC ASPHALT

1 DECK

Timber boards should be well seasoned and not less than 19mm nominal thickness. Tongued and grooved boarding is essential, with boards closely clamped and securely nailed. Plain edge boards should not be used.

2 THERMAL DESIGN

To reduce the risk of condensation the cavity space formed between the deck and suspended ceiling should be ventilated to the outside air. Openings equivalent to 4% of the plan roof area are suggested.

If cavity ventilation is likely to be restricted it is advisable to change the design to a warm roof construction.

TIMBER BOARDED DECK

COLD ROOF

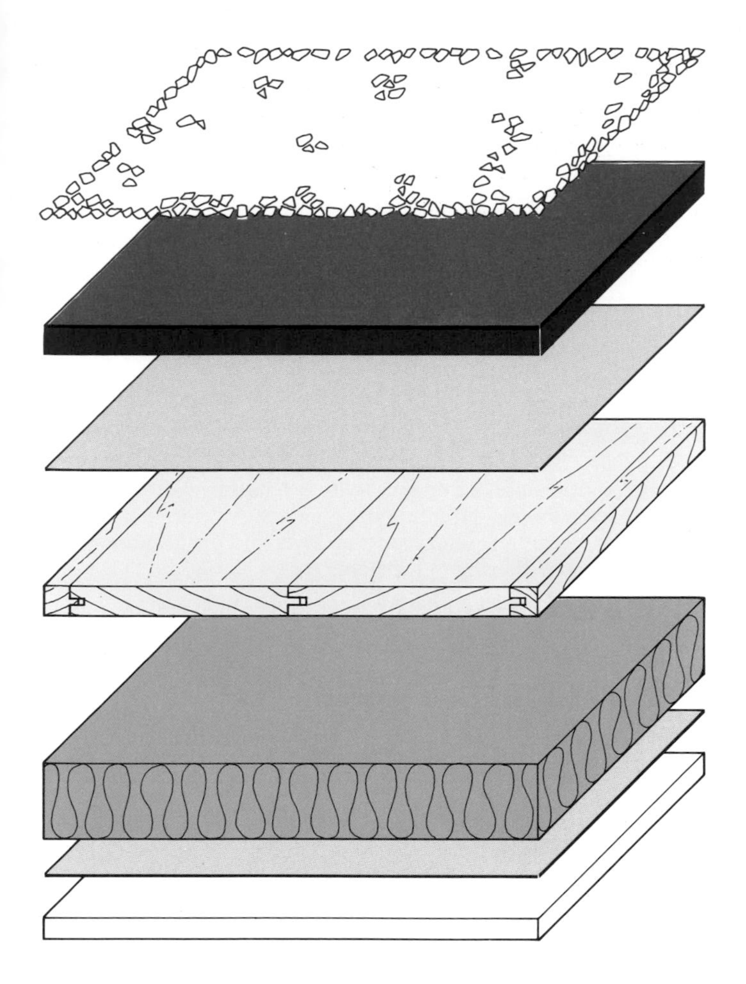

Layer of 10mm stone chippings
in bitumen based adhesive compound

20mm two-coat mastic asphalt
BS 988T recommended

Separating layer of loose-laid sheathing felt

Timber boarded deck laid to falls (Note 1)

Cavity ventilated to outside air (Note 2)

Insulation

Vapour check where required

Suspended ceiling

1 DECK

Woodwool slabs must be type SB to BS 1105 and the joints of all slabs should be taped according to manufacturers' instructions. In the case of prefelted slabs, taping will be necessary to complete the temporary waterproof covering. Where a plain finish slab is used, a sand and cement screed or slurry should be applied to provide a suitable surface for the roofing specification.

2 UNDERLAY

A fully bonded layer of BS 747 type 2B asbestos base or type 3B glass fibre base felt may be applied to provide waterproofing cover to prescreeded and plain finish slabs before the main roofing specification is applied.

3 VAPOUR CHECK AND VAPOUR BARRIER

The requirements for a vapour check or vapour barrier with woodwool constructions are indicated in Table 1.10, Section 1.3 Vapour Barrier Design Guide.

3.1 Vapour check
The recommended specification for a vapour check is a single layer of BS 747 type 2B asbestos base or type 3B glass fibre base felt bonded in bitumen. The facing to prefelted woodwool slabs will also function as a vapour check if taped.

3.2 Vapour barrier
For a vapour barrier, a two layer system is recommended.
First layer: BS 747 type 3G glass fibre base perforated felt partially bonded.
Second layer: High performance roofing bonded in bitumen.

4 THERMAL INSULATION

Mastic asphalt requires continuous firm support. Most grades of mineral wool and glass fibre must be overlaid with a firm heat resistant board such as wood fibreboard, perlite board or cork. Heat sensitive boards such as expanded polystyrene or polyurethane foam must be overlaid in a similar fashion.

The following thicknesses (mm) of insulation are required to achieve specified U-values for the total roof construction, calculated from the thermal values listed in Table 4 Appendix A.
Deck: 50mm Woodwool slab

Insulation	U-value (W/m^2 °C)				
	0.7	0.6	0.5	0.4	0.3
Wood fibreboard	35	46	63	88	130
Cork board	29	39	53	74	109
Glass fibre roofboard	24	32	43	60	88
Mineral wool slab	24	32	43	60	88
Cellular glass slab†	31	42	57	79	117
Perlite board	35	46	63	88	130
Expanded polystyrene*	15	23	34	51	80
Polyisocyanurate roofboard	15	21	28	39	57

† Cellular glass slabs are mopped with hot bitumen after laying and two layers of non-bituminised paper are therefore required between the sheathing felt separating layer and the insulation to prevent the felt adhering to the slabs.

* U-value allows for additional 13mm wood fibreboard overlay

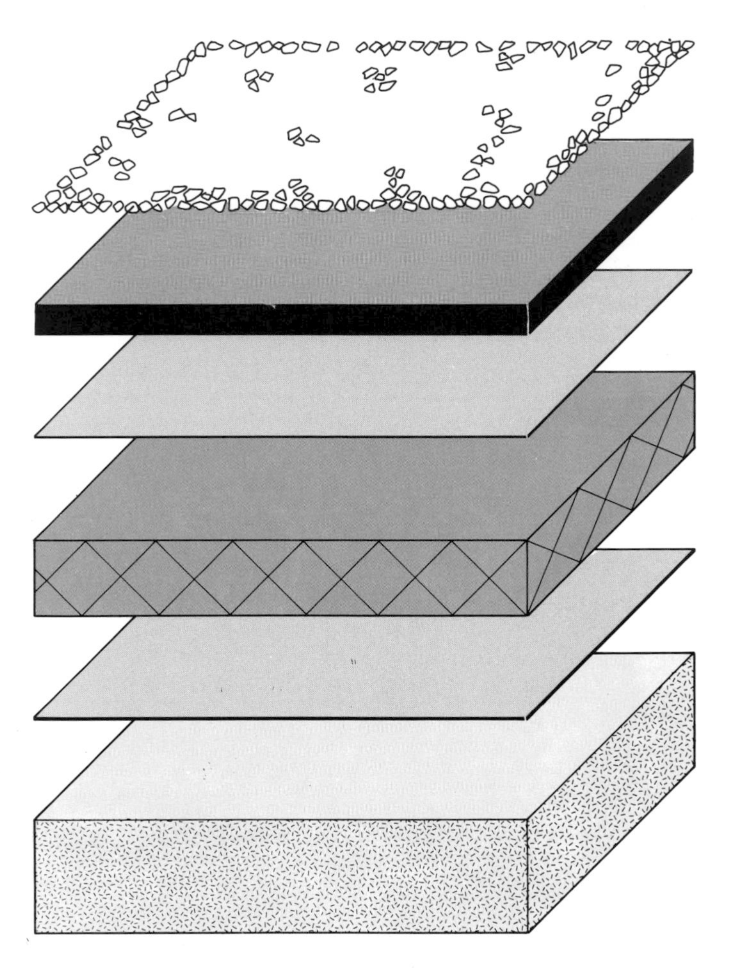

Layer of 10mm stone chippings in bitumen based adhesive compound

20mm two coat mastic asphalt BS 988T recommended

Separating layer of loose-laid sheathing felt

Insulation board bonded in hot bitumen (Note 4)

Underlay, vapour check or vapour barrier as required (Notes 2 and 3)

Woodwool deck slabs (Note 1)

INSULATION
Wood fibreboard
Cork board
Glass fibre roofboard
Mineral wool slab
Cellular glass slab
Perlite board
Expanded polystyrene (HD grade, prefelted)
Polyisocyanurate roofboard

1 DECK

Woodwool slabs must be type SB to BS 1105 and the joints of all slabs should be taped according to manufacturers' instructions. Where a plain finish slab is used, a sand and cement screed or slurry should be applied to provide a suitable surface for the roofing specification.

2 SEPARATING LAYER

When prefelted woodwool slabs are used or bituminous taping for closing slab joints, a loose-laid non-bituminised paper underlay to the sheathing felt separating layer is required to prevent the sheathing felt adhering to the pre-felted deck surface.

3 THERMAL DESIGN

Standard thicknesses of woodwool slab provide the following U-values for the total construction, calculated from the thermal values listed in Table 4 Appendix A.

Thickness of woodwool (mm)	50	75	100	125
U-value (W/m^2 °C)	1.34	0.99	0.78	0.65

The effects of adding a ceiling or insulation at ceiling level are discussed in Section 1.2 Thermal Design.

WOODWOOL DECK

NO INSULATION ABOVE DECK

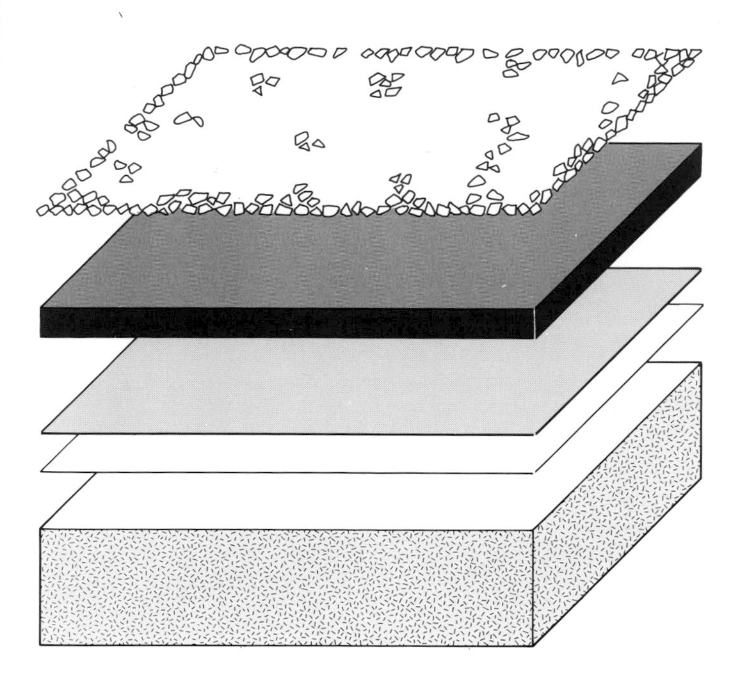

Layer of 10mm stone chippings
in bitumen based adhesive compound

20mm two-coat mastic asphalt
BS 988T recommended

Separating layer of loose-laid sheathing felt with non-bituminised
paper underlay as required (Note 2)

Woodwool deck slabs (Note 1)

1 DECK

Metal decking is usually fixed by a specialist contractor as part of a single responsibility roofing operation comprising deck, thermal insulation and waterproofing.

For high humidity conditions, aluminium decks are recommended. If steel decks are used under these conditions, a protective coating to both sides of the deck should be specified.

2 UNDERLAY

Depending on the thickness and type of insulation board and the deck profile, an underlay may be required to support the board over the deck troughs. The specialist contractor will advise on the need for an underlay and will recommend a suitable material. material.

An underlay is always required beneath expanded polystyrene to improve the bitumen bond of insulation to deck, and may also be required beneath insulations of low laminar strength to act as the critical layer for wind attachment, see Section 1.7 Wind Attachment Design Guide.

3 VAPOUR CHECK AND VAPOUR BARRIER

The requirements for a vapour check or vapour barrier with metal deck constructions are indicated in Table 1.11, Section 1.3 Vapour Barrier Design Guide.

3.1 Vapour check
The recommended specification for a vapour check is a single layer high performance roofing or a hessian or glass fibre reinforced felt bonded in bitumen to the top flats of the deck.

3.2 Vapour barrier
For a vapour barrier a two layer system is recommended.
First layer: High performance roofing bonded in bitumen to the top flats of the deck.
Second layer: BS 747 type 2B asbestos base or type 3B glass fibre base felt bonded in bitumen.

4 THERMAL INSULATION

Strong moisture resistant insulations such as isocyanurate foam and cork provide suitable support for the mastic asphalt. Most grades of mineral wool and glass fibre must be overlaid with a firm heat resistant board such as wood fibreboard, perlite board or cork. Heat sensitive boards such as expanded polystyrene or polyurethane foam must also be overlaid.

The following thicknesses (mm) of insulation are required to achieve specified U-values for the total roof construction, calculated from the thermal values listed in Table 4 Appendix A.

Deck: steel or aluminium decking

Insulation	U-value (W/m^2 °C) 0.7	0.6	0.5	0.4	0.3
Wood fibreboard	59	71	88	113	154
Cork board	50	60	74	95	130
Glass fibre roofboard	40	48	60	77	105
Mineral wool slab	40	48	60	77	105
Cellular glass slab†	53	64	79	102	139
Perlite board	59	71	88	113	154
Expanded polystyrene*	32	40	51	68	96
Polyisocyanurate roofboard	26	32	39	50	68

† Cellular glass slabs are mopped with hot bitumen after laying and two layers of non-bituminised paper are therefore required between the sheathing felt separating layer and the insulation to prevent the felt adhering to the slabs.

* U-value allows for additional 13mm wood fibreboard overlay

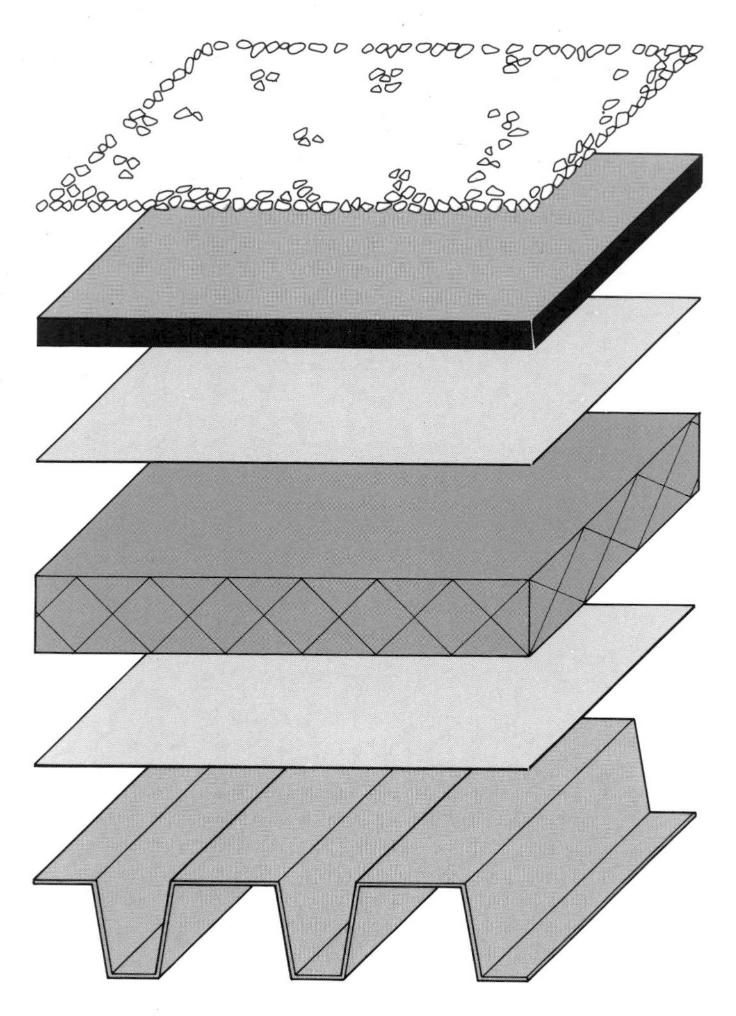

INSULATION
Wood fibreboard
Cork board
Glass fibre roofboard
Mineral wool slab
Cellular glass slab
Perlite board
Expanded polystyrene (HD grade, prefelted)
Polyisocyanurate roofboard

SECTION 5 DETAIL DESIGN

5.1 BUILT-UP ROOFING DETAIL DESIGN

INTRODUCTION

The formation of roof waterproofing details calls for a combination of the skills of the designer and the roofer. There are many different approaches to detail design and it is not possible to claim any one approach as the only correct method.

There are, however, fundamental principles of detail design which apply to all flat roofs, irrespective of the waterproofing material. These are described in the following pages.

At all skirtings and upstands, the waterproofing must be carried at least 150mm, and preferably between 150mm and 250mm, above the level of the finished roof, to protect against rain splash-back.

At open edges, the waterproofing must be turned down at least 50mm. The only reliable alternative to this is to incorporate a parapet and coping.

These minimum dimensions should never be overlooked or compromised because of other conflicting demands such as appearance or cost.

SIDE OF ROLL AND END OF ROLL

When considering built-up roofing details, there are two separate cases to be examined: the end of roll formation of details and the side of roll formation. At the end of roll vertical sections of the detail are formed by a continuation of the roll of roofing. At the side of roll the roofing will usually be finished on the flat and the vertical sections of the detail will be inserted as extra pieces.

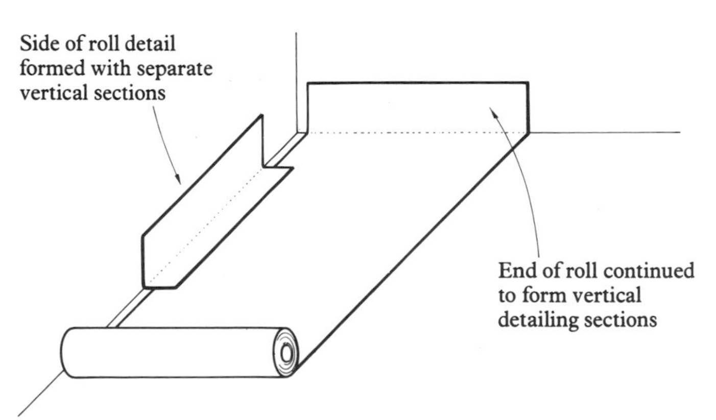

The end of roll detail is therefore easier to form, and for this reason roofers normally run the length of rolls along the shortest dimension of the roof. This gives them the maximum opportunity for carrying out the details at the end of roll.

Most catalogues and codes of practice show details at the end of roll formation only and this can give a misleading impression of the final detail at the side of roll. The application techniques for side and end of roll are best illustrated by considering the example of a skirting to a wall.

The positioning of the layers of felt will depend on the number of layers specified and on whether the specification is fully or partially bonded.

Fully bonded waterproofing
If the deck or insulation is one which will accept a fully bonded system of either 2 layers or 3 layers, it will be possible to carry the successive layers of roofing straight up the skirting at the end of roll condition.

It is usual to complete the upstand with staggered layers of roofing to prevent a thick ledge forming at the top of the upstand. The staggered formation is also more stable against slipping and slumping and the thinning of the waterproofing at the top of the upstand causes no problems.

End of roll condition

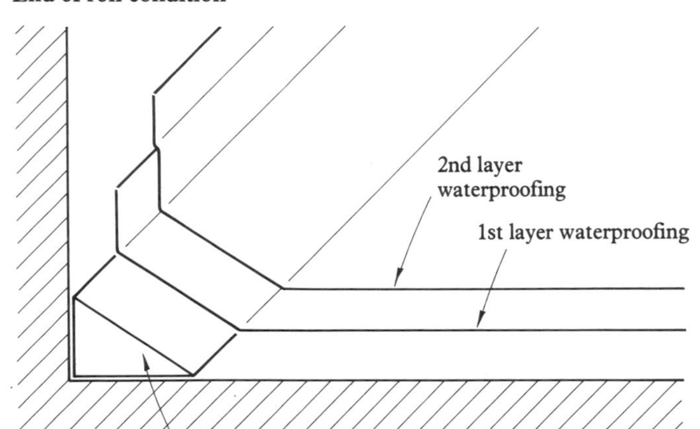

2 layer specification

1st layer
2nd layer
3rd layer

3 layer specification

The layers of waterproofing at the side of roll will finish on the horizontal, and it will be necessary to complete the layers through the detail with separate pieces.

Side of roll condition

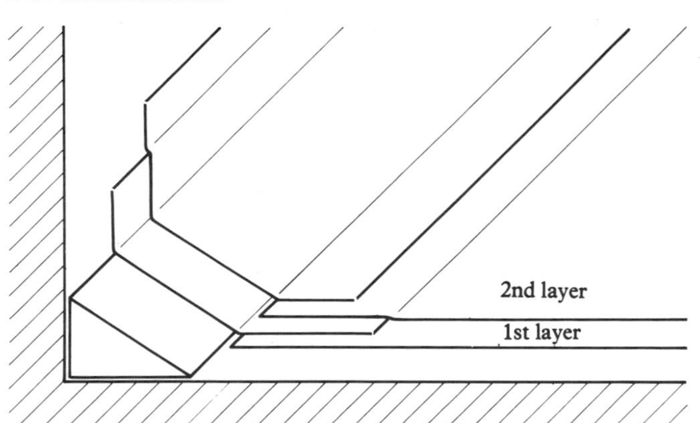

2 layer specification

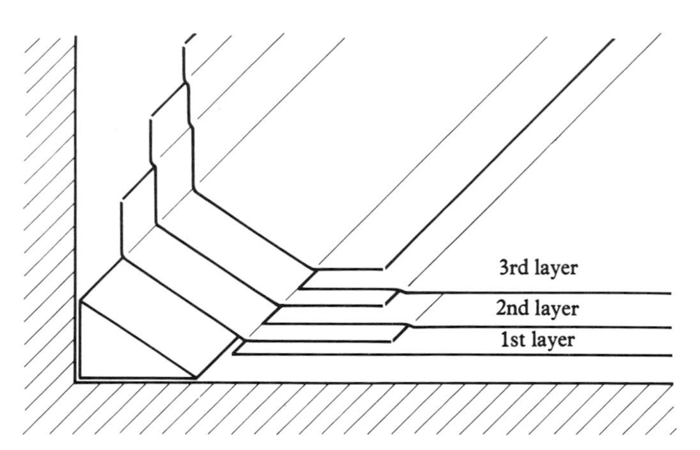

3 layer specification

Partially bonded waterproofing
If the deck or insulation requires a partial bond system, this will probably be achieved by a layer of BS 747 type 3G perforated gritted roofing.
To prevent wind damage, the edges of the roof must be closed off against the entry of air to the underside of the waterproofing and it is therefore necessary to fully bond all the layers of the detail work. Codes of practice suggest that the 3G should be stopped back from the roof edge to allow a 450mm wide fully bonded strip to be formed but this is seldom carried out and has not proved necessary. It is better to have the maximum area part-bonded provided the detail work is fully bonded and effectively prevents the entry of air to the underside of the waterproofing.
3G is only suitable for application to horizontal areas and is always stopped at the base of the skirting.

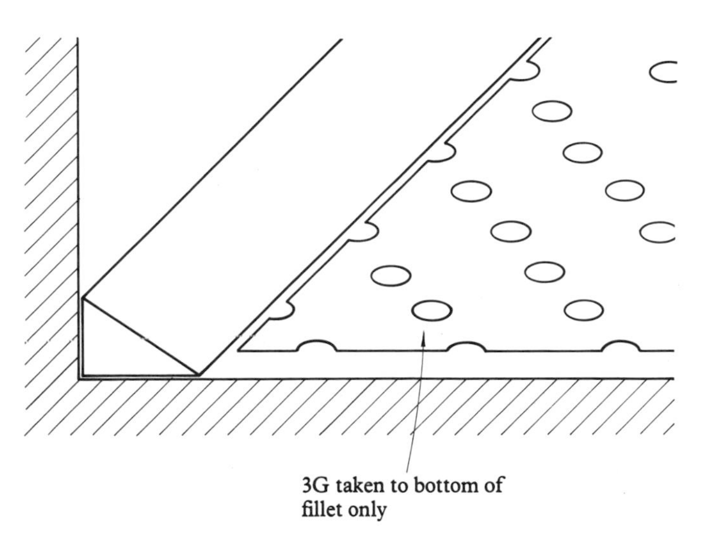

If the succeeding layers are to be two layers of high performance polyester base felt, the 3G underlay is only added to form a means of attachment by a partial bond. It is not an essential part of the waterproofing and it is not necessary to extend the first layer through the upstand, as the two polyester felts are sufficient.

If, however, the waterproofing specification is of 3 layers of BS 747 roofing, it is important that the first layer is extended through the detail, otherwise the upstand would comprise only two layers of low strength roofing and this would be insufficient for good performance. The first layer should be completed with material to match the remainder of the specification, probably BS 747 type 3B.

End of roll condition

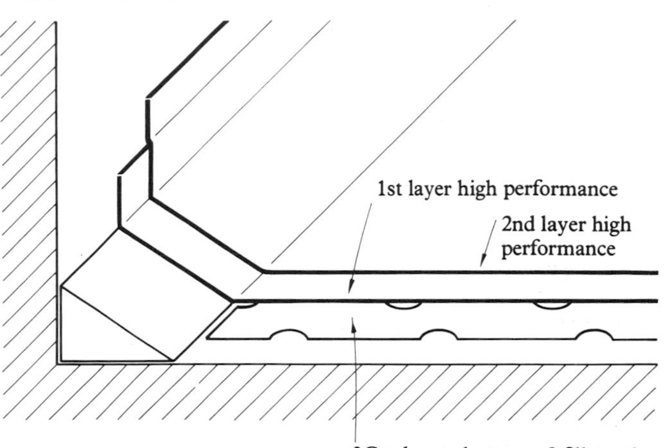

Partially bonded high performance specification

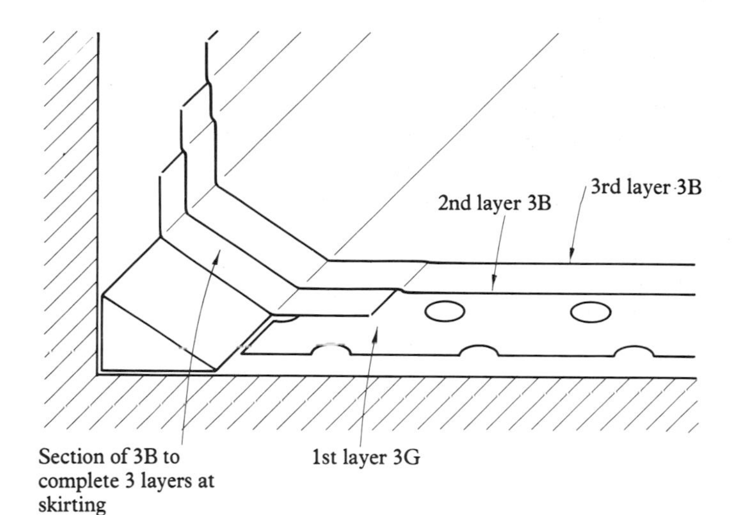

3 layer BS 747 specification

At the side of roll condition, all layers of waterproofing are finished on the horizontal and the continuation through the detail is formed with separate pieces.

Side of roll condition

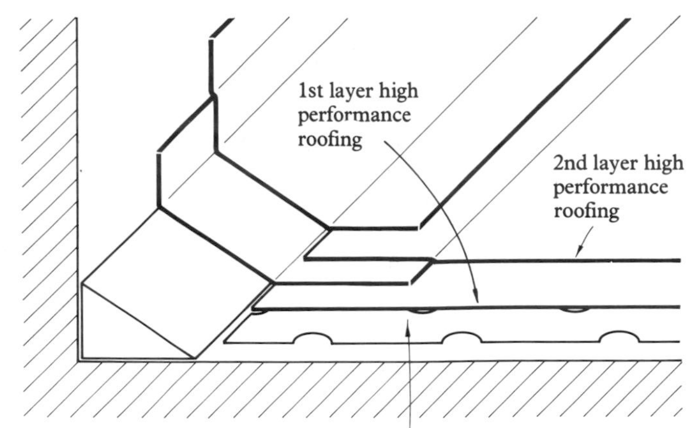

Partially bonded high performance specification

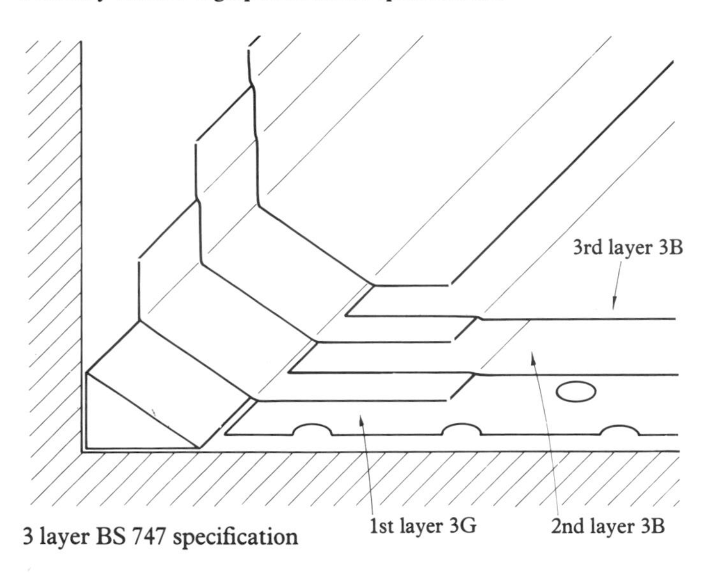

3 layer BS 747 specification

FILLETS

A fillet is normally incorporated at the base angle of the upstand to take out the sharp corner and provide a satisfactory base for the upstand material. If a fillet is not provided, a void would frequently be formed between the waterproofing and the corner of the structure. The waterproofing would then be unsupported and likely to split if there is traffic in the area, particularly if the upstand material has become hard and brittle after several years of service, and if there is differential movement between the roof and wall.

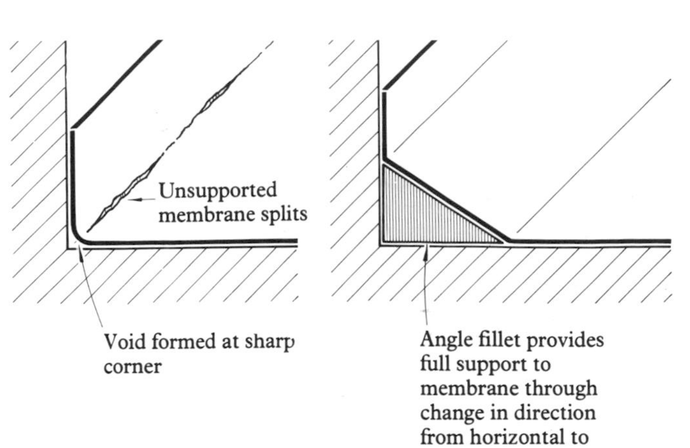

The traditional materials for fillets, sand and cement or timber, are now largely replaced by wood fibreboard, or sometimes high density polyurethane. Sand and cement fillets are generally only used between concrete decks and brick or concrete walls. Wood fibreboard fillets are preferred because they present a dry and stable sloping face for the waterproofing. A 50mm by 75mm fillet is probably the best size, and may be used either way to give a choice of height.

SKIRTINGS

Skirting at junction of brick wall and concrete deck
If there is no insulation on the roof deck, a fillet is introduced immediately and is bonded to the deck and to the wall with a continuous mopping of bitumen. The bitumen bonding also protects the fillet from dampness in the construction.

If insulation board is to be included and a vapour barrier, vapour check or underlay is required, this layer should be taken through the detail to extend far enough above the fillet to enable the waterproofing to complete an envelope to the insulation. For the side of roll condition, it will be necessary to complete the layer with separate inserts for the vertical areas. It is not strictly necessary to carry out this operation for a vapour check or an underlay, but in practice it is easier to treat all layers as if they were vapour barriers. This will avoid any on-site confusion regarding the purpose of the layer and its method of application.

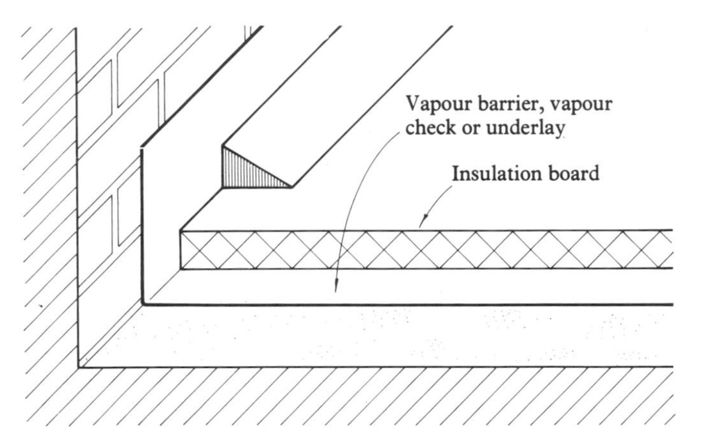

Single layer vapour barrier

A vapour barrier should be applied with precautions to ensure that where necessary it is isolated from building movement. This will sometimes lead to the use of a two-layer vapour barrier. The first layer will be partially bonded, generally by using BS 747 type 3G roofing, and for both side of roll and end of roll this layer should be finished on the flat. The second layer of the vapour barrier should continue through the detail and extend far enough above the fillet to allow the envelope to be completed.

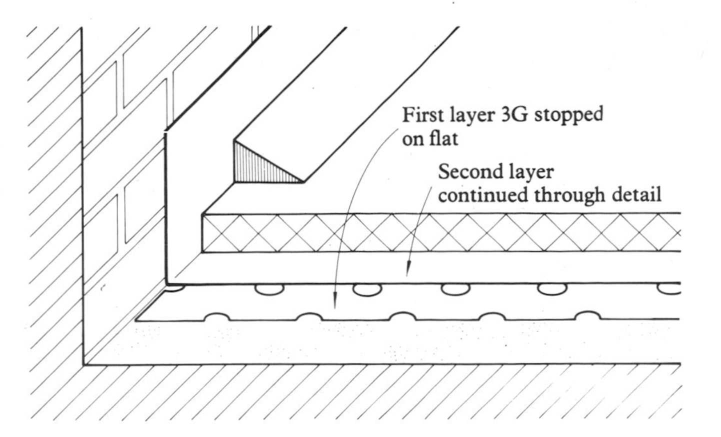

Two layer vapour barrier

The metal cover flashing is normally positioned immediately under the damp proof course, and set a minimum 30mm into the wall. It is good practice to build it into the wall under the dpc during construction.

With the majority of built-up roofing specifications, the top of the upstand will be thinned by the staggered formation of the materials and a ledge will not be formed at the top of the flashing. If the full thickness of waterproofing is maintained at the top of the upstand or if the thickness is increased by the addition of insulation or a freestanding kerb, a ledge may be formed which might hold water. To avoid this, the flashing should enter the wall one course higher than the top of the upstand. It will be necessary to take this into account when setting the level for the dpc at design stage.

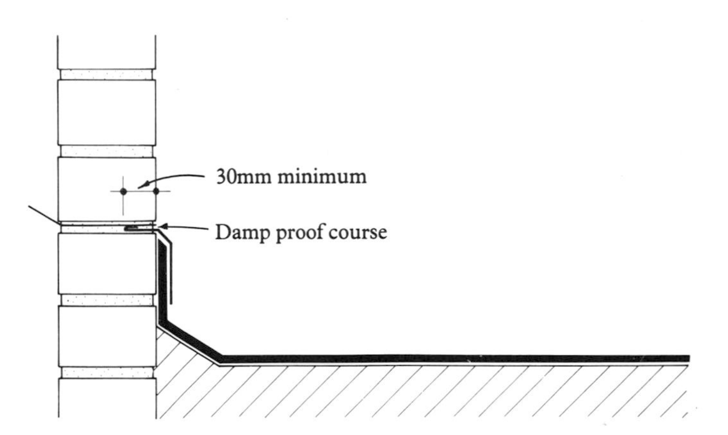

The formation of the upstand waterproofing will then follow the principles previously described for side of roll and end of roll conditions.

A protective surfacing at skirtings is normally achieved by substituting a mineral surfaced or metal foil faced roofing for the final layer. This extends on to the horizontal area for approximately 75mm and is continued up to the level of a cover flashing which is set underneath the damp proof course in the brickwork.

The chippings may be stopped at the bottom of the fillet or carried to the top.

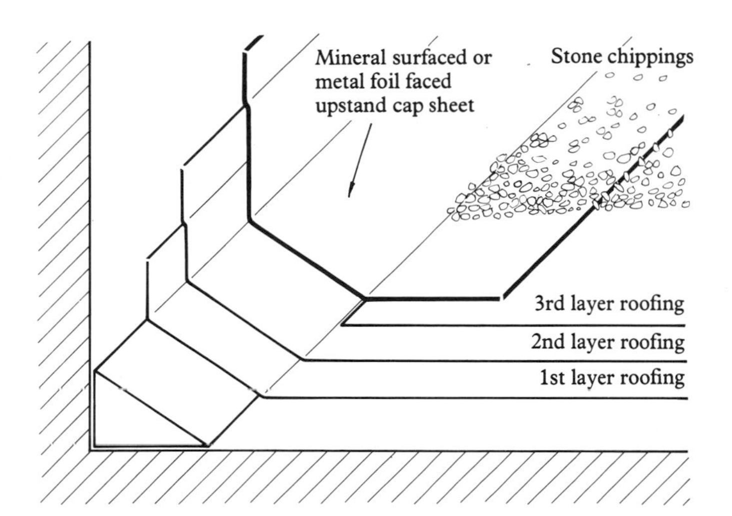

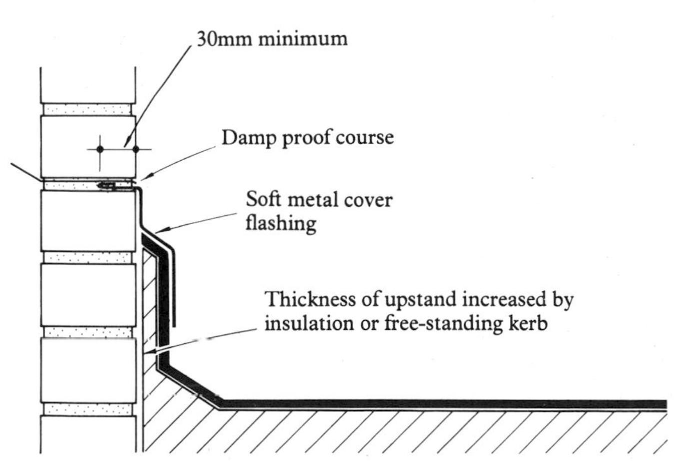

An alternative method of protecting the membrane is to form the upstand from the same cap sheet material used on the flat areas of the roof, carry the chipping finish to the top of the fillet, and dress the metal cover flashing down to the top of the fillet. A 50mm x 75mm fillet has a shallow angle that allows the chippings to be taken to the top of the fillet and remain stable when bonded in bitumen.

Lead is the traditional and most successful flashing material, but is too often ruled out on the grounds of cost. Lead weathers well in most localities and allows for movement in relation to the upstand which it covers. It is easily dressed down over the finished waterproofing and turned back for inspection and remedial work. Copper or super-purity aluminium are used as alternatives, but they cannot rival lead for the combined virtues of ductility, weathering and corrosion resistance.

If a metal flashing is to be installed after completion of the walls, a chase will have to be cut beneath the dpc and the cover flashing held with lead wedges and sand and cement pointing.

If a metal cover flashing is not included, the high performance mineral surfaced or metal faced roofing is turned into the chase and held in position by sand and cement pointing.

A similar arrangement using separate mineral surfaced or metal faced flashing pieces applies when the waterproofing is taken over a low parapet wall to form a capping, or is dressed over an eaves check curb.

The instructions or recommendations of the manufacturers of the roofing materials should always be consulted for special requirements.

Skirting at junction of metal decking and brick wall
At the junction between a metal deck on a steel frame and a brick wall which is independent of the steel frame, differential movement between the roof and the wall can be expected and an independent upstand will be required.

The independent upstand can be formed by metal angles attached to the deck.

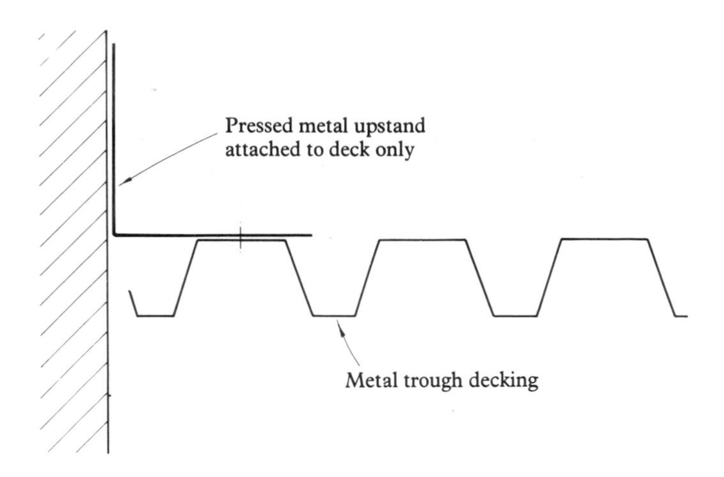

If the roof specification includes an underlay, a vapour check or a vapour barrier, this should be taken through to the top of the metal upstand. At the side of roll condition it will be necessary to complete the layer through the detail with separate pieces.

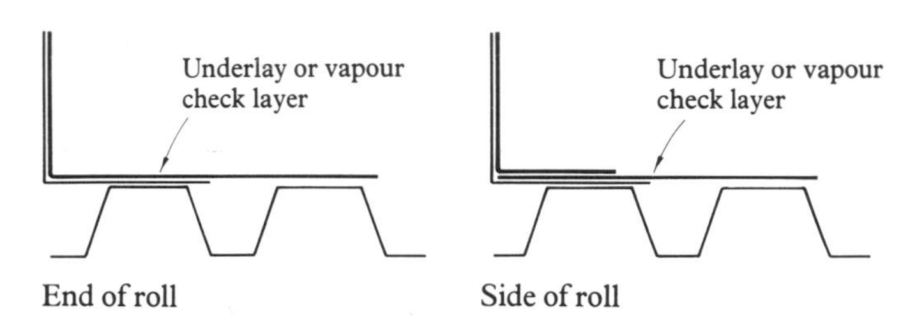

If a vapour barrier is specified, this will generally be a two layer system with the first layer a high performance material and the second layer a BS 747 felt. Under these circumstances, it should only be necessary to continue the first layer through the detail at both side and end of roll.

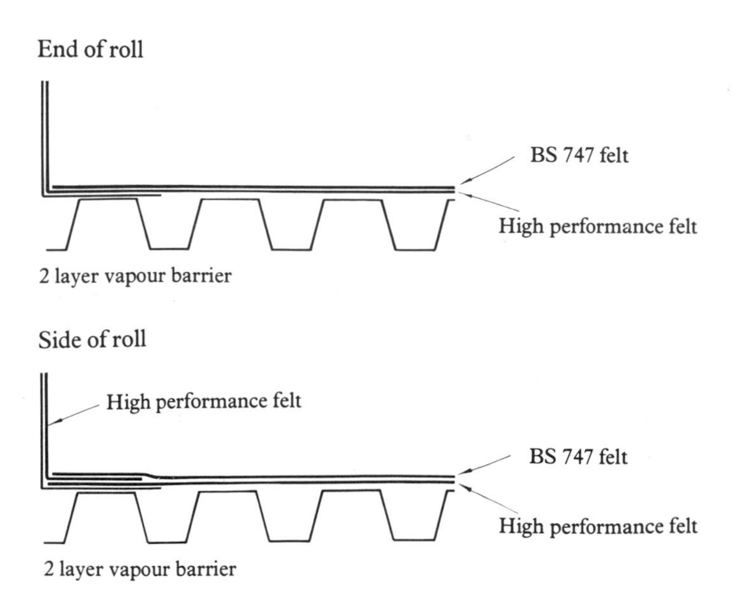

The insulation may be set down from the top edge of the metal upstand by about 25mm. This will allow the waterproofing to complete an envelope around the insulation and provide protection from rain splash. The envelope should always be completed if a vapour barrier is specified and the top edge is exposed to the internal environment.

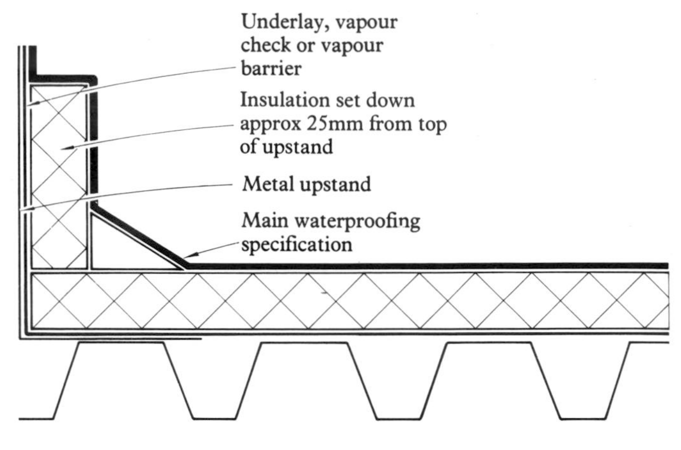

An alternative is to bring the vapour barrier layer over the top and onto the front face of the insulation which, in this case, will be continued to the top of the metal upstand.

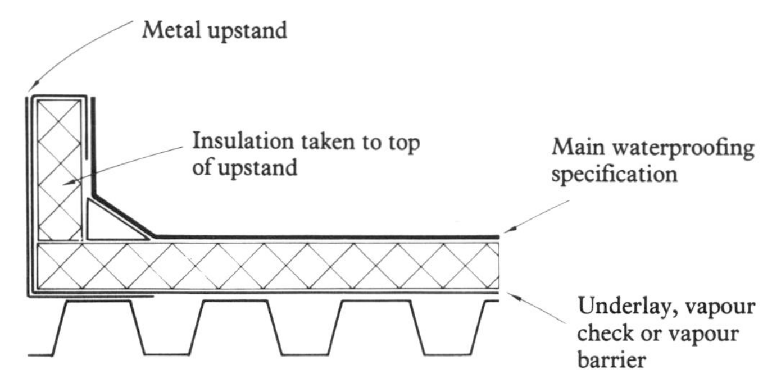

If the insulation board selected for the main horizontal areas of the roof requires a partially bonded first layer, it is advisable to change the insulation at the upstand to one which will accept a full bitumen bond. The waterproofing should not be applied direct to the metal upstand due to the risk of splitting at joints in the metal.

A metal cover flashing is essential to allow for the movement between roof and wall, and this should enter the wall one course higher than the top of the upstand, so that the flashing can be dressed more easily over the upstand without forming a ledge which might hold water.

Fixed upstands and kerbs
The majority of skirtings on metal deck roofs are formed by metal upstands fixed to the deck and to the structural frame or cladding members. The formation of skirtings is exactly the same as for independent upstands described above.

The principles so far discussed for skirtings apply equally well to ventilators and rooflights and to similar items set on kerbs above the general level of the roof. The termination of the waterproofing and the flashing arrangements will, however, vary according to the nature of the kerb.

Timber and concrete kerbs
The first layer of the waterproofing specification will be bitumen bonded to the majority of upstands but should be nailed to large areas of timber to avoid blistering and the effects of movement. The waterproofing should be taken over the top of the kerb. The ventilator or rooflight must include an integral flashing with a 50mm turndown to provide reasonable protection to the upstand and 50mm clearance for the roofing, to allow repair work to be carried out without disturbing the flashing. A flashing which is deeper or which projects less than this should not be incorporated in the design unless it is easily removable.

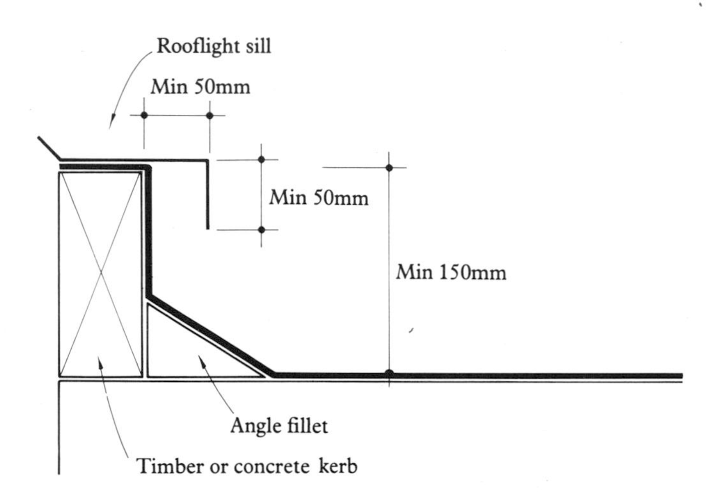

The waterproofing at the corners of the kerb is formed by allowing each individual upstand piece to project past the corner when it is positioned. The projecting part is then cut into flaps, leaving one flap for each flat surface which forms the detail. These are then turned round the corner and bonded in bitumen.

The process is repeated for each upstand piece. For the final mineral surface cap sheet, the first piece is cut as before but the second piece is cut on the mitre. This results in a substantial thickening of the specification at the corners and the clearance of the flashing must be sufficient to avoid cutting into the waterproofing.

Metal kerbs
Metal kerbs are frequently installed on a metal deck to support ventilators and rooflights. The kerbs should be designed to incorporate a projecting flashing which should allow for the thickness of the insulation, and a further 50mm clearance for the waterproofing. The flashing should be approximately 50mm deep to allow a good cover to the upstand and plenty of working space for the roofer. The waterproofing cannot be turned over the top of the kerb and secured. It is therefore not desirable to have such kerbs higher than 150mm above the level of the roof. If substantially higher upstands are essential it will be necessary to add a timber batten at the top in place of the insulation to provide a position for nailing the waterproofing at the top of the upstand.

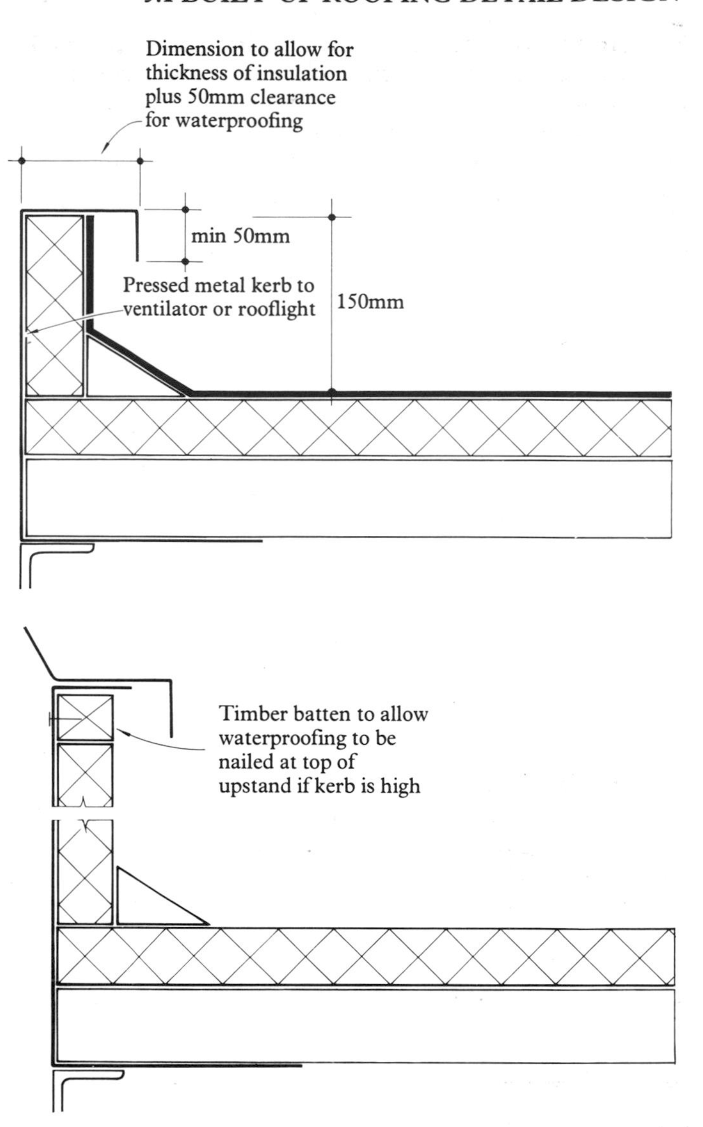

Skirtings to thresholds and cills
Designers sometimes miscalculate the levels of the roof, and in an effort to produce as much height of glazing and doors as possible can leave an upstand which is far too small. Thresholds and cills leading on to a flat roof should allow a 150mm upstand. A lead cover flashing under a threshold cill and throating forms the most satisfactory detail.

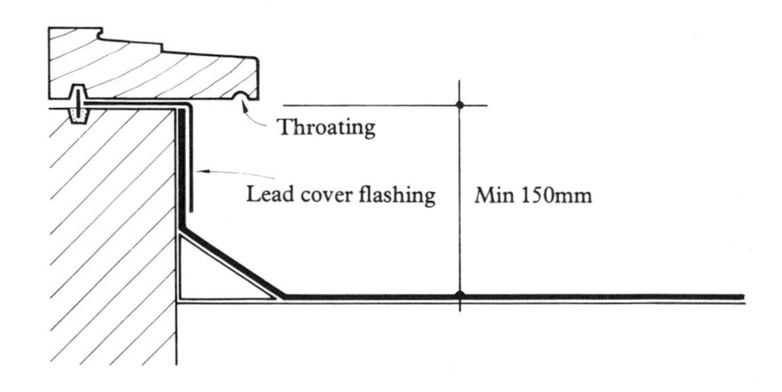

Skirtings to cold pipes
Upstands to pipes are a form of skirting but the technique of application generally requires star cutting of the underlays to form the base of the upstands, and collars to form the upstand itself.

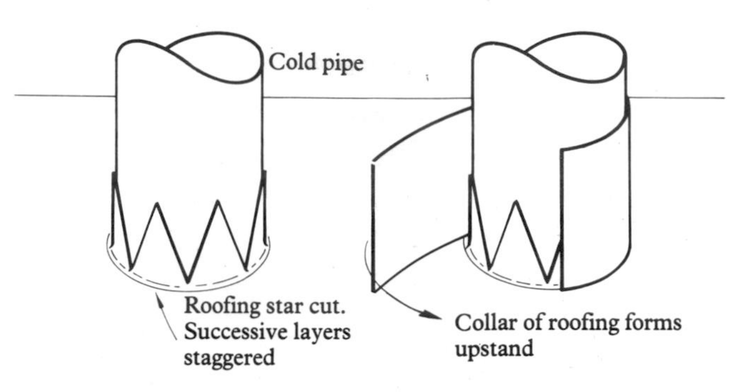

The principle of forming star cuts and collars applies to the treatment of a vapour barrier layer around penetrations but for an underlay or vapour check beneath insulation, it will be sufficient to cut the felt and insulation close to the projection. The bonding bitumen used will fill the voids and will provide sufficient protection.

If a cold metal pipe is firm in relation to the roof deck, there should be no need to form an independent upstand. It is common practice for a cold pipe to be left with no further flashing arrangement, leaving the seal of the waterproofing to the pipe as the only safeguard.
This is generally satisfactory but an independent flashing or cowl secured to the pipe is preferred.
A pipe sleeve may be necessary to form an independent upstand if differential movement is expected between the pipe and the deck. A pipe sleeve may also be necessary if the pipe is fitted after the roofing has been completed, and for pipes of asbestos cement.

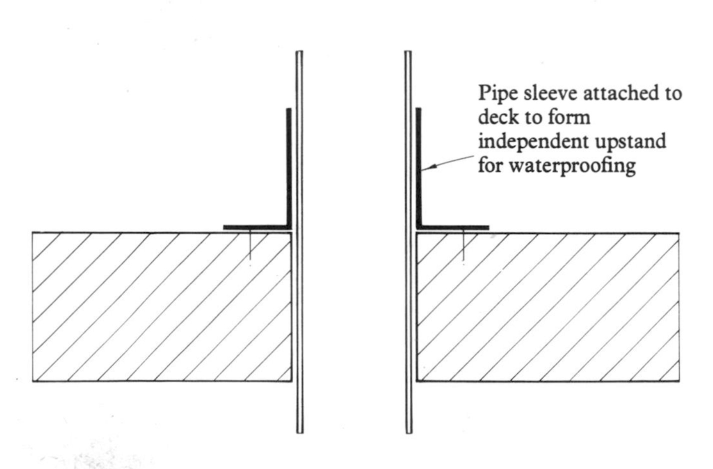

Badly planned pipe penetrations make successful waterproofing almost impossible

An independent flashing on a cold vent pipe may be formed by strapping and caulking a short metal apron into position. Proprietary pipe sleeves are available with integral weathering collars which prove very satisfactory.

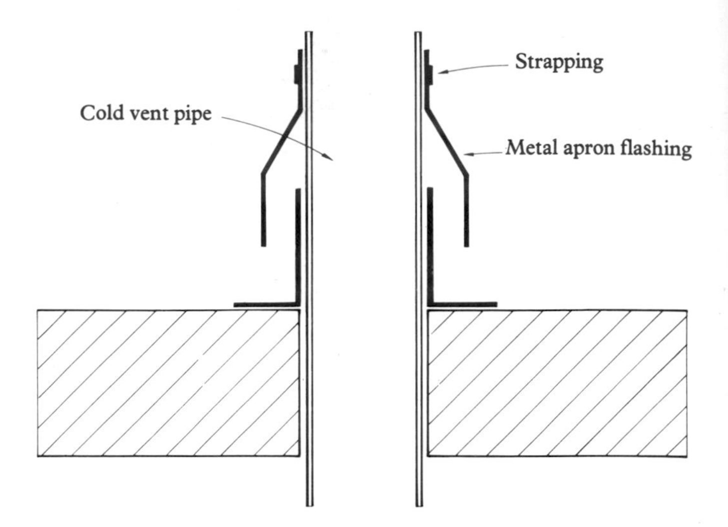

Skirtings to hot pipes and flues
Bitumen roofing should not be applied direct to hot pipes or flues. Hot pipes must have an independent sleeve with a separating air space or insulation between the sleeve and the pipe. A 25mm air gap will suffice for pipe temperatures up to approximately 100°C. Above this temperature it is necessary to add insulation. As a rough guide, a 25mm air gap and 50mm mineral wool insulation will be needed for temperatures up to 200°C.

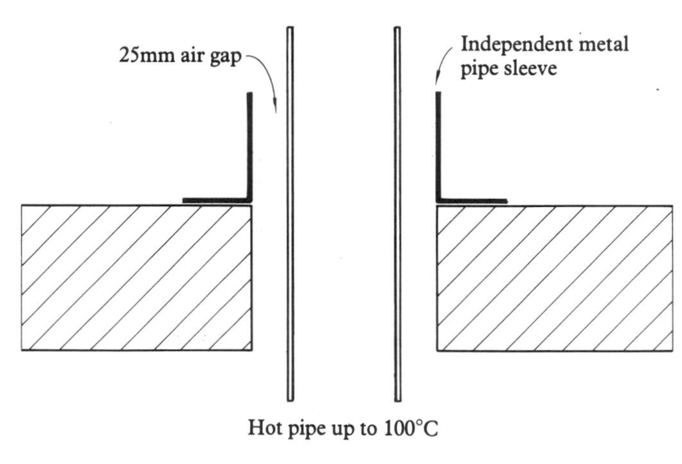

Hot pipe up to 100°C

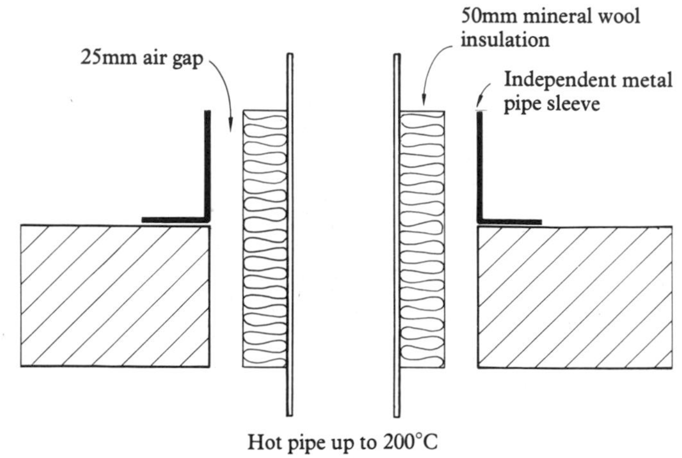

Hot pipe up to 200°C

Large industrial flues require kerbs constructed of concrete or insulated metal. Ventilation should be allowed between the flue and the kerbs. The insulation must be non-combustible and the flue or the flashing should be removable to allow access for repair work.

Where temperatures are so high that the use of combustible material could form a fire risk, the roof structure in the immediate vicinity of the flues should be constructed of heavy plate steel or concrete supported back to the main structure.

EAVES AND VERGES

Where the waterproofing is taken over the edge of a roof, either as a drip into an external gutter, or to form a weathering edge to a kerb, there should be a minimum 50mm turndown of the waterproofing.

On the gable ends of a roof, some form of edge upstand or check kerb should be provided to prevent blown water spilling over the edges. The height of check kerb required to prevent overspill will depend on the degree of exposure of the roof, the roof pitch and the shape of the upstand. A height of 50-75mm would normally be sufficient, and an angle of approximately 45° for the upstand face will avoid sharp changes in direction of the waterproofing.

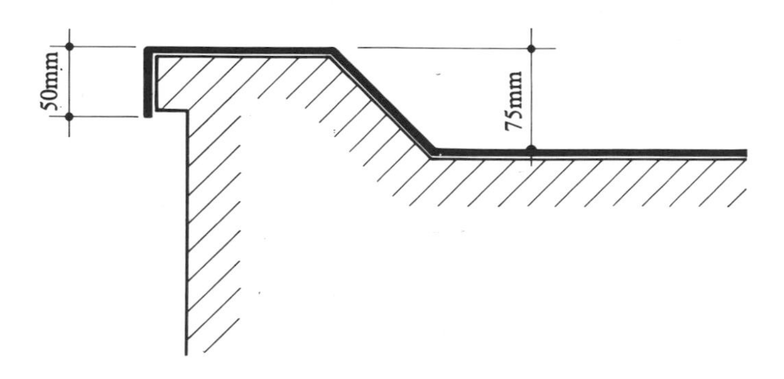

Welted drips

Welted drips of mineral surfaced roofing are the traditional and reliable edge detail for the eaves and verges. Designers frequently discount their use on grounds of appearance or fashion, but the welted drip is favoured by many roofers as the most satisfactory way of forming a drip into an external gutter, or to weather the edge of a kerb.

Welted drips are traditionally formed on timber battens to throw the rainwater clear from the fascia. Strips of roofing are cut from the roll and folded into shape to form the minimum 50mm downturn. The roofing is then nailed to the timber batten, bonded to form a drip and returned on to the roof surface to be interleaved with the waterproofing.

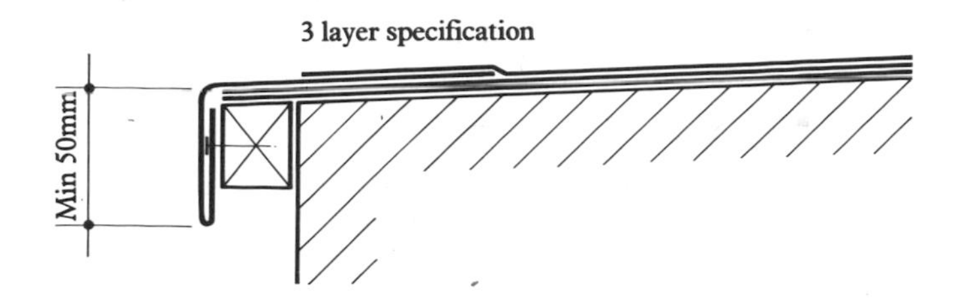

An alternative method of forming drips is to pre-bond the welt separately and then to bond the welt to either a timber batten or metal box closure.

The drip can be made from BS 747 type 2E mineral surfaced roofing but a high performance mineral surfaced roofing is preferable.

The materials used for the drip can be prone to blistering, top pitting and granule loss, which can mar appearance but will not detract from the waterproofing performance of the detail. Some mineral surfaced materials are also prone to cracking at the bends, but this is not usually noticeable on a finished roof and again performance is not likely to be affected.

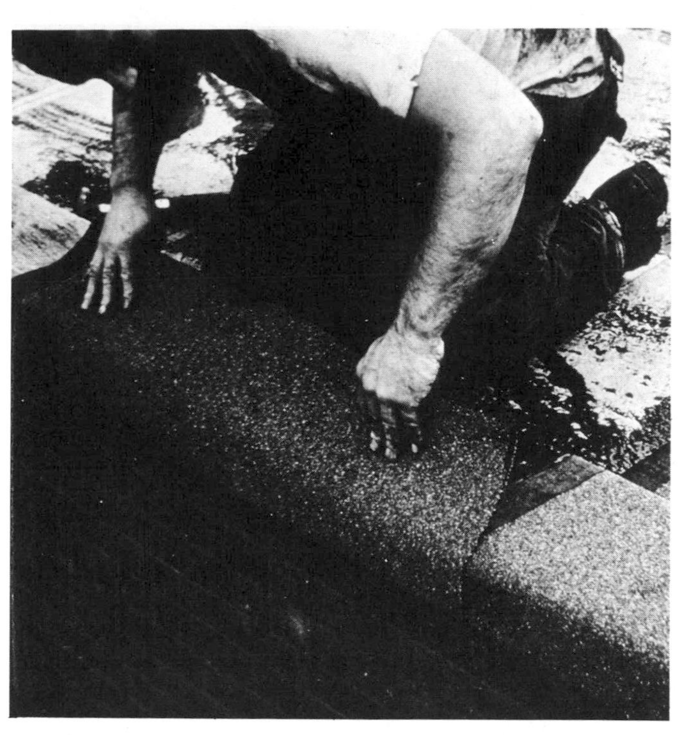

Forming a welted drip detail with mineral surfaced roofing

Some of the new high performance mineral surfaced roofings will maintain a better appearance and those with a polymer modified bitumen have a better resistance to cracking than BS 747 roofings. There can be some difficulty, however, with the thicker high performance materials in folding the material to form the lower edge.

The formation of the drip requires craft skills for the hot bitumen bonding of the folded materials and the shaping of the edges of the roofing strip to accommodate the side laps without unsightly bulges at the roof edge. There are a number of ways of shaping the edges of the strips of roofing to avoid this but an increased thickening will occur along the roof edge. Small amounts of water may be held back after rainfall, but this will not have a detrimental effect on the performance of the detail.

On concrete or brick kerbs, timber plates should be incorporated with a 25mm minimum overhang. The detail then follows normal practice with the welted drip formed after the upstand detail.

Where insulation extends to the roof edge, it will be necessary to provide firm support for the waterproofing and to protect the vulnerable roof edge of the insulation from damage by ladders and other maintenance traffic. This can be achieved in a variety of ways. For example, a timber fascia can be extended upwards to the correct height for the insulation to butt behind.

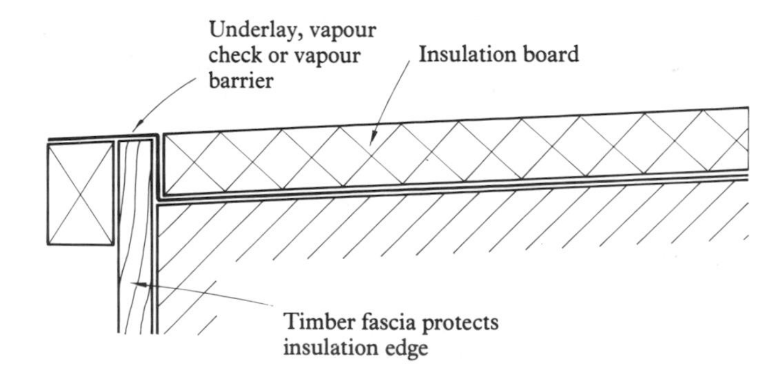

If an underlay, vapour check or vapour barrier is required, this should be taken up and over the edge to allow the waterproofing to complete an envelope around the insulation.

Metal trims

The most common form of edge trim is the single piece extruded aluminium trim. Many proprietary forms are available, and whilst there are differences in the design they all work on the same butt jointing principle. They are designed for use on kerbs and are not usually suitable for a drip into a gutter due to the raised portion of the extrusion which encloses the leading edge of the membrane.

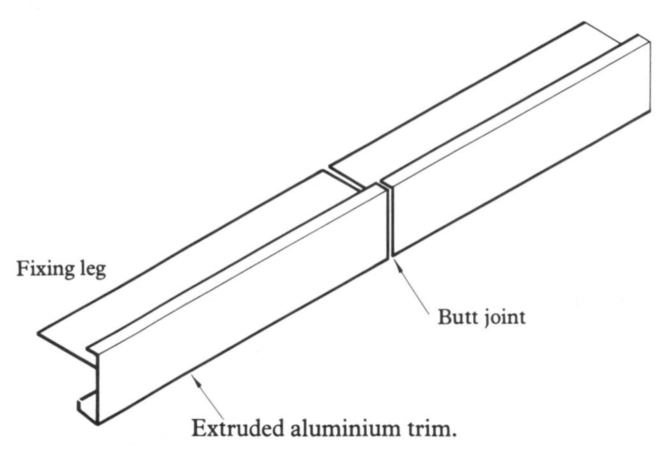

Grounds must be provided for the trim and these grounds must themselves be firmly fixed to the structure. The grounds may be timber plates, timber inserts, or metal closures and kerbs in conjunction with metal deck. It is possible to fix direct to good quality concrete with cartridge driven pins or by drilling and plugging, but in this case the trim must have a large back leg to allow the fixing to be sufficiently far back from the edge to prevent spalling. 50mm back from the roof edge is generally sufficient but site trials will also be necessary to establish the most satisfactory fixing method.

Metal trims should only be fixed directly to brickwork when the bricks, their bond and the fixing of the trim are proved sound by site trials. Otherwise, a timber plate should be firmly strapped to the top of the wall to provide a fixing. In general, brickwork does not provide a good fixing for metal trims and a traditional parapet and coping is preferable.

The waterproofing of the trim system depends on an efficient seal of the roofing to the horizontal leg of the trim and this should be primed and completely dry before application of the roofing. Before fixing the metal trim, the first layer of the waterpoofing specification is taken across the kerb to project about 25mm. This acts as a final line of defence to deflect any small amount of rainwater which might pass through the butt joint system and ensures that the water falls clear of the outer face of the building.

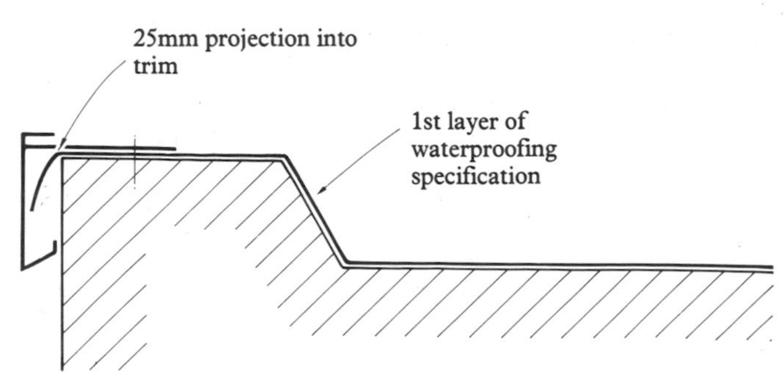

The trim should incorporate metal butt-straps, which are close fitting to the underside of the joint. The straps may be supplied loose or fixed at one end of the trim. If supplied loose, the end fixing of the trim should pass through the end of the butt strap.

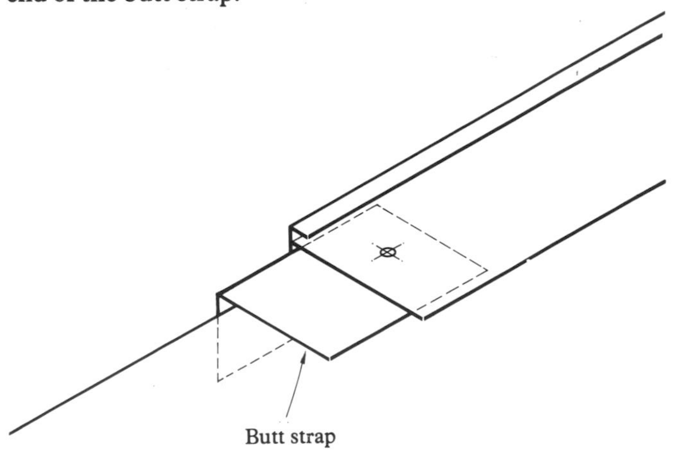

The edge trim of a roof is always vulnerable to wind damage, and a good principle is to fasten the trims with screw fixings at 300mm centres with additional fixings in positions of extreme exposure. Fixings may be placed through both sides of the butt-strap. This practice would appear to prevent movement at the joints, but in practice there is enough slip at the fixing point to prevent a problem arising.

Metal trims were originally used with BS 747 felts or BS 743 hessian based felts, and these would generally show distortion and some splitting above the joints of the trims as a result of expansion and contraction of the metal.

The expansion and contraction of metal trim edges can lead to splitting over the joints.

It has been found that a reinforcing pad of high performance roofing, in particular an elastic material such as a bitumen polymer roofing, forms a satisfactory sheathing pad and eliminates most of the distortion and splitting. These pads should be 150mm long and are bonded in hot bitumen above the trim joint.

High performance roofings generally behave well when used with metal trims, but again it is a sound principle to add an elastic sheathing pad to reduce distortions.

Trims of 3 metres long are economic and form the best straight line in practice. Shorter trims will move less, but the fatiguing action of the joints caused by temperature changes is still likely to lead to some distortion or splitting unless a reinforcing pad is incorporated.

The temperature range for self-finished light coloured aluminium trims in service is likely to be from -10°C to 60°C, and the free expansion and contraction of 3m lengths of aluminium trim over this range is 5mm. In practice, the movement of the trim will be restricted by the fixings, but if no allowance is made for expansion at the joints there is the danger of buckling of the face of the trim. A gap of 2mm proves to be sufficient in service.

Pressed aluminium trims of 1.2mm thickness are simple and effective and are also useful at an open edge where there is no kerb, as there is no interruption to the flow of water with this type of trim. The principle of application of a pressed metal trim is exactly the same as for extruded trims. The top layer of roofing may be set back 25mm from the edge of the trim provided the butt straps and projecting underlays are properly fixed.

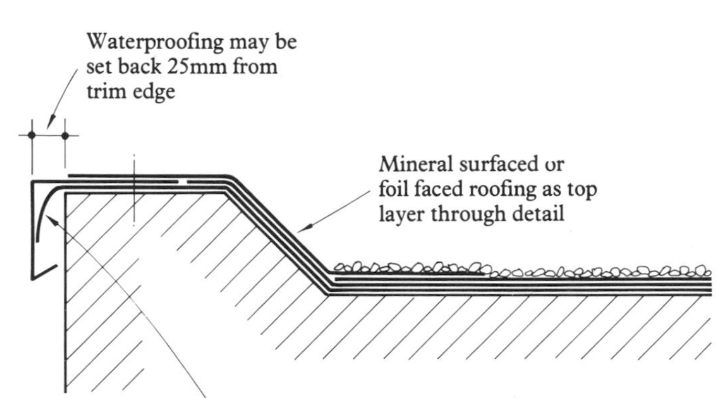

The pressed aluminium system may also be used to achieve deep fascias. Front fixings will be required when the fascia is deeper than 100mm and a firm continuous support backing should be provided. Typically, this detail may be used to form the eaves and verges of metal deck roofs.

Galvanised steel or coated steel trims are not recommended as the bitumen and heat of application may harm the coating. The flat ledge at the edge of the trim would also be vulnerable to corrosion. Where coated steel fascias are required, they should be designed as part of a capping which covers the roofing membrane so that no question of bonding to the steel surface arises.

Lead and copper trims and fascia flashings
Lead trims are not widely used today although they normally prove satisfactory. They form the best and perhaps the only satisfactory means of cloaking slates or tiles at the top of a mansard roof. The principles of application are similar to those for aluminium trims, including the need for sandwiching the lead between the layers of built-up roofing.

The back edge of lead is not normally welted for built-up roofing. The lead must be fixed at close centres and tightly dressed to the slates or tiles. Lead movement can split the membrane and the maximum length of lead should therefore be in the order of 1.5 metres.

Cappings
A common detail for buildings with metal decking and profiled cladding is to take the vertical cladding up above roof level to form a dwarf parapet. The fascia trim will probably match the colour of the cladding and be of similar material. This detail combines the trim detail with a flashing detail, and the conventional principles apply. The 50mm turndown is required as usual but it is not always necessary to allow a 50mm working space for the membrane as the fascia is easily removed for inspection and remedial work.

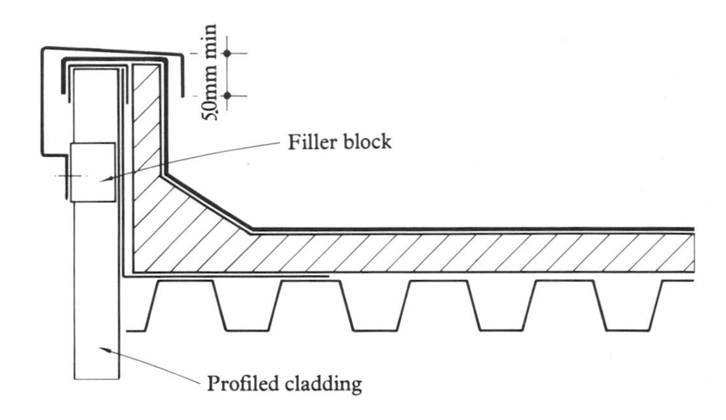

The top of the flashing should have a slope or haunching to prevent standing water on the top, which would find any weakness at the butt joints or fixing points. The water may also act as a lens for the sun and can lead to premature breakdown of the coating.

Cappings are particularly vulnerable to the wind and will require fixing on the front face and the top surface, both at approximately 300mm centres.

GUTTERS

Lined gutters
As mentioned in Section 1.1 Falls and Drainage, lined gutters are best avoided on a flat roof. When it is considered essential to incorporate gutters, the formation of the membrane in the gutter is equivalent to a narrow roof with kerbs on either side installed with an end of roll formation.

There are differing schools of thought on the subject of the finishing or surfacing of the gutter.

In wide gutters it is best to protect the waterproofing in the sole with stone chippings and to treat the sides of the gutter as skirtings, using a mineral surfaced felt to form the final layer. If the gutter is relatively narrow, the mineral surfaced felt can be used to line the whole of the gutter.

The exclusion of a surfacing in an attempt to increase the flow of water is not recommended as the separation and blistering of a poorly protected gutter will cause more of a restriction to flow of water than either a chipping or mineral surface. The gutter is the most vulnerable part of the roof and it makes little sense to provide the least protection to this part of the membrane.

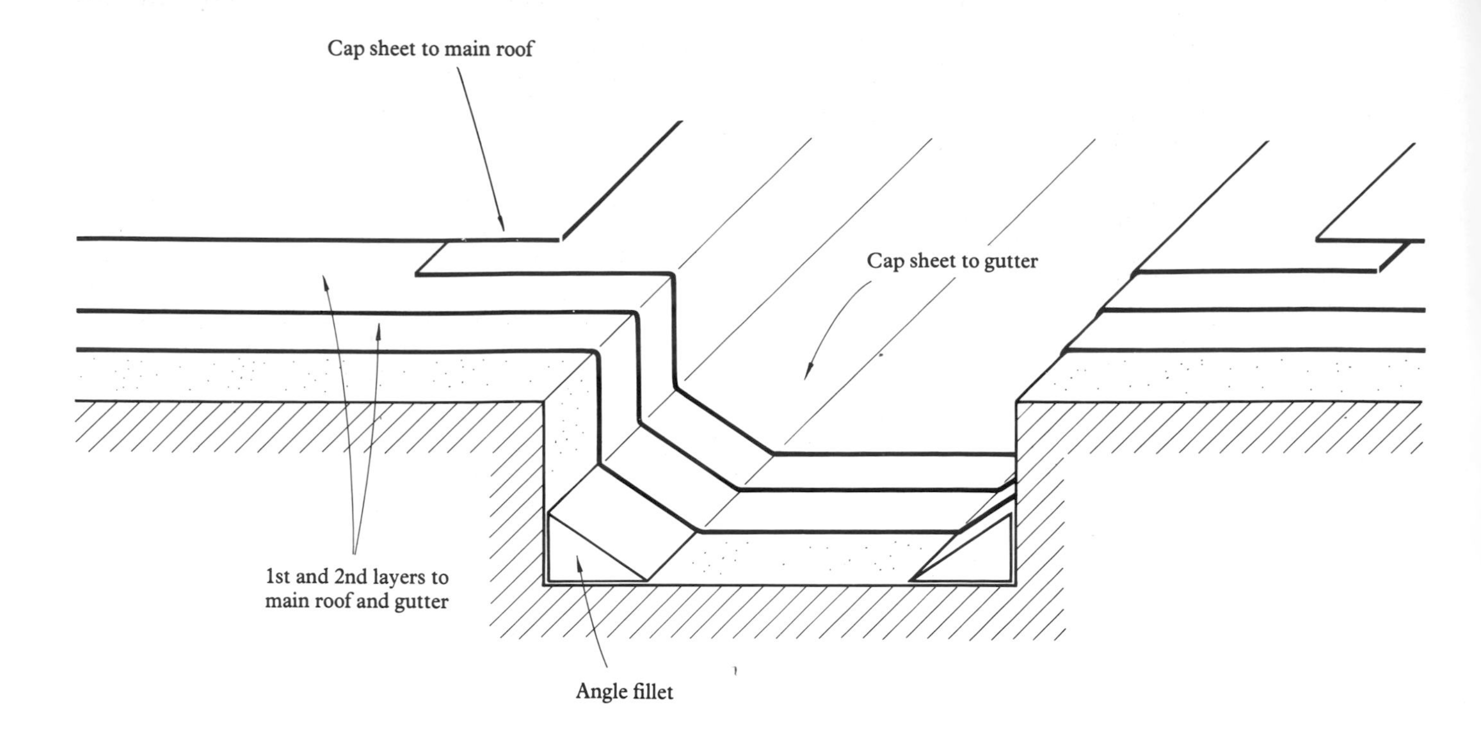

The roofing in the sole of the gutter will follow the normal specification for flat areas, including partial bonding using BS 747 type 3G roofing as necessary. The sides of the gutter will always be fully bonded and are therefore more vulnerable to blistering if the insulation is one which would normally require a partial bond. To reduce the amount of blistering, the insulation for the vertical section is often changed to a material which is suitable for full bonding, such as wood fibreboard or cork.

Mineral surfaced roofing normally has a 50mm margin or selvedge at the side of roll, with the mineral granules replaced by a sand finish to give the best contact surface for bonding side laps. It is desirable to start the mineral surfaced roofing at the outlet points so that the lap is in the direction of the flow of water. This pattern of laying is not always possible but in practice laps formed with the selvedge will not be at risk regardless of direction.

However if the main area of the roof is formed from mineral surfaced roofing, it is essential that the gutters are fully lined before the mineral surfaced cap sheet is applied to the adjacent main roof area, so that the main area cap sheet overlaps the gutter cap sheet in the direction of flow.

With this particular overlap situation, the laps are formed at the ends of the mineral surfaced roofing rather than the selvedge sides of roll. The mineral surfacing of the lower sheet is therefore sandwiched into the lap and it is difficult to ensure that all interstices between the mineral granules are filled with bitumen. If the lap is formed against the flow of rainwater, there is an increased possibility that water would drive into the leading edge and cause delamination or blistering.

With a stone chipping finish, most self-finished felts can be safely applied with the laps against the direction of falls.

Valley gutters

Valley gutters may be formed with a separate lining in much the same way as a box gutter, or by the continuation of the roofing from one slope through the valley and up the other slope for a short distance. This will then be overlapped by the main area roofing applied separately to the second slope.

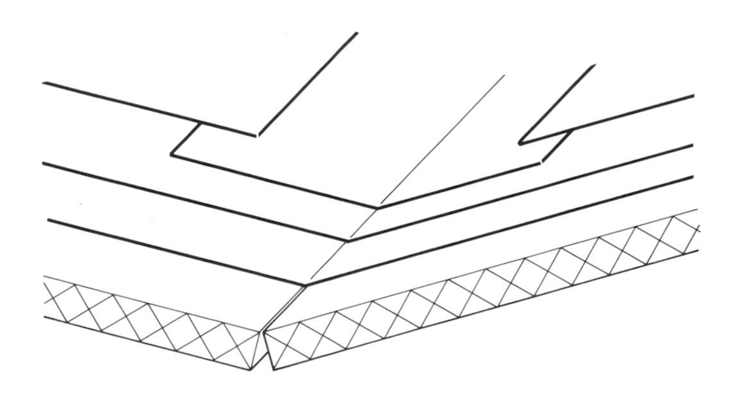

Valley gutter formed with separate lining

The intersection of the valley results in a straight line joint of insulation, which is likely to show up differential movement between the two roof slopes. The specification should therefore be strengthened by substituting a high performance layer for one of the valley pieces unless the main specification already includes high performance materials.

RIDGES

At a ridge the layers of roofing will be taken over on to each side, and a separate ridge cap formed as a final operation.

If the slope of the roof is such that nailing of the roofing is required to prevent slippage, the timber battens should be fixed to the deck and the insulation butted up to them. The roofing is then close nailed at 50mm centres, with the ridge cap protecting the heads of the nails.

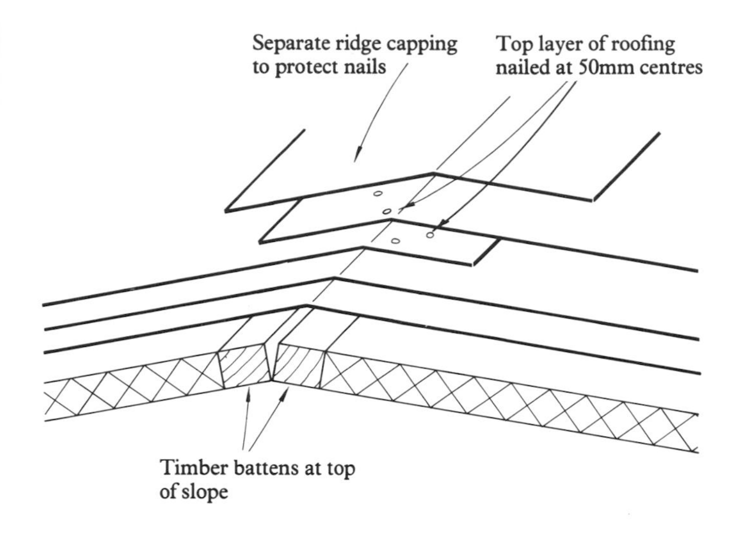

RAINWATER OUTLETS

Lead outlets

The traditional outlet still used to a limited extent is the lead drop outlet, which, although crude and inefficient in comparison to proprietary outlets, generally proves effective.

The lead flange and lead pipe are welded together and the joint is not always entirely dependable. Roofers tend to turn a collar of roofing down into the outlet pipe and this can reduce the effective diameter of an outlet by as much as 25mm. The efficiency of lead outlets can be improved by making them with a tapered side, but this is a difficult shape for leadwork and proprietary outlets are generally preferred.

The flange of a lead outlet should be primed and interleaved between the layers of roofing. BS 747 felts should be reinforced with an additional layer or a local change to a high performance roofing. Where necessary, turn downs are formed by star cutting the material and the cut portions are turned into the pipe with a wiping of hot bitumen. Wire mesh cages are inserted as necessary to act as gravel guards.

Proprietary outlets

A number of cast iron, aluminium or plastic outlets are available. They usually allow for a turndown of roofing into the bell-mouthed portion of the outlet, but sometimes the flange is rather small for the formation of an effective seal: 100mm of flange is normally needed for the bonding of the membrane. In many cases, the outlet is set in position before the application of the waterproofing and it is not possible to interleave the flanges between the layers of roofing. If this is the case, it is even more important to have a good flange area for the bond and an allowance for the membrane to be turned down to form a collar.

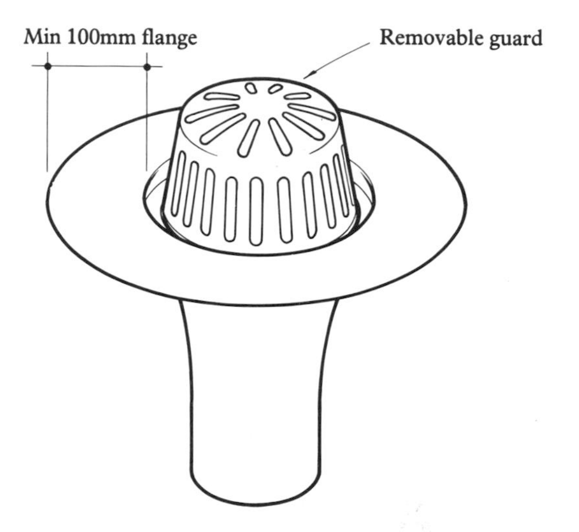

Outlets should be carefully positioned with close attention to levels at the design stage. A number of problems arise with outlets as a result of careless bedding on site or careless application of concrete or screeds after the outlet has been positioned. This can lead to wrongly positioned outlets with badly contaminated flanges and throats, all of which make it difficult for the roofer to form an effective bond.

Attempts are sometimes made to set down the level of outlets to prevent a thickened portion at the flange of the outlet which holds back water and allows the formation of a small pool around the outlets. Such attempts are not usually successful. It is preferable to allow some thickening at the outlet to ensure the formation of a satisfactory seal. The connection between the downpipe and the outlet should make provision for thermal expansion of the downpipe.

Where a stone chipping finish is applied to the roofing, particular care should be taken to bond the chippings firmly around the outlet. Some roofers and designers prefer to incorporate a pad of mineral surfaced roofing around the outlet with the chippings set back to the edge of the pad. The idea is to allow a catchment area for loose chippings, and a surround to the outlet which is more likely to prove self-cleansing. Only high performance felts or glass base felts should be used for the mineral surfaced pad. Chippings will provide better protection to the membrane, however, and it is recommended that chippings be taken right up to the outlet and are well bonded.

EXPANSION JOINTS

Expansion joints in the waterproofing are only required where an expansion joint is allowed in the roof structure. Twin kerb expansion joints are always recommended.

The formation of loops or folds in the roofing will not provide a satisfactory allowance for movement as the majority of roofings, even high performance, should not be subjected to flexural movement.

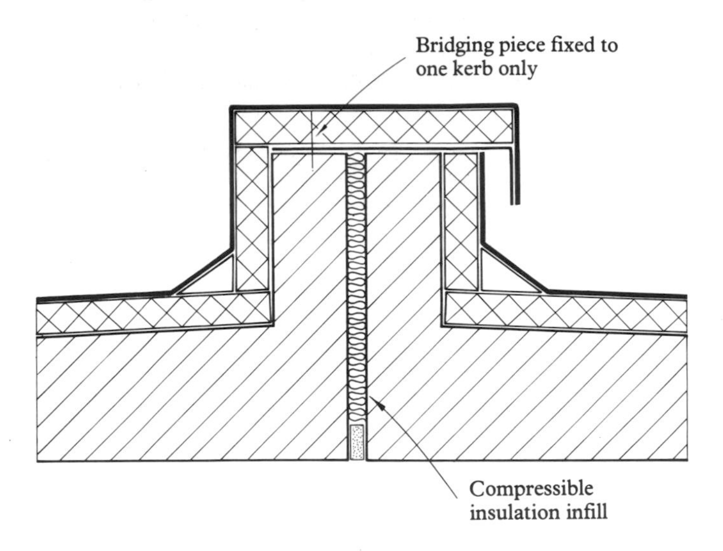

A bridging piece fixed only to one kerb allows for differential movement between the kerbs. Insulation taken up and over the kerb will prevent a cold spot. Compressible insulation material may be placed within the gap to reduce draughts and further reduce the incidence of cold spots.

The ends of expansion joints will normally project above the level of the kerb which forms the roof edge to allow boxed ends which make full allowance for the anticipated movement. Attempts to conceal the expansion joint on the roof edge will normally entail a fully bonded end which will not allow for movement. This is sometimes accepted in the belief that significant expansion and contraction will seldom take place, and occasional tearing will be treated as a maintenance item. Many expansion joints with fixed ends perform satisfactorily in service, and no doubt this indicates that these joints exhibit little movement in practice. At the ends of expansion joints against a higher level abutment, the joint should be taken up the vertical upstand without change of detail, and careful design may be necessary for provision of counter flashings. Again, this detail is sometimes formed with no allowance for movement in the belief that there may be little movement in reality and occassional repair is an acceptable risk.

Proprietary expansion joints are available which are generally successful, and special units and intersections are available to simplify the detailing. Expansion joints which are designed to take up movement by the flexure of elastomeric strips are draughtfree and are easier to form into a satisfactory joint at the end and at intersections. These joints are usually set on kerbs to keep them out of water.

Copper V expansion joints are sometimes used, but they are stiff and require very firm attachment. They should be set on kerbs and the fixing of the kerbs to the deck must be extremely firm as the forces imposed by the copper joint are substantial. Copper expansion joints have not always behaved satisfactorily. They can suffer fatigue failure and the quality of workmanship required is of a high standard and may include the need to provide brazed joints between lengths.

All expansion joints should be formed above the roof level and will prevent drainage across them. The roof drainage system must be designed to take into account the position of any expansion joints and if necessary outlets should be provided adjacent to the expansion joints.

ROOF TOP ADDITIONS

Building services
Services can be taken across the roof and supported by the roof if the compression strength of the insulation is sufficient. The load should be applied through pads which are removable for inspection and repair of the membrane. 300mm x 300mm x 25mm thick concrete paving slabs form a suitable pad. Most high performance roofings will accept loads in the order of 50kg transmitted through the pad, provided the deck and insulation can also accept the load without distortion or overloading. Membranes composed of BS 747 felts will accept the load in compression, but they are more likely to crack at the edge of the pad through minor movement and local stresses, and a further pad of high performance roofing approximately 600mm x 600mm square will be a suitable precaution.

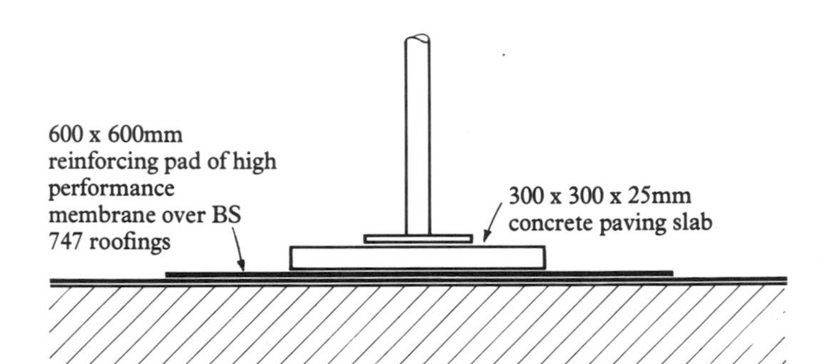

Ducting or pipework should be kept at least 300mm above the surface of the roof with minimum 300mm vertical sections leading to it. This allows the formation of satisfactory collars and flashings and gives room beneath the horizontal runs for repairs or re-roofing.

If the provision of services across a roof concentrates traffic into local areas, extra protective surfacing will be required to accept the extra traffic in that area.

Handrail standards should be attached to the structural frame and should preferably be set on kerbs. The waterproofing is made good as described for cold pipes.

Tanks and housings

Tanks, housings, condensers, ventilation units and similar items of roof top equipment should be installed on separate structural kerbs. It may be that a waterproofing is necessary under the tanks or ventilation units as a precaution against overflow or leakage of the casings or housings, but this should be separated from the waterproofing operation on the main roof areas. If this is not possible, there should be at least 900mm clearance between the underside of the unit and the roof surface to allow inspection and repair of the membrane beneath.

Cleaning rails

Cleaning rails and tracks for wheeled window-cleaning rigs must be included in the early stages of design; supports should be set on structural stools which are in turn attached to the structural frame. The membrane will be turned up to form a skirting around the stools and covered by a metal flashing inserted between the rail and the stool. Bituminous flashings are not suited to the compression and lateral forces which are associated with the support and fixing of the rail. Metal flashings should allow access to the membrane for inspection and repair without removal of the rail. If the load is light and lateral forces negligible, tracks can be provided as raised paved areas near to the edge of the roof but not forming part of the roof edge detail.

Aerials and signs

Aerials and signs should be supported on pads or frames designed for the purpose and trades who erect them should be instructed not to make fixings through the roofing without prior discussion with a roofing contractor, who should agree a satisfactory detail and make good afterwards. Large signs attract substantial wind forces and should be fixed to stools or frames tied back to the structural frame, and with proper provision for making good where the waterproofing is penetrated.

Lightning conductors

Short runs of lightning conductors may be laid loose on the surface of the roof, but if secure fixings are required special flanged brackets should be set into the roof and made good by a roofer. Alternatively clips can be fixed to concrete pads resting on the roof to avoid penetrating the waterproofing.

It is not generally satisfactory to attempt to bond lightning conductors to the roof with bitumen or to locate them with bonded strips of roofing materials.

TYPICAL DETAILS

The principles of detail design described above are further illustrated in the following section in the form of typical details for warm roof constructions.

The details are grouped according to the structural deck material in the same sequence as the typical specifications of section 3.4.

The principles of detail design are identical for timber and woodwool decks. To avoid unnecessary repetition, these decks have not been separately illustrated. Details illustrated for plywood deck constructin are therefore equally applicable to woodwool, chipboard and timber boarded decks.

In all cases, typical adjacent structures have been included to enable the principles of detail construction and waterproofing to be shown, but it is not intended to illustrate details of building construction.

BUILT-UP ROOFING CONCRETE DECK

Skirting to brick parapet

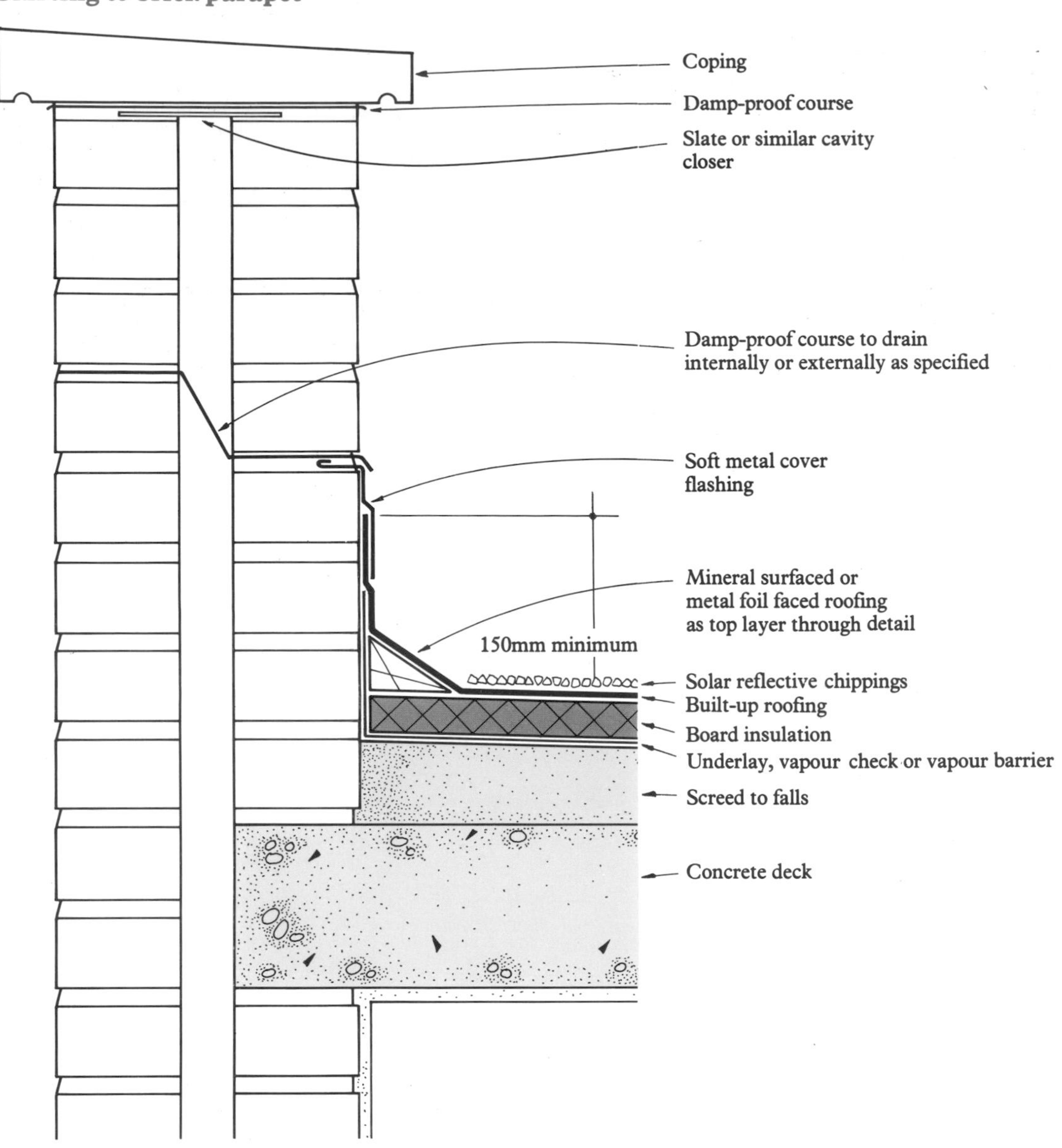

Check kerb and metal trim edge

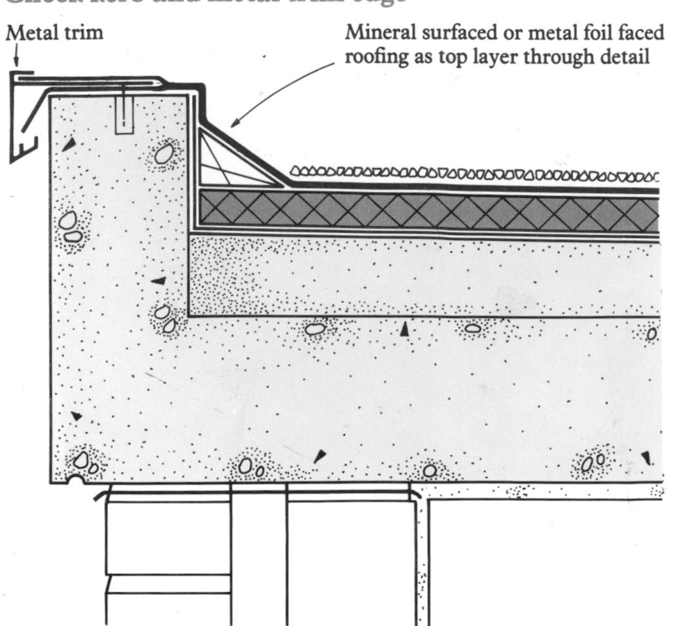

Check kerb and welted drip

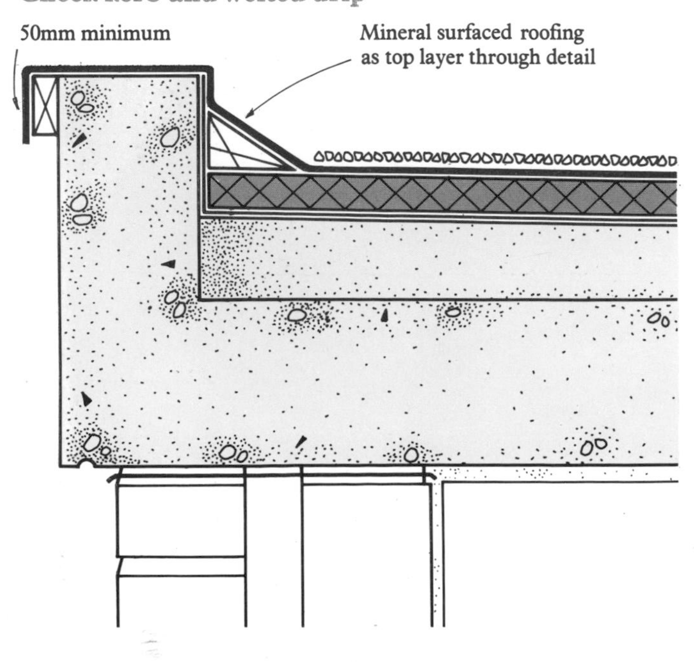

Welted drip to external gutter

Mineral surfaced roofing as top layer through detail

Skirting to metal rooflight or ventilator kerb

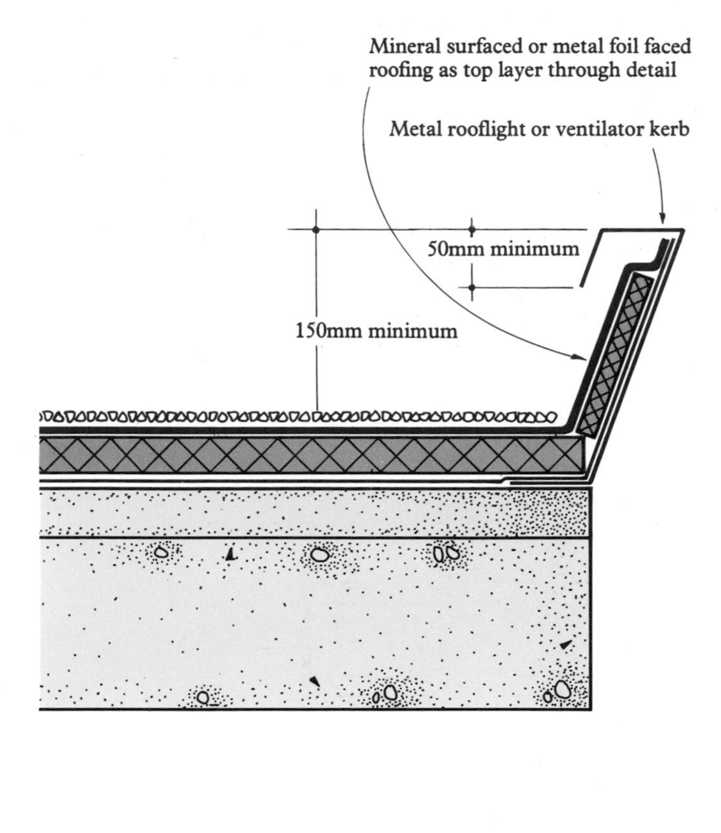

Roof outlet: clamping cone type

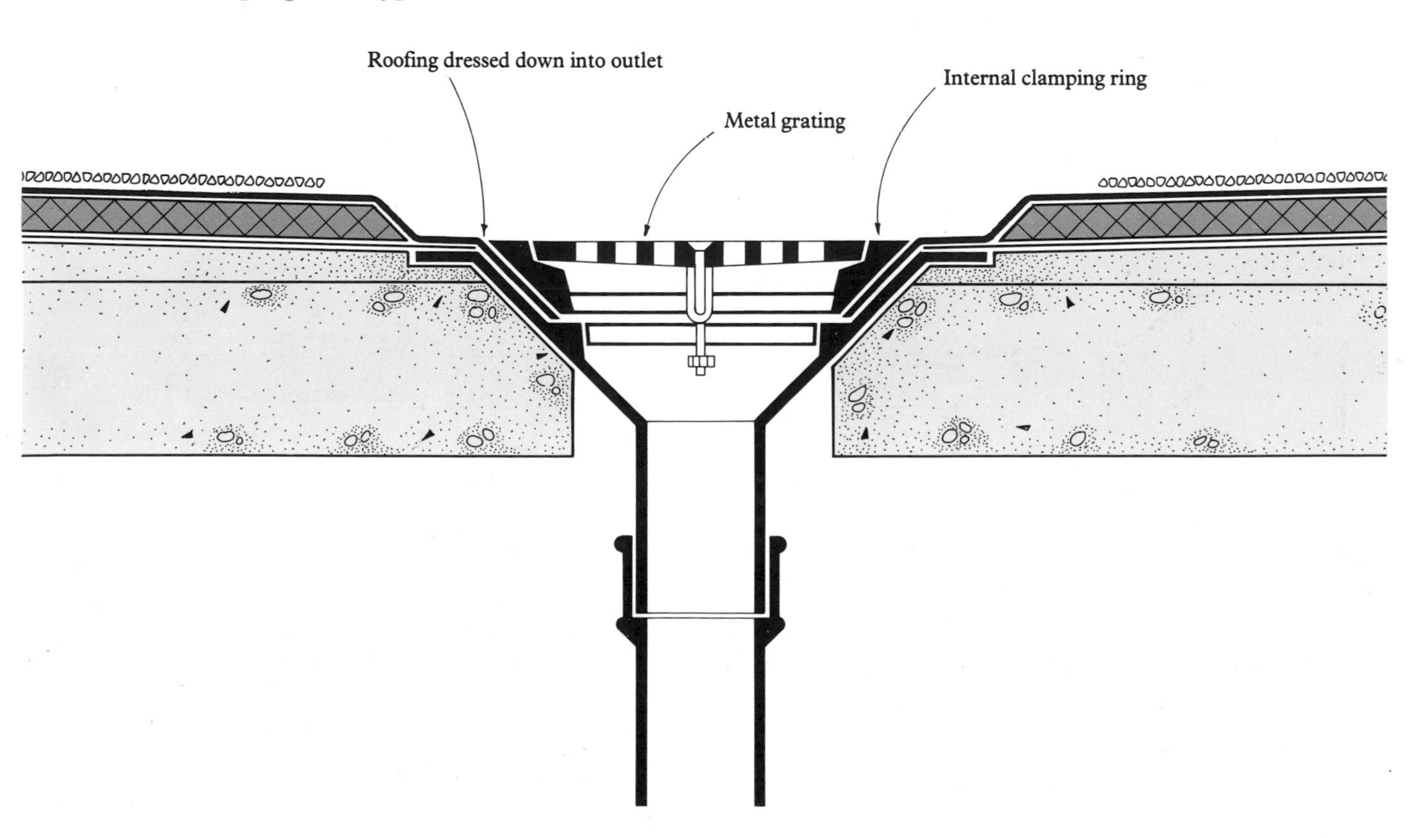

Twin-kerb expansion joint

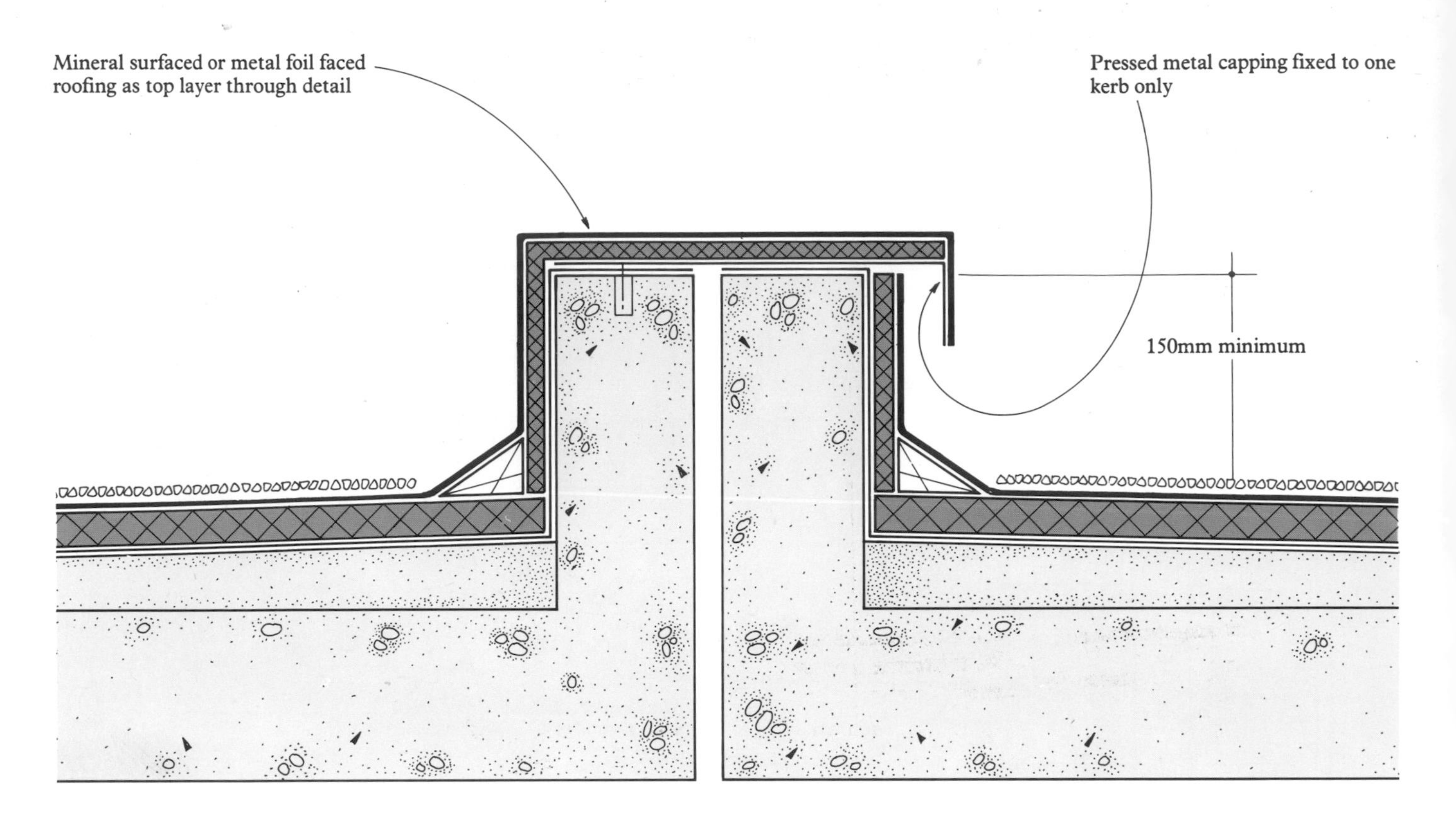

Skirting to cold pipe

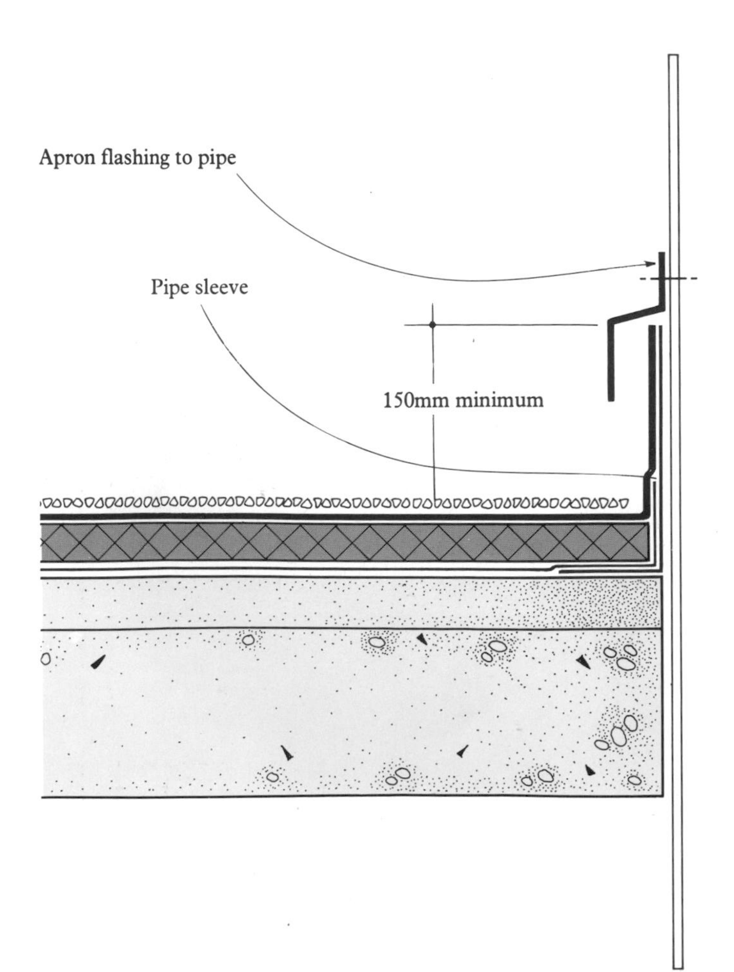

Skirting to hot pipe

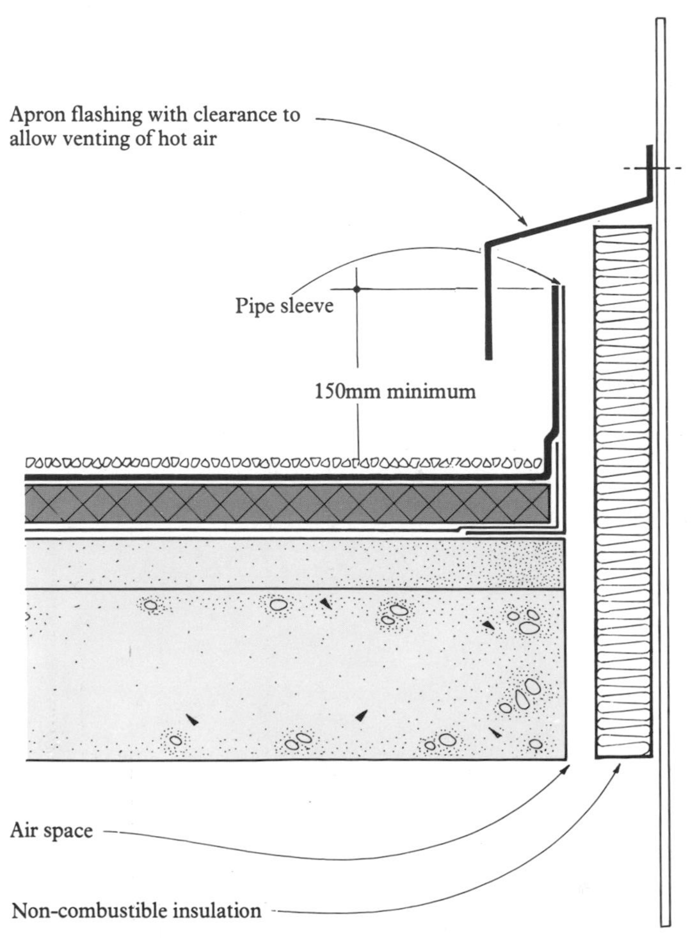

BUILT-UP ROOFING
TIMBER/WOODWOOL DECK

Plywood deck illustrated

Skirting to brick parapet

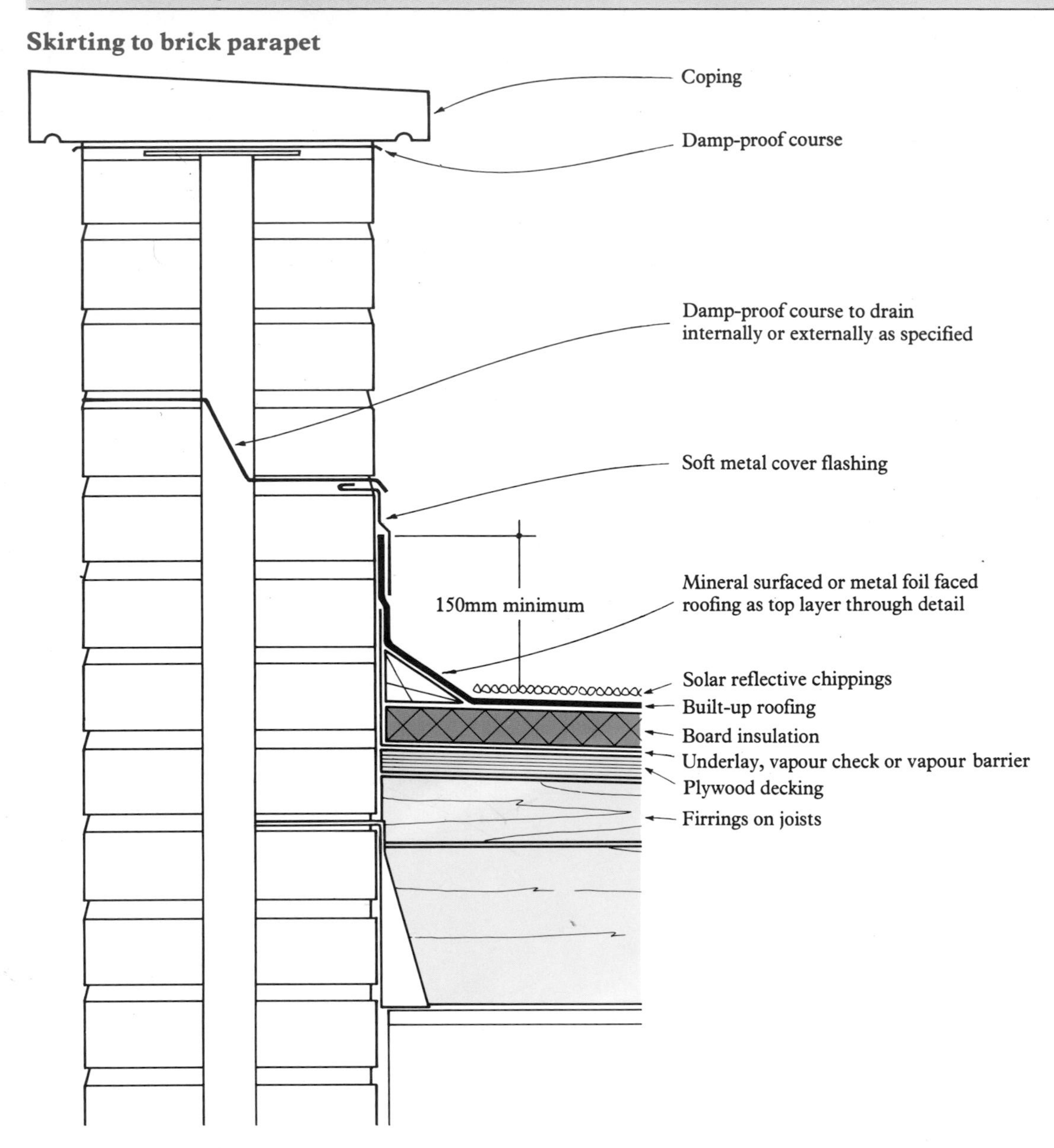

Check kerb and metal trim edge

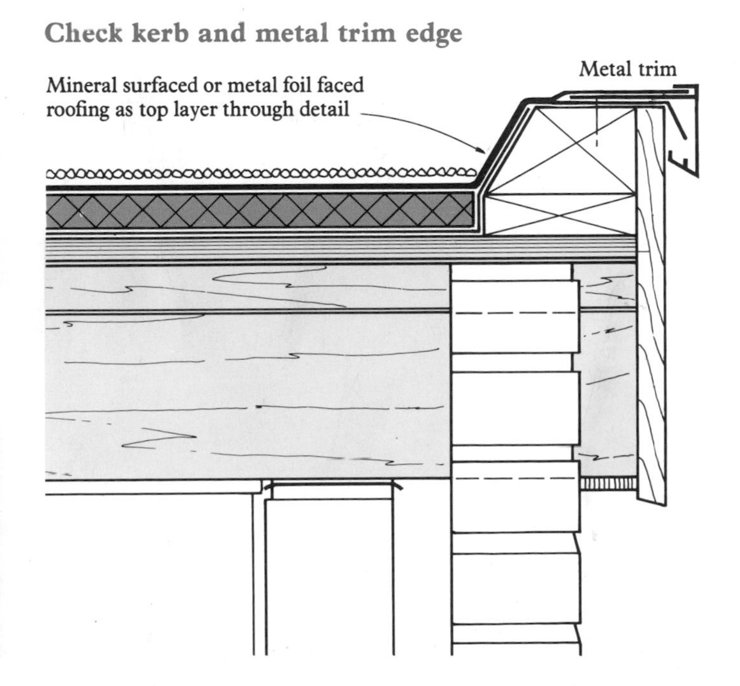

Check kerb and welted drip

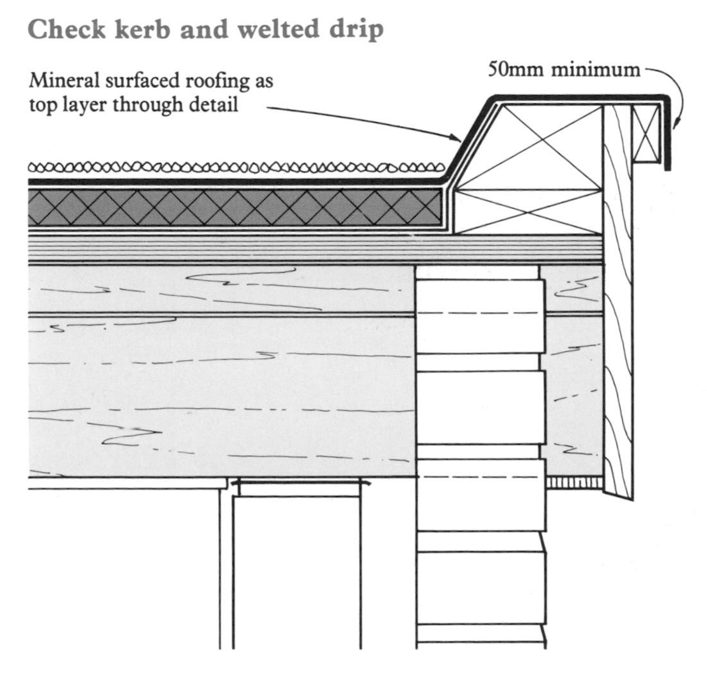

Welted drip to external gutter

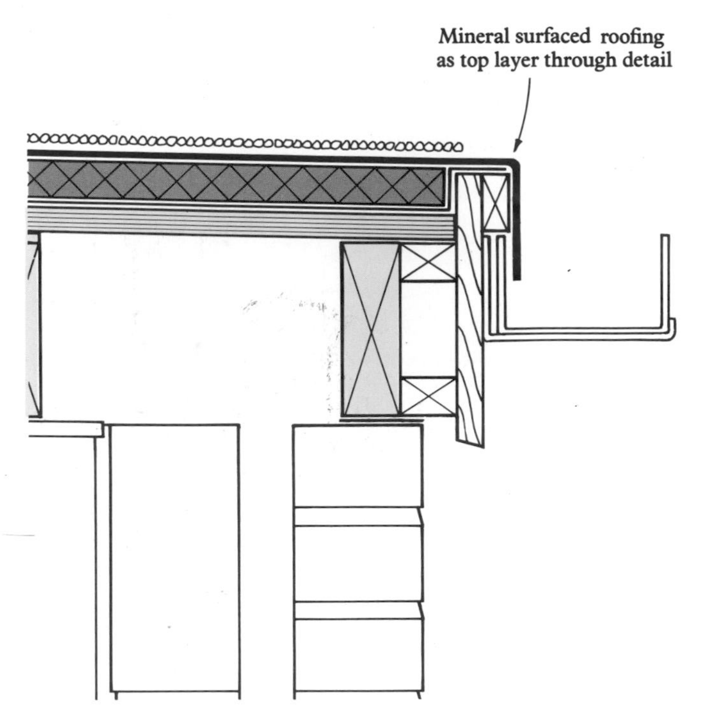

Roof outlet: proprietary flanged type

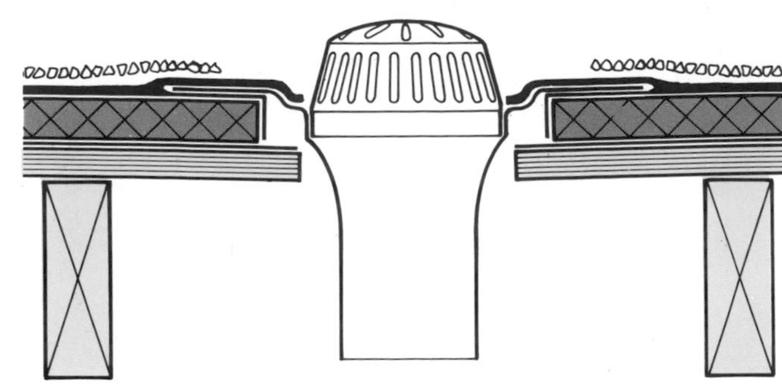

Skirting to metal rooflight or ventilator kerb

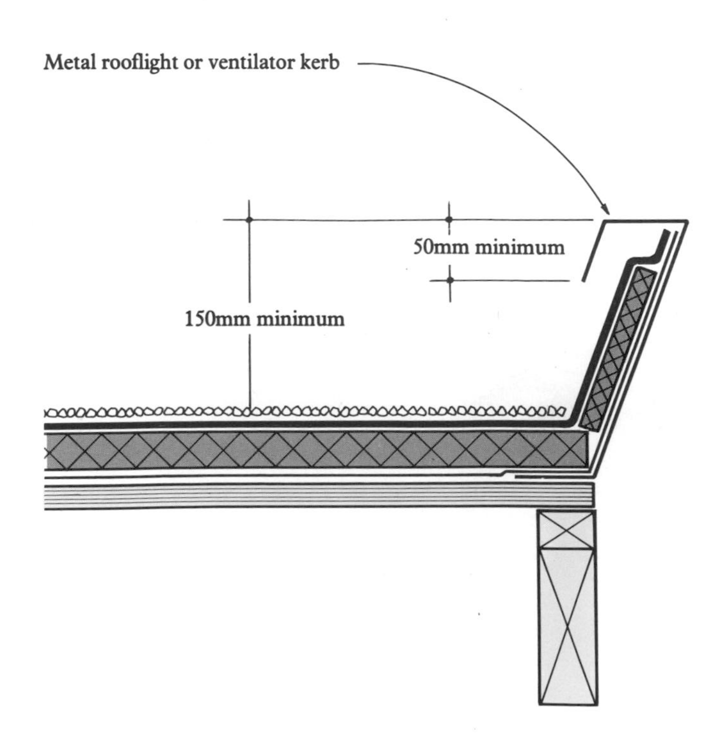

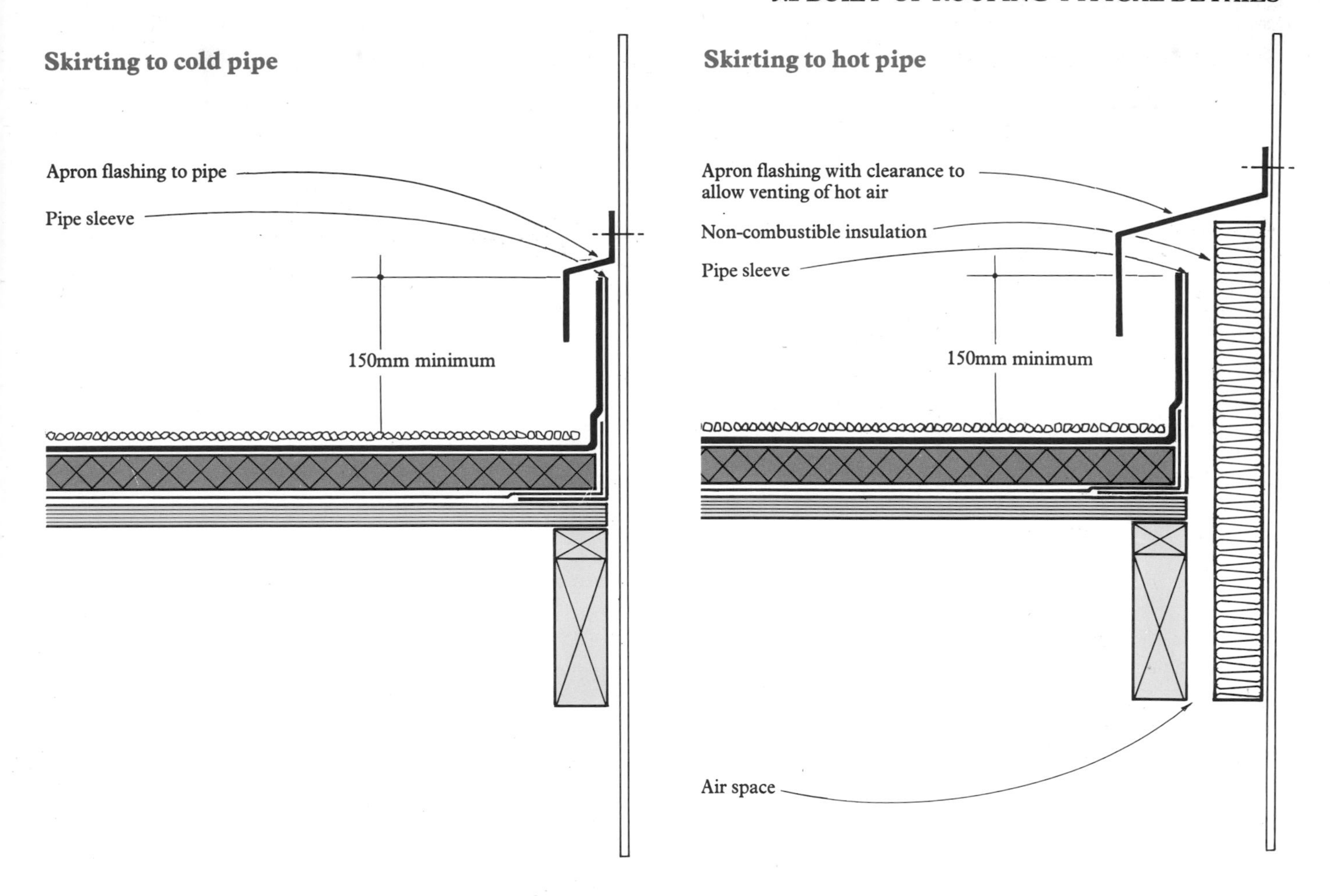
Skirting to cold pipe
Apron flashing to pipe
Pipe sleeve
150mm minimum
Skirting to hot pipe
Apron flashing with clearance to allow venting of hot air
Non-combustible insulation
Pipe sleeve
150mm minimum
Air space

BUILT-UP ROOFING METAL DECKING

Skirting to brick parapet

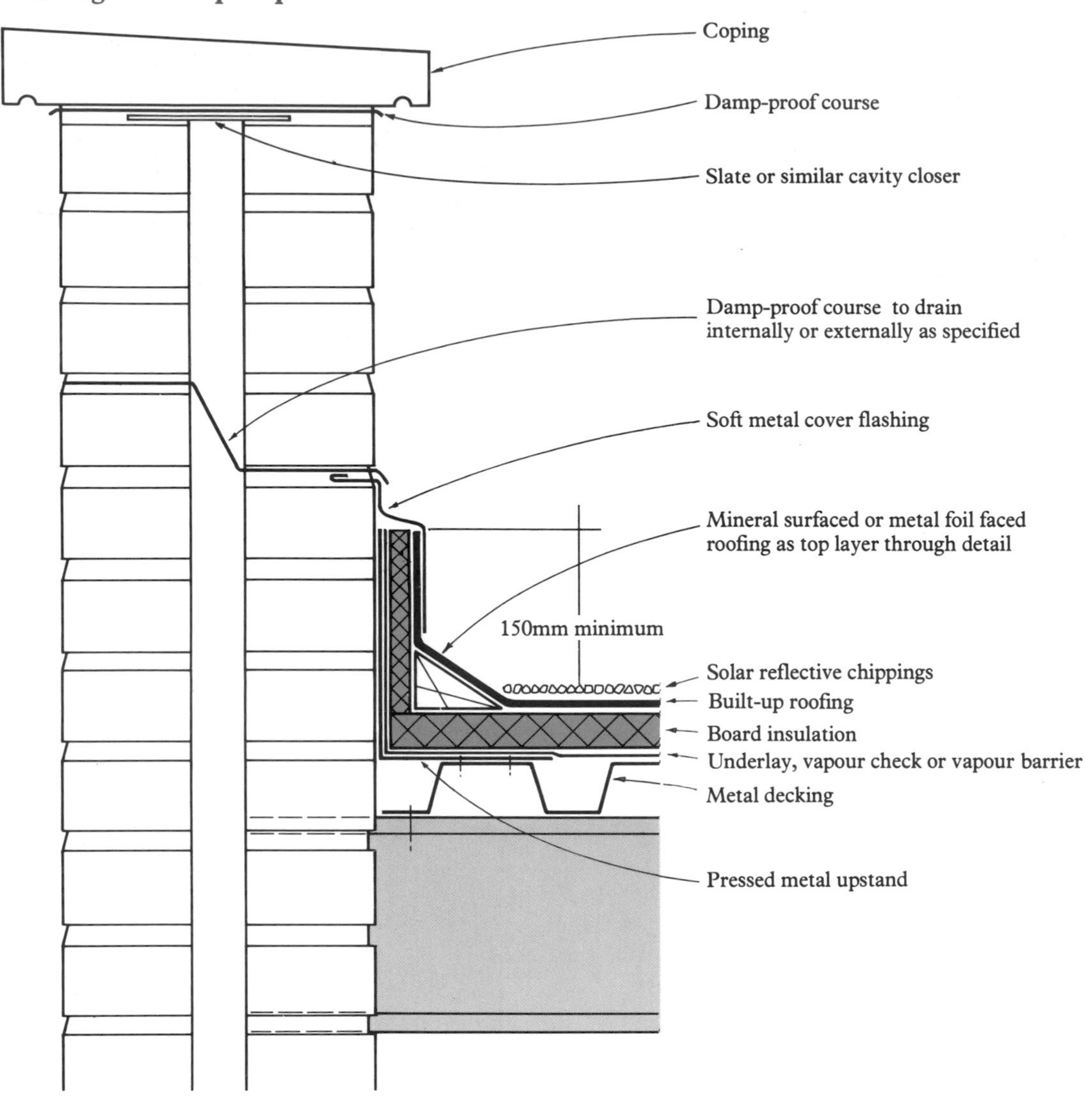

Skirting to cladding parapet

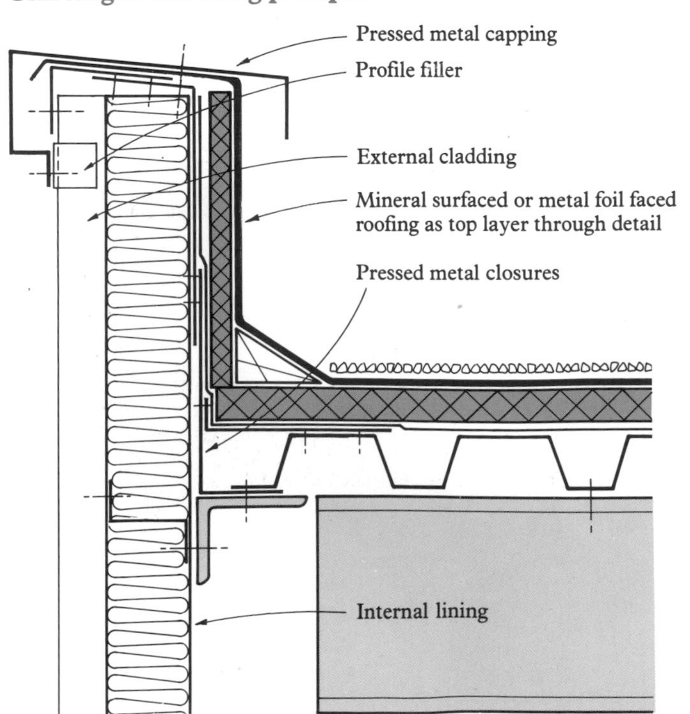

Skirting to cladding abutment

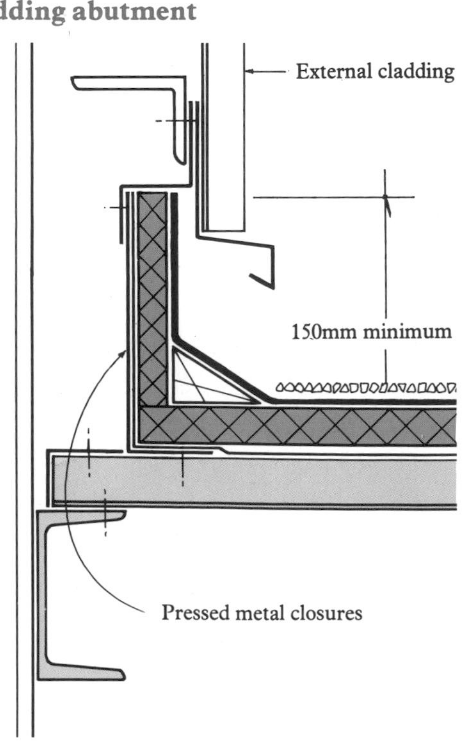

Metal check kerb

Pressed metal capping

Mineral surfaced or metal foil faced roofing as top layer through detail

Pressed metal closures

Welted drip to external gutter

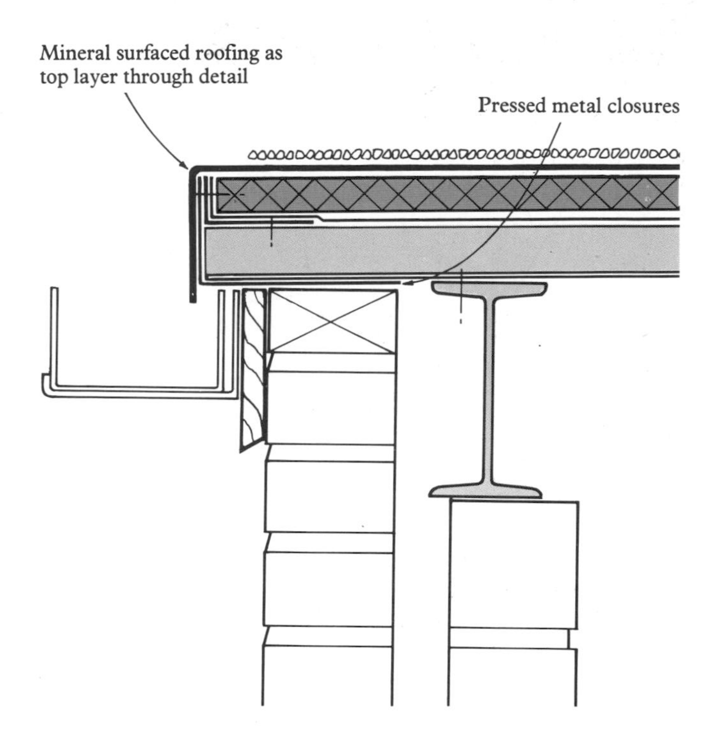

Roof outlet: proprietary flanged type

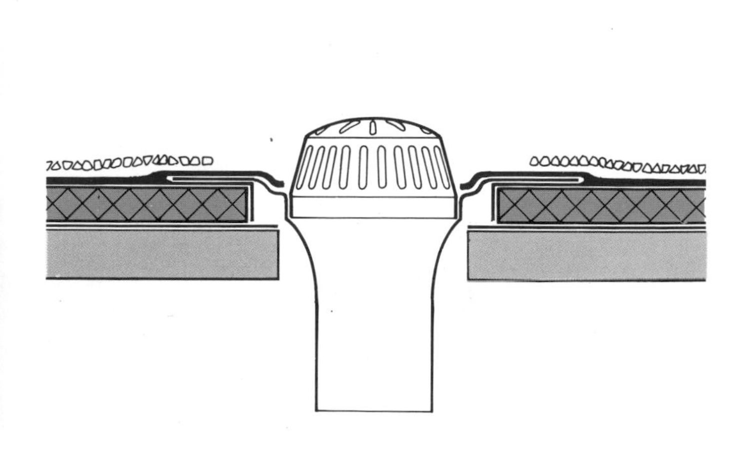

Skirting to metal rooflight or ventilator kerb

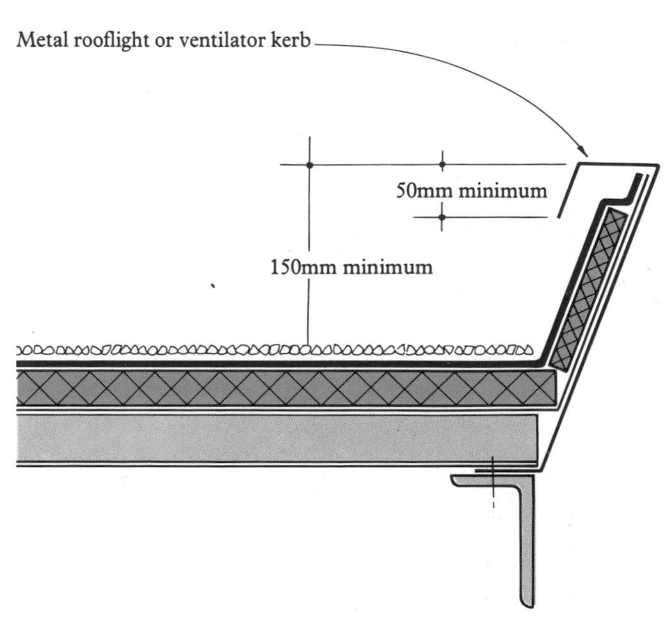

Twin kerb expansion joint

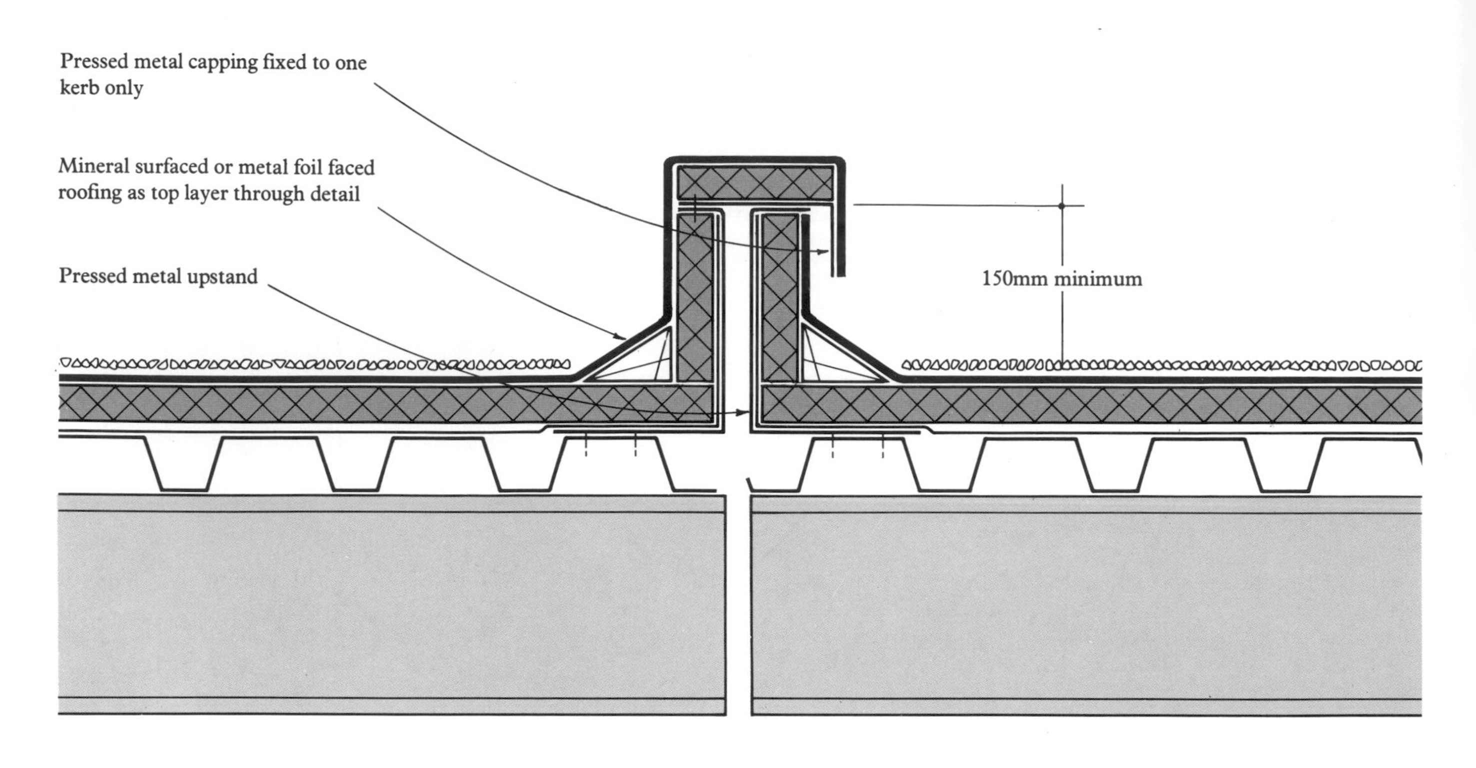

Skirting to cold pipe

Skirting to hot pipe

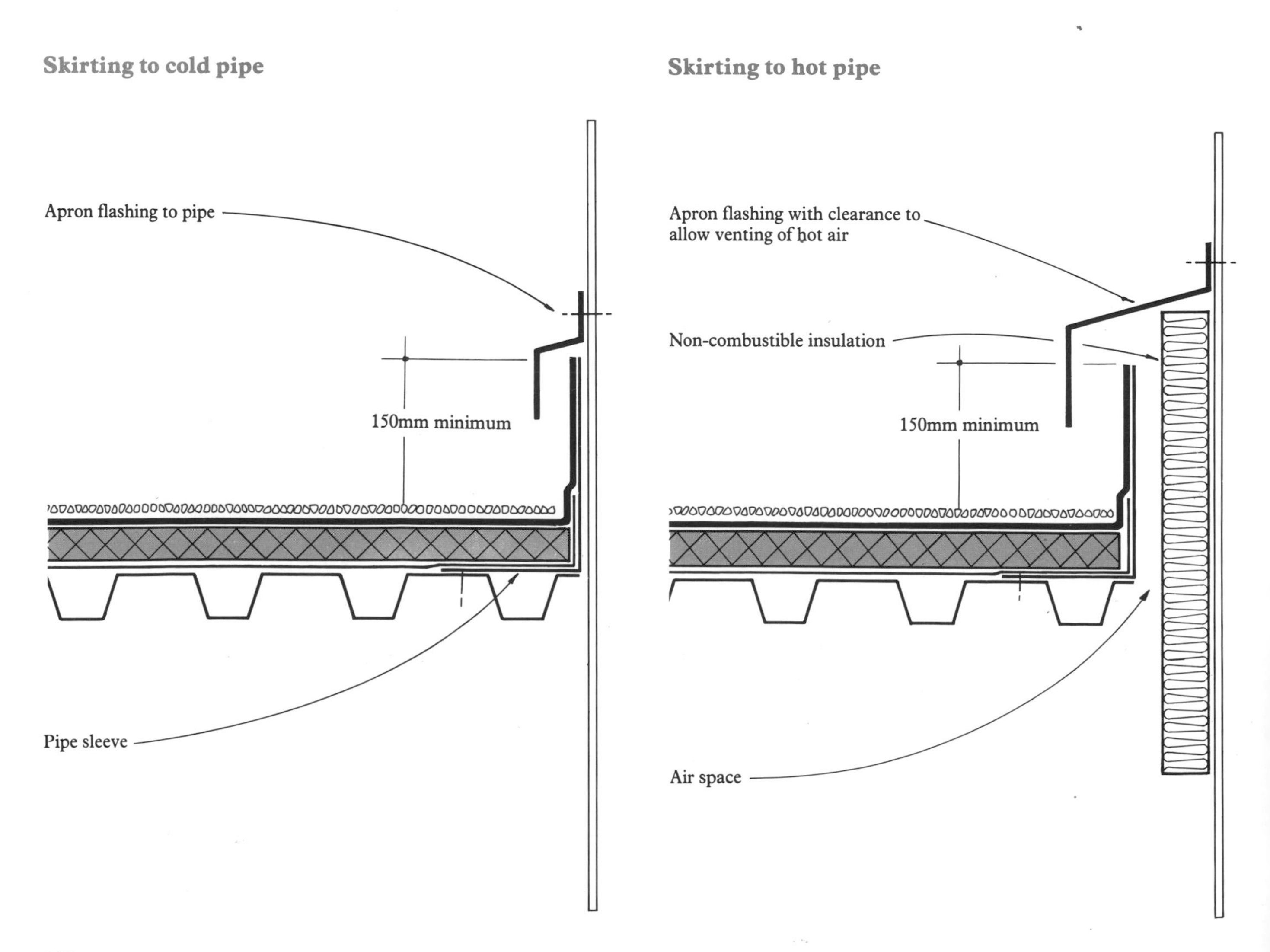

5.3 MASTIC ASPHALT DETAIL DESIGN

INTRODUCTION

The fundamental principles of allowing a minimum 150mm upstand for the waterproofing at skirtings and a 50mm turndown at open roof edges apply equally for asphalt as for built-up roofing, including the need where necessary to form independent upstands where differential movement between the roof and wall is expected.

With asphalt detailing however, the paradox also arises that although for the main horizontal areas of the roof the asphalt is invariably applied with a separating layer of sheathing felt, for the detail work there is always the need for full adhesion of the asphalt to the vertical face at upstands and open roof edge details.

ADHESION AT DETAILS

The techniques for achieving adhesion at detail work will depend on the substrate.

Applications to concrete, brick or sand/cement rendering will be bonded direct after suitable preparation of the surface.

If it is not possible to achieve a surface which will allow reliable adhesion for direct applied asphalt, it will be necessary to secure expanded metal lathing to the surface to provide a key.

Lightweight concrete blocks, sand lime bricks, timber, plywood and chipboard are unsatisfactory surfaces for the direct application of asphalt and a prior application of expanded metal lathing on sheathing felt is always necessary.

Vertical metal surfaces other than small pipes will not allow sufficient adhesion and again expanded metal lathing will be required, although in this case the separating layer of sheathing felt is not necessary.

There must be satisfactory adhesion of the asphalt to vertical surfaces to restrain movement and reduce blistering and slumping

The expanded metal lathing must be fixed with nails or screws at 150mm centres in both directions. The lathing must be firm and unyielding to remain permanently in position and withstand distortion from the heat of the asphalt. Expanded metal lathing is manufactured with a vaned or cupped shape which must point upwards to provide a satisfactory key for the asphalt.

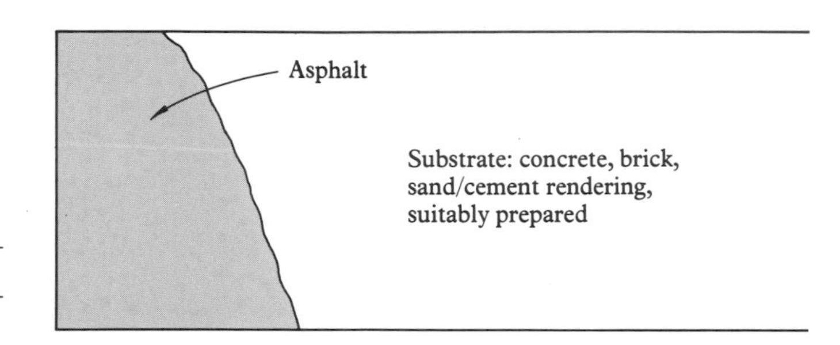

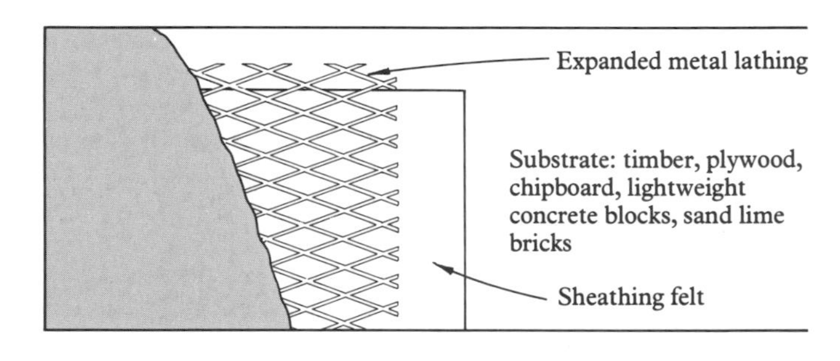

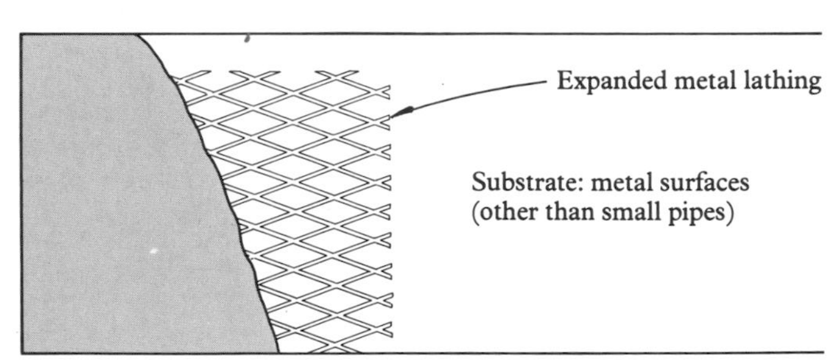

In recent years the traditional practices for achieving direct adhesion to brick or concrete wall faces have been changed by the availability of reliable high bond primers. Most asphalters now apply the high bond primer to all vertical brick or concrete surfaces as a matter of routine. This overcomes the majority of difficulties which might otherwise arise from unsatisfactory conditions such as smooth surfaces, surface laitence or mould oil. Using a high bond primer will normally avoid the need for the more traditional methods of surface preparation. These include roughening or wire brushing or the application of a rough surfaced rendered skin.

The traditional craft skills call for the first layer of asphalt to brick or concrete to be applied with a steel trowel. This entails the application of small dabs of asphalt pressed well home to the prepared surface to ensure good adhesion. The asphalter will frequently add a small amount of bitumen to each bucket of asphalt. The bitumen then spreads over the top surface of the asphalt in the bucket and as each dab of asphalt is removed with the trowel it receives a coating of the bitumen which acts as a primer and helps the adhesion. This method is extremely labour intensive, and it is now generally accepted that an equally good adhesion will be obtained by a float application of mastic on a high bond primer. This is a faster and more suitable method for modern building programmes.

Another traditional practice now discontinued is the raking of joints of vertical brickwork to form a key for the asphalt. This has proved unnecessary and is now thought undesirable as it introduces stress lines which show in the surface of the asphalt.

THICKNESS AT DETAILS

Asphalt roofing will generally be carried out to the following nominal thicknesses:

Main horizontal areas: 20mm in two coats.

Vertical work at details direct to a suitable surface: 13mm in two coats.

Vertical work on expanded metal lathing: 20mm in three coats.

Vertical work exceeding 300mm in height on any surface: 20mm in three coats.

Two-coat work on expanded metal lathing would not be satisfactory as it is not easy to bring the asphalt to a satisfactory finish giving full cover to the lathing. Three-coat work is necessary including a thin first coat of asphalt to form the key to the lathing. This is followed when cool by a second coat which builds up thickness but leaves a rough finish. The third then completes the full thickness and can be brought up to a first class finish.

With two-coat work to surfaces such as brick or concrete treated with a high bond primer, the first coat will form the bond and key to the surface and will take out irregularities. This first coat will be relatively thin, about 4 or 5mm and it will be allowed to cool to a stable consistency before the application of the second coat which will bring the detail work to the nominal thickness of 13mm.

It is not possible to achieve a satisfactory standard of surface finish with two-coat work if the height of the work exceeds 300mm. This is known as 'shown vertical' work and is carried out in three layers with the additional layer bringing the nominal thickness to 20mm.

Three-coat work may also be required to old brickwork, using the first coat as a dubbing out layer to establish a suitable base.

ASPHALT FILLETS

Horizontal asphalt work is carried out as a separate application from the vertical detail work. It is not possible to form a fully efficient junction between the vertical and horizontal without the separate operation of adding a fillet. This adds a substantial thickness of asphalt to reinforce the junction between horizontal and vertical and fuse the surfaces together.

To make sure of a fully efficient fusion, all solid angle fillets must be formed in two-coat work. All dirt, sand and foreign material is first brushed away and a poultice of hot asphalt is then applied to the angle. This is left for a few minutes to soften the horizontal and vertical asphalt at their intersection. The hot poultice is then removed and the area is scraped and cleaned before fresh asphalt is applied to the warmed junction and dressed in with a filleting tool. A second coat is applied immediately and is cut in with a filleting tool.

The pressure of the filletting tool on the face of the fillet will cut into the softened faces of the horizontal and vertical asphalt to leave an indentation of perhaps 2mm or 3mm. This is good practice and is often taken as a sign of a well formed fillet. The reduction of thickness caused by the indentation is normal and not a cause for concern.

The fillet should be formed with a minimum face of 40mm. Some authorities call for a 50mm minimum face but this is difficult to achieve in practice due to the larger quantity of asphalt tending to slump and lose its shape.

Solar reflective chippings may be taken to the base of the fillet or set back by approximately 75mm so that an easy visual check can be made on the condition of the junction. If the chippings are set back and a solar reflective paint finish is applied to the asphalt detailing work the paint should be continued onto the flat to protect those horizontal areas not finished with chippings.

SKIRTINGS

Skirtings direct to brickwork and concrete
In addition to obtaining full adhesion between the asphalt and concrete or brickwork, it is essential that the asphalt at the top of the skirting is turned into a chase or over a kerb. This secures the top of the asphalt and prevents it from falling or slumping away from the wall. The chase must be cut at least 25mm deep and 25mm on face and the bottom edge of the chase should be knocked off to leave a rough sloping face. This forms the ideal shape for keying the asphalt, and allows a full thickness of asphalt to be applied at the arris.

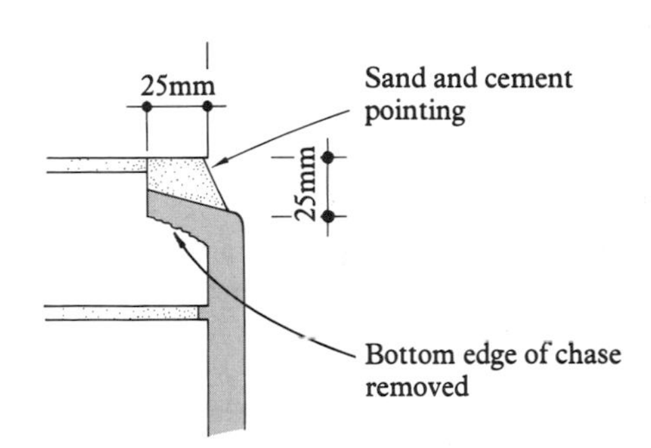

The chase should not be cut immediately under the damp proof course because of the danger of damaging the dpc and the bulky nature of the detail which would result.

Both first and second coats of asphalt to the skirting must be turned into the chase or over the kerb. The asphalt in the chase is splayed back at an angle to allow a sand and cement pointing which is essential. It is not good practice to fill the chase with asphalt and omit the pointing.

The preferred flashing arrangement at skirtings to brickwork is a metal cover flashing inserted directly beneath the damp proof course which is one brick course higher than the top of the asphalt.

The cover flashing provides additional security by protecting the brick course below the damp proof course and prevents a flow of water to the top of the skirting. The cover flashing may be omitted but the top of the asphalt skirting will then be subjected to the flow of water down the brick face, and it can be expected that more maintenance will be required.

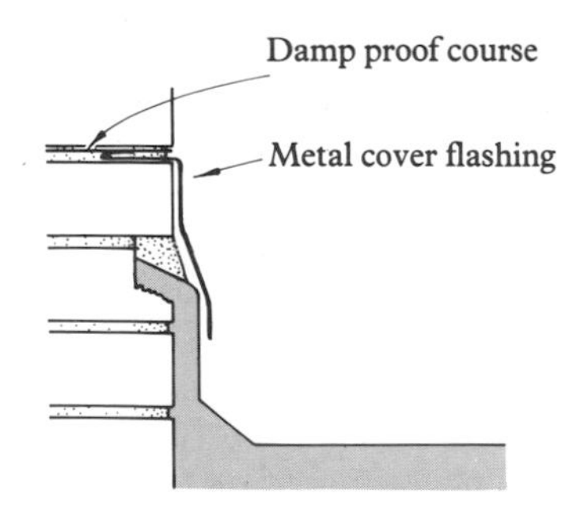

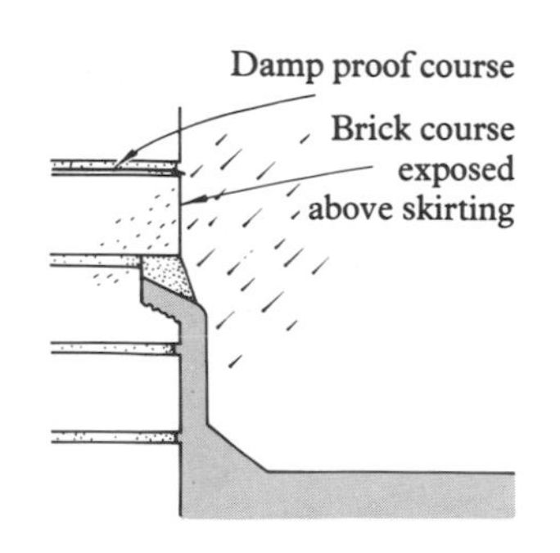

Independent upstands
The transition from total freedom of movement of the asphalt for the main horizontal areas of the roof to total restraint at skirtings is likely to give rise to locally increased stresses on the line of transition between separation and attachment. It is clearly undesirable to risk further stresses as a result of differential movement between the wall and the roof and it is advisable to err on the side of including independent upstands in the design.

Independent upstands are generally formed on free-standing timber kerbs nailed or spiked to the roof deck. If the shape of the kerb prevents successful nailing, it may be necessary to divide the kerb into two parts, fixing the first part as a base plate and the second plate to the first.

On metal decks it is difficult to fix a timber kerb and a free-standing metal upstand will normally be used with a timber facing, sheathing felt and expanded metal lathing.

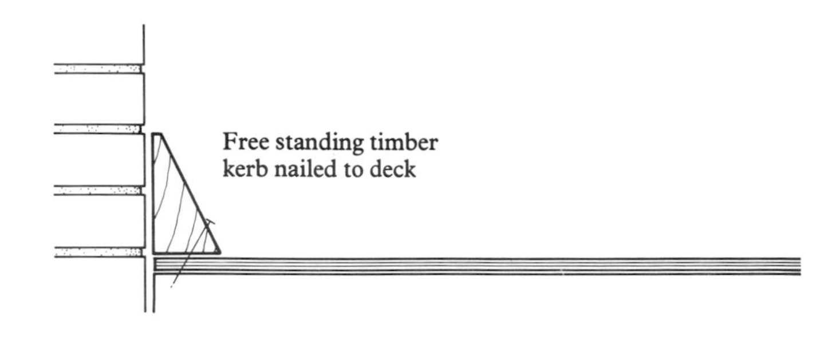

Forming asphalt fillets

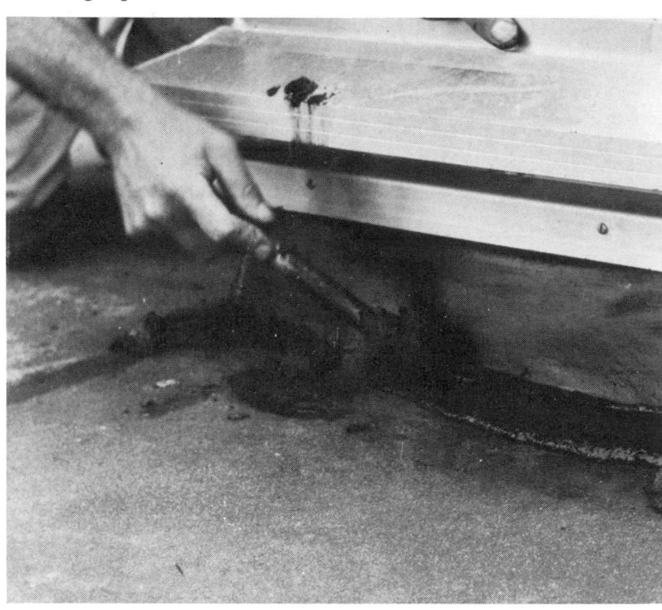

Hot poultice of asphalt is removed and the warmed junction is scraped and cleaned.

Second coat of the fillet is applied immediately

Fresh asphalt is applied to the warmed junction and dressed in with a filletting tool

The finished detail with a solid asphalt fillet

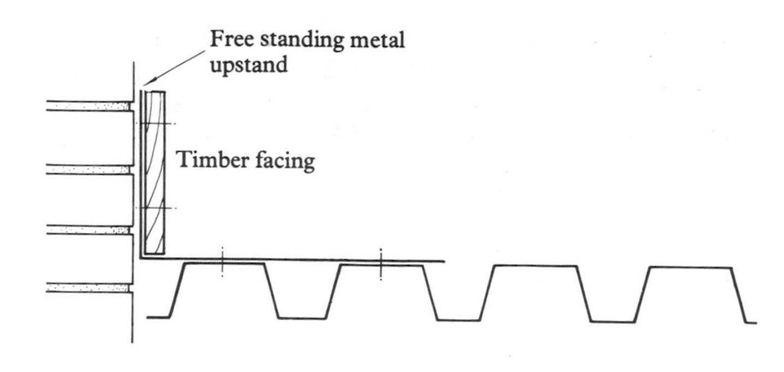

Skirting support

It is essential to provide firm support at the base of all asphalt skirtings whether they are formed on independent upstands or direct to the vertical wall.

This calls for special treatment if the roofing specification includes a board insulation, to avoid problems which may arise from the compressible nature of the insulation, the difficulty of cutting the individual insulation boards tight to the vertical face, and also the possibility of a foam insulant being damaged by the heat of the asphalt.

Skirtings formed direct to a brick or concrete wall face will require additional support if an insulation board is used. This may be achieved by allowing a controlled gap in the order of 25mm wide between the edge of the insulation and the wall face. During the application of the first coat of asphalt to the main areas, the gap may then be filled with asphalt to form a supporting leg for the fillet, provided the insulation materials are not adversely affected by the heat of the asphalt. If the insulation is heat sensitive, the gap may be filled with sand and cement with the water content kept to a minimum to prevent the entrapment of water against the insulation. This mix is sometimes referred to as earth-damp sand and cement.

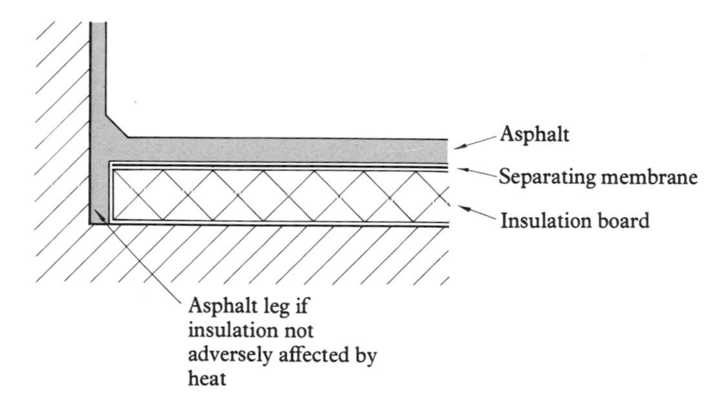

In particular, expanded polystyrene would melt and cavitate from the heat of asphalt and a sand and cement infill will always be necessary. Expanded polystyrene also requires an overlay of fibreboard, cork or a similar temperature-resistant insulation and in this case both insulations are set back to leave a gap which is subsequently filled with sand and cement.

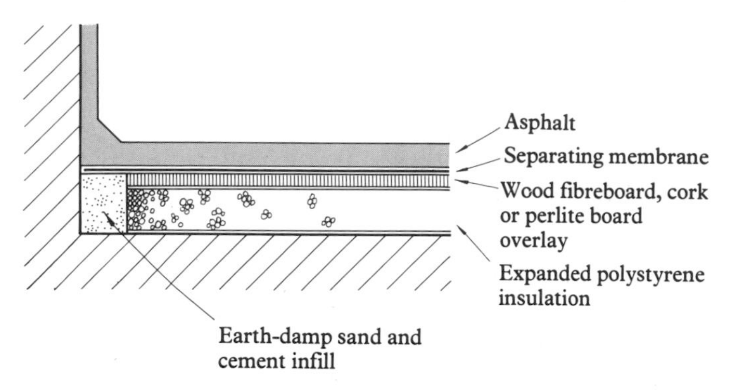

If the skirting is formed on timber or any surface which calls for the use of expanded metal lathing, the lathing can be continued onto the horizontal for approximately 75mm to provide reinforcement for the mastic in the vicinity of the skirting. In this case it will suffice for the insulation to be cut close to the vertical face and the small gap which is inevitable in practice will cause no problem. The reinforcing of the lathing will also prevent problems arising from the compressible nature of the insulation.

Alternatively, the insulation board may be stepped back from the edge and the gap filled to form a supporting leg as previously described.

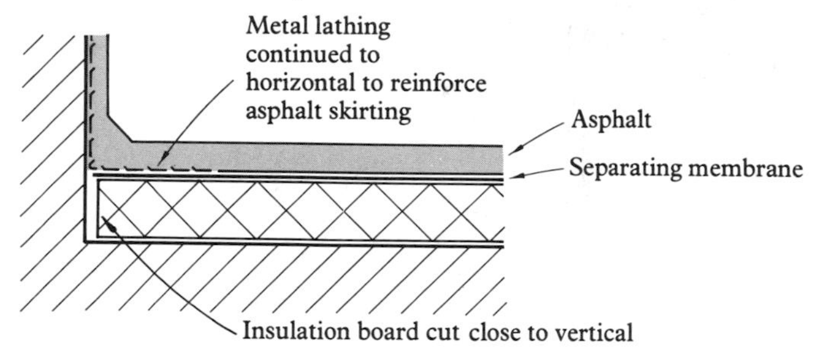

Vapour barriers and vapour checks

Roofing felt vapour barriers and vapour checks should be continued up the wall face in such a way that the asphalt membrane will make contact with them. The contact area should be kept small as it is not desirable to use bitumen felt as a base for the direct application of asphalt. Alternatively the felt can be turned back onto the surface of the board for a short distance before the separating membrane is applied. This detail may cause the sheathing felt to act as a separating layer between the asphalt and the bitumen felt but no significant vapour leakage should arise.

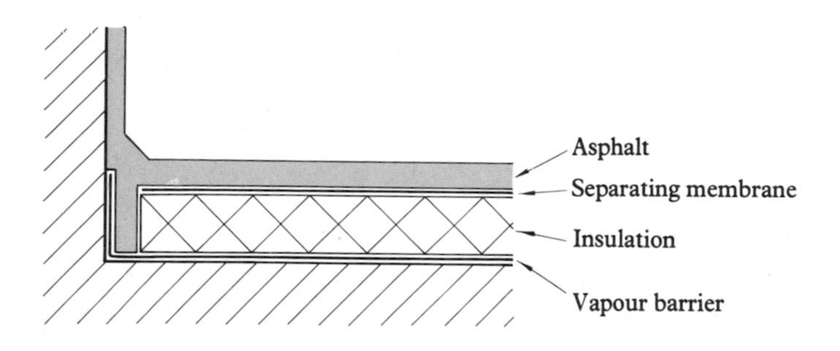

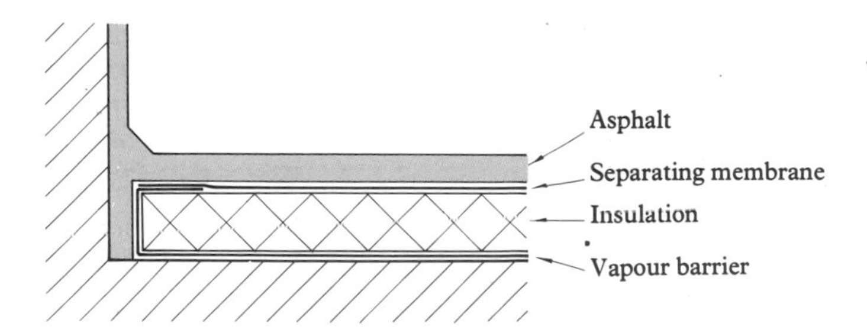

Mastic asphalt vapour barriers or vapour checks are occasionally specified, generally in connection with a special need to provide early waterproofing cover or a level surface to avoid cavitation beneath a tapered insulation board.

A glass tissue separating membrane is first applied to the deck, followed by a single-coat asphalt to a 13mm nominal thickness. A thin coat of vertical asphalt is used to complete an envelope to the insulation at skirtings.

For heat sensitive insulations such as expanded polystyrene the detail will be associated with a sand and cement infill and a wood fibreboard or similar overlay as described above.

Skirtings with paving grade asphalt

Before the application of paving grade asphalt, it is necessary to complete the roofing grade including skirtings, but without surfacing or sand rubbing. The paving grade is then applied to the flat areas only and finished tight to the roofing grade skirting but with no vertical element in the paving. A further roofing grade fillet is then applied in the normal way.

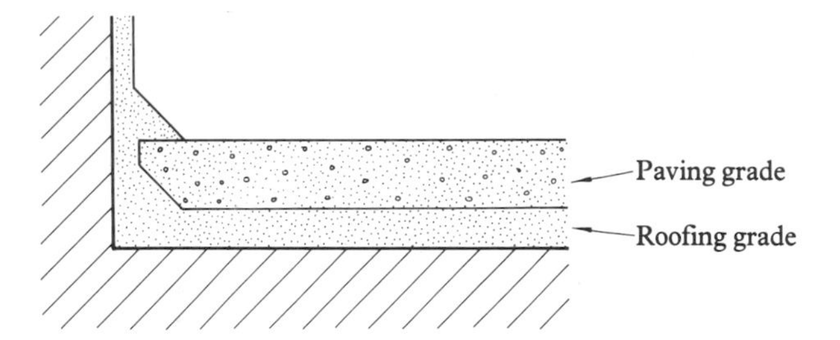

Skirtings to ventilators, rooflights etc

Upstand details to ventilators and rooflights are formed in precisely the same way as skirtings but with the top of the asphalt secured by turning it onto the horizontal face of a timber kerb or into a chase formed in the top edge of a concrete or brick kerb.

Sheathing felt and expanded metal lathing are applied to the vertical face of timber and are dressed over the kerb and securely fixed.

Suitable grounds for fixing the ventilator or rooflight top should be provided and the asphalt butted against these grounds. The asphalt should not be continued to form a base for the rooflight or ventilator as it is not a suitable material to provide the necessary support and is prone to damage from the fixings.

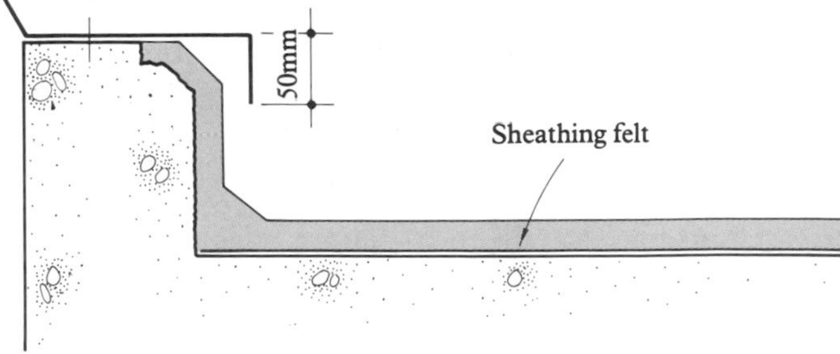

Concrete or brick kerb

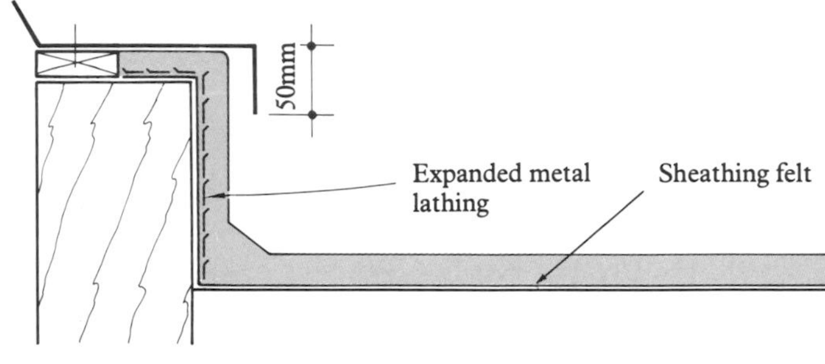

Timber kerb

With upstands to metal kerbs, it will not normally be possible to turn the asphalt over the top of the kerb. The asphalt will finish on the vertical and a key will be provided by a proprietary kerb with grooved surfaces or by spot welded expanded metal lathing. If spot welding is not practical, the expanded metal lathing may be fixed to a timber facing in conjunction with sheathing felt.

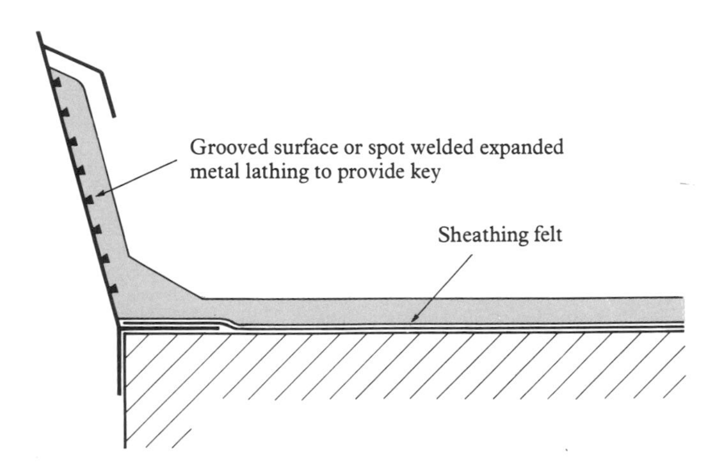

Proprietary metal rooflight kerb

Skirtings to cold pipes and hand rails

Flanged metal collars, preferably of lead, should be provided for pipes which are to pass through the roof. A welt will be required at the edge of the flange to provide a key for the asphalt and thereafter the asphalt skirting is formed against the collar in exactly the same way as normal skirtings with high bond primer and fillets. An apron flashing should then be installed to the pipe to protect the top edge of the asphalt.

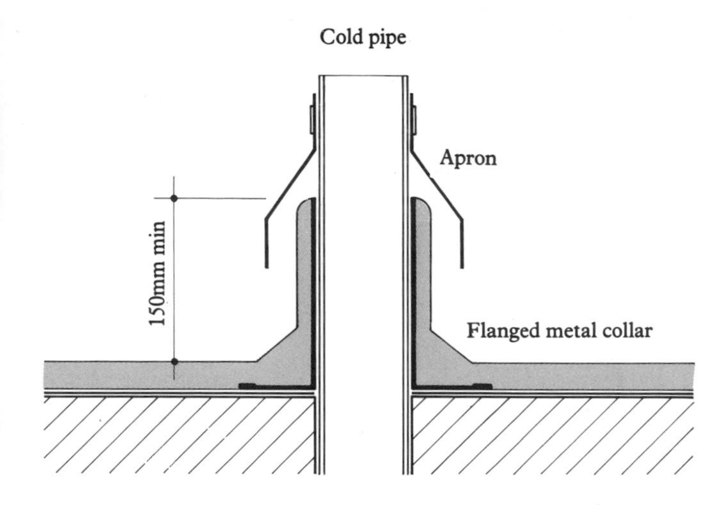

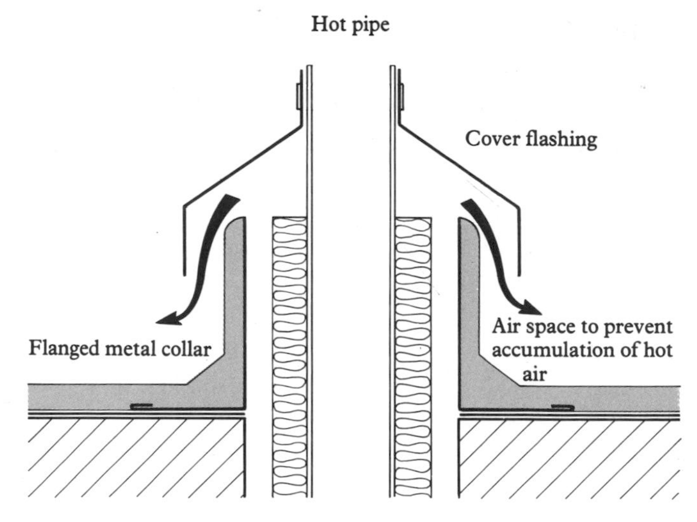

Small cast iron pipes which are in position before the waterproofing is carried out can receive an asphalt skirting direct. A high bond primer is necessary unless the pipe has a bitumen faced protection which will act as the primer. Large pipes may require expanded metal lathing wired in position to provide a key for the asphalt. Hand rails are usually set on a kerb approximately 75mm high, in which case a skirting of 75mm around the hand rail will complete an effective 150mm skirting height above roof level. An independent apron flashing should then be installed onto the hand rail. If this flashing is omitted the top edge of the asphalt may separate from the pipe in the course of time and this could lead to leakage. An apron flashing should only be omitted when the consequences of leakage are not severe and the need for occasional maintenance of the item is accepted.

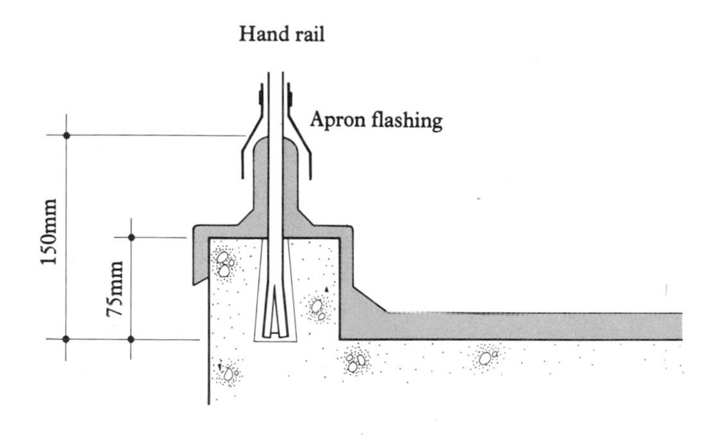

Certain manufacturers of vent pipes make special fittings available to provide a protection and weathering to the top edge of the asphalt. These are an excellent arrangement and are to be recommended.

Skirtings to hot pipes and flues
Asphalt skirtings to hot pipes follow the same procedures as for cold pipes with flanged metal collars but with an air space or insulation or a combination of the two to reduce the temperature of the asphalt skirting. A larger apron flashing will be required to cover the insulation and asphalt and a ventilation space between the asphalt and the flashing to prevent the accumulation of hot air around the pipe.

The insulation and ventilation should be designed to ensure that the asphalt is not taken significantly above 70°C.

GUTTERS

The asphalt to the sole of a gutter is applied on sheathing felt and the formation of a gutter will include sides which are conventional skirtings jointed by an arris to the horizontal main areas. The gutter will form a heat trap and particular care should be taken with the detailing and application of gutter sides to ensure good adhesion to safeguard against blistering or slumping.

The sides of the gutter should be uninsulated or the insulation should be covered by a timber facing with sheathing felt and expanded metal lathing applied before the asphalt.

EXPANSION JOINTS

Twin kerb expansion joints are recommended with a metal cap flashing fixed to one kerb only, or a capping system held by cleats or spring clips. In either case suitable grounds should be provided to avoid fixing through the asphalt. The boxing in of the ends to complete the waterproofing but still allow movement is sometimes difficult as mentioned in the section describing built-up roofing. Special detail design will be required unless a measure of risk of increased maintenance and repair is accepted.

Flush surface expansion joints are always difficult to form and the design of the structure should avoid the occurrence of expansion joints of this type if at all possible. Proprietary expansion joints are available and specialist application is necessary if expansion joints are to run across areas covered with paving grade and subject to wheeled traffic.

Eaves and verges must provide firm restraint to the edge of the asphalt which must either be turned down at the outer face of the building or finished by a metal apron.

Asphalt aprons
Asphalt aprons form a reliable roof edge detail much favoured by asphalters. A chamfered batten is placed as a guide and thickness gauge against which the lower undercut edge of the apron is formed. The eaves corner will be formed as an arris in the asphalt joining the vertical to the horizontal.

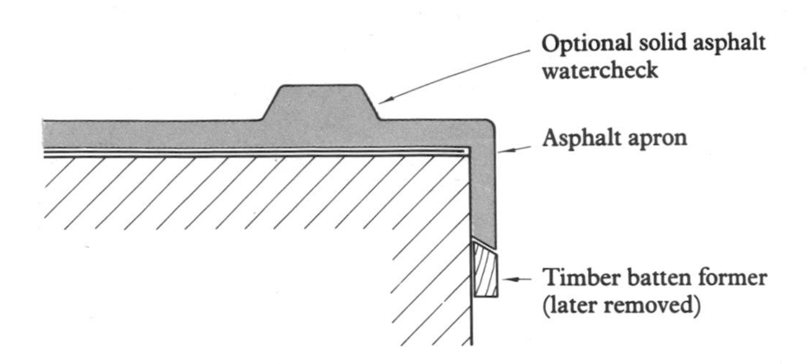

Check kerbs may be formed in an extra thickness of asphalt near to the roof edge, but most commonly a check kerb will be formed in the structure and the asphalt will be taken over and finished by an apron or metal trim detail. The vertical face to receive the asphalt apron will be prepared in exactly the same way as skirtings, with primer, sheathing felt and expanded metal lathing as necessary.

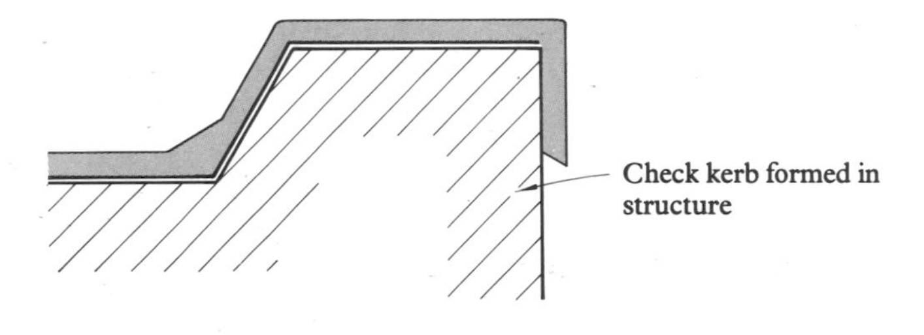

Lead aprons
A traditional eaves detail is a lead apron to external gutter, but problems have arisen from cracking of the asphalt at the joints and over the inside edge of the lead. The cracking is caused by the differential movement between the lead and the asphalt and this can occur more frequently if the asphalt is thinned by the design of the detail. Asphalt will not bond sufficiently to the horizontal lead unless a welt is formed at the internal edge. The asphalt will key to the welted edge and this will help to maintain a bond and waterproof seal to the entire horizontal face of the lead. The welted edge will significantly reduce the thickness of the asphalt however, and provide a line of weakness for cracking if the detail is not designed to allow the lead to be placed in a recess slightly below the level of the main areas of roof. The asphalt can then be applied at full thickness throughout the detail.

Individual lengths of lead should be limited to 1.5 metres.

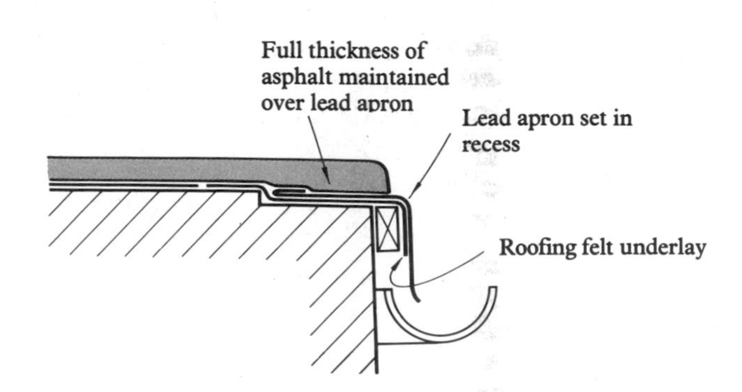

A roofing felt underlay to the lead flashing is generally provided as an isolating layer. The lead should be well nailed at close centres to restrain movement and secure it aginst wind forces.

Aluminium trims
Extruded aluminium trims are widely available for use with asphalt. Some trims are designed on the same principle as lead aprons with a welted or ribbed formation to provide a key, and the trim is set into a recess to allow the application of full thickness asphalt. Experience has shown a tendency for the bond to the aluminium to be ineffective or for cracking to take place at the trim joints or inside edge of the trim due to the movement of the aluminium.

Aluminium trims are available with the extrusion designed to act more as a permanent formwork to hold a traditional asphalt apron concealed behind the aluminium fascia and these trims do not depend on the formation of a bond between asphalt and aluminium. As the horizontal leg of the trim is relatively thin, there is no need to set it down into a recess. The turn down of the asphalt behind the vertical face of the trim provides a key and holds the main areas of asphalt against the forces of contraction. A felt underlay is not required with this form of trim.

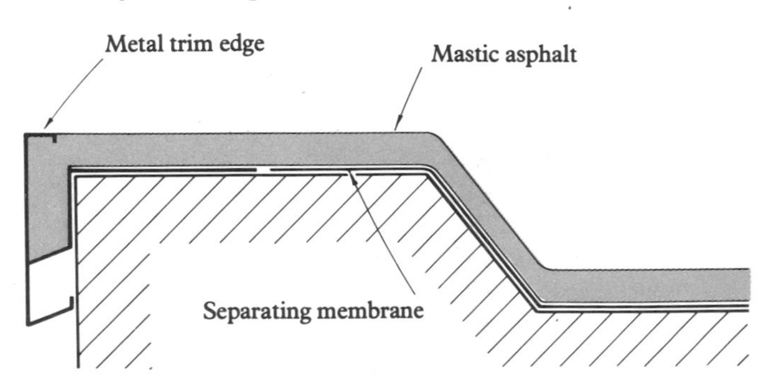

Asphalt cappings to parapets
Asphalt cappings to parapet walls combine the detailing principles for skirtings, eaves and verges previously described. The asphalt is continued up and over the parapet to be finished by an asphalt apron or a metal trim edge. Finishing the waterproofing on the flat is never recommended.

Asphalt cappings are most applicable to low parapets of solid construction and where differential movement between the roof and parapet structure is unlikely to stress the waterproof covering.

The preparation of the horizontal parapet surface to receive the asphalt is the same as that required for the vertical. Vertical asphalt is generally applied to a nominal thickness of 13mm in two-coat work if the parapet height is under 300mm and three-coat work to 20mm if the parapet height is over 300mm. The horizontal and vertical asphalt surfaces are applied as separate operations. As with asphalt fillets, a poultice of hot asphalt is applied to the previously formed vertical or horizontal surface to ensure a successful fusion at the arris.

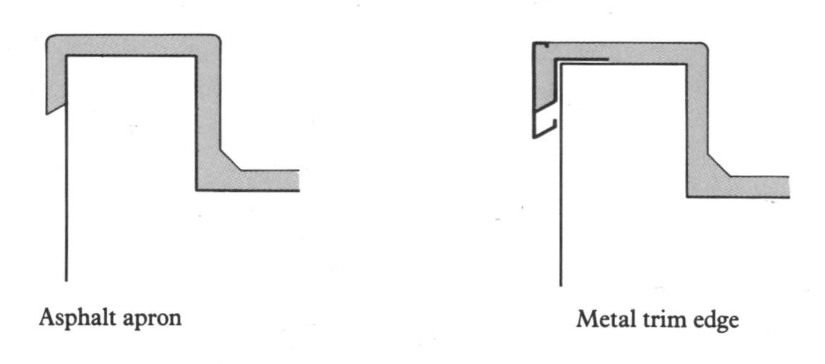

RAINWATER OUTLETS

Outlets are available with flat flanges formed of lead, aluminium or plastic with welted edges or ribs to provide a key and ensure good adhesion. The outlet should be recessed to ensure that a full thickness of asphalt is maintained.

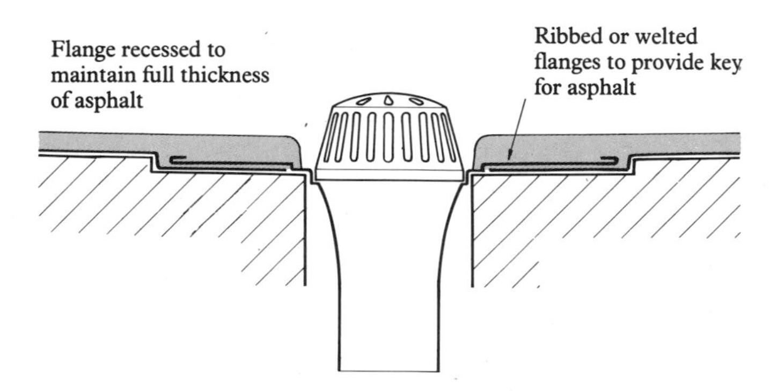

Proprietary outlets which incorporate turn-downs are preferred, as these provide a key for the asphalt. Certain proprietary outlets are also fitted with compression rings which are tightened against the inner surface of the asphalt and help to provide a stable formation at thc outlet.

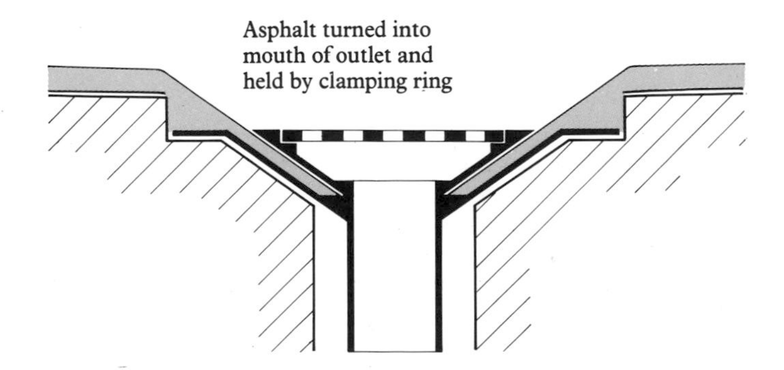

Outlets should be carefully positioned with close attention to levels at the design stage. The connection between the downpipe and the outlet should make provision for thermal expansion of the downpipe.

TYPICAL DETAILS

The principles of detail design described above are further illustrated in the following section in the form of typical details for warm roof constructions.

The details are grouped according to the structural deck material in the same sequence as the typical specifications of section 4.4

The principles of detail design are identical for timber and woodwool decks. To avoid unneccessary repetition, these decks have not been separately illustrated. Details illustrated for plywood deck construction are therefore equally applicable to woodwool, chipboard and timber boarded decks.

In all cases, typical adjacent structures have been included to enable the principles of detail construction and waterproofing to be shown, but it is not intended to illustrate details of building construction

MASTIC ASPHALT CONCRETE DECK

Skirting to brick parapet

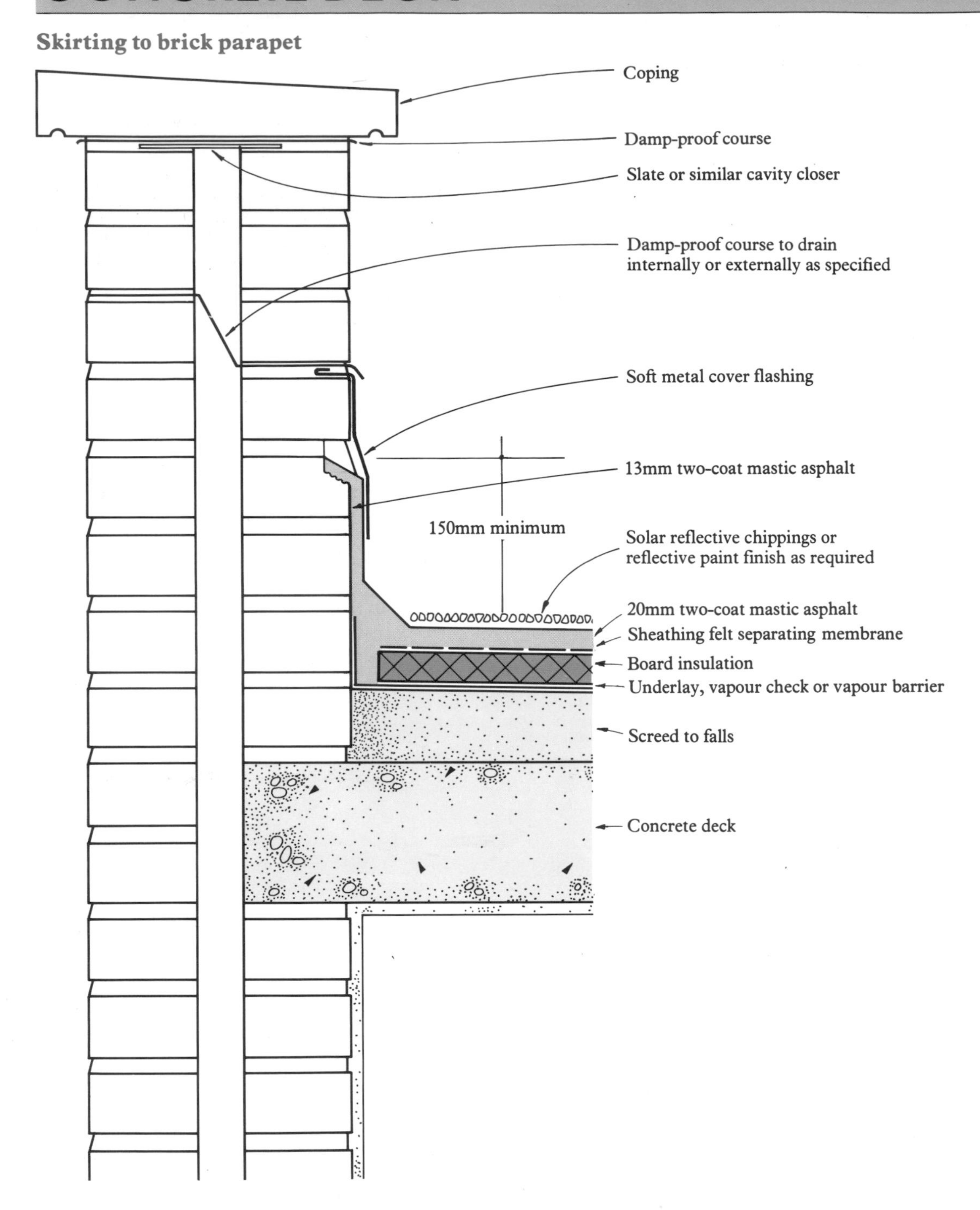

Check kerb and metal trim edge

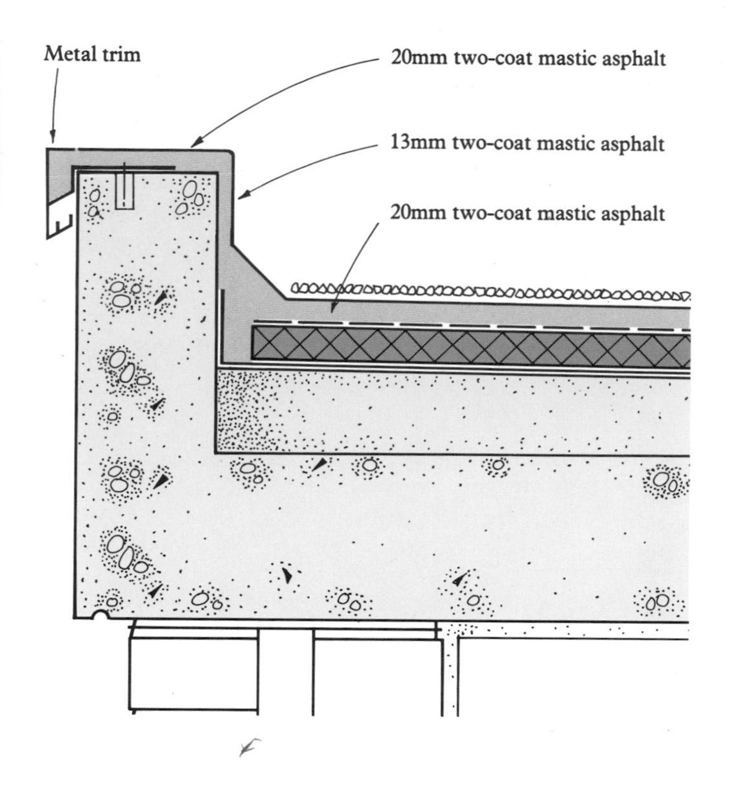

Asphalt apron and drip to check kerb

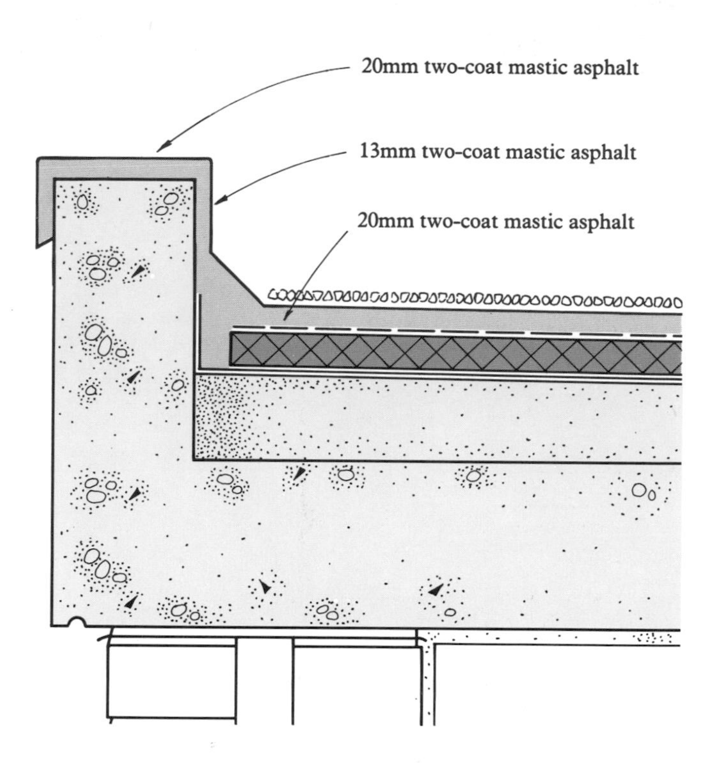

Asphalt apron and drip to external gutter

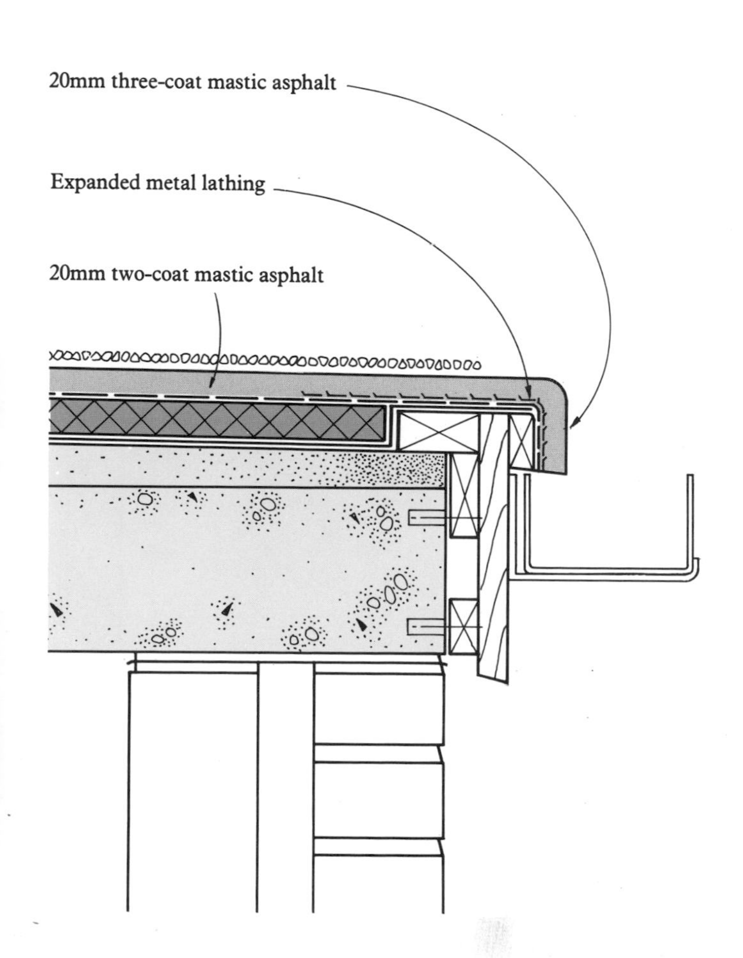

Skirting to metal rooflight or ventilator kerb

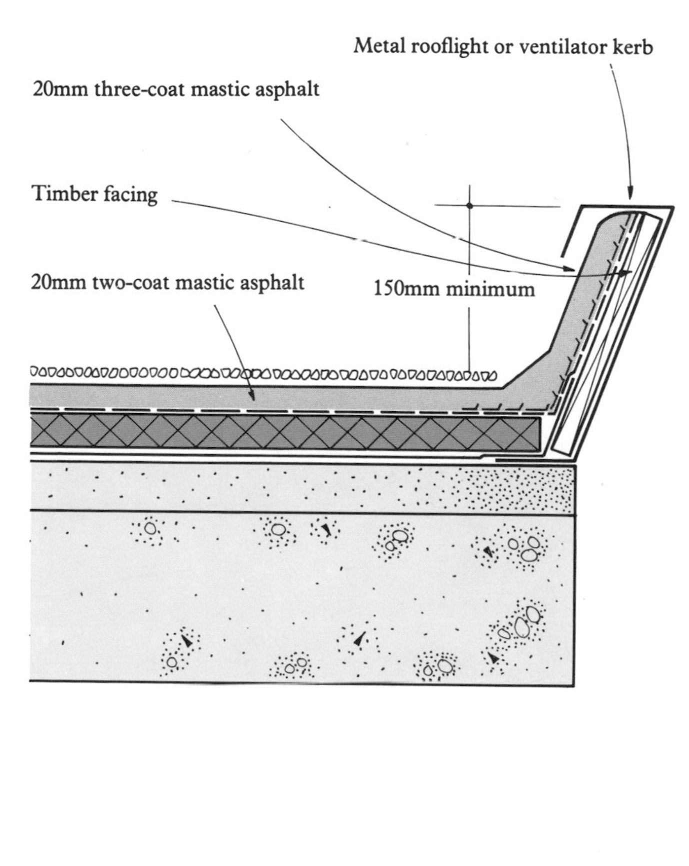

Roof outlet: clamping cone type

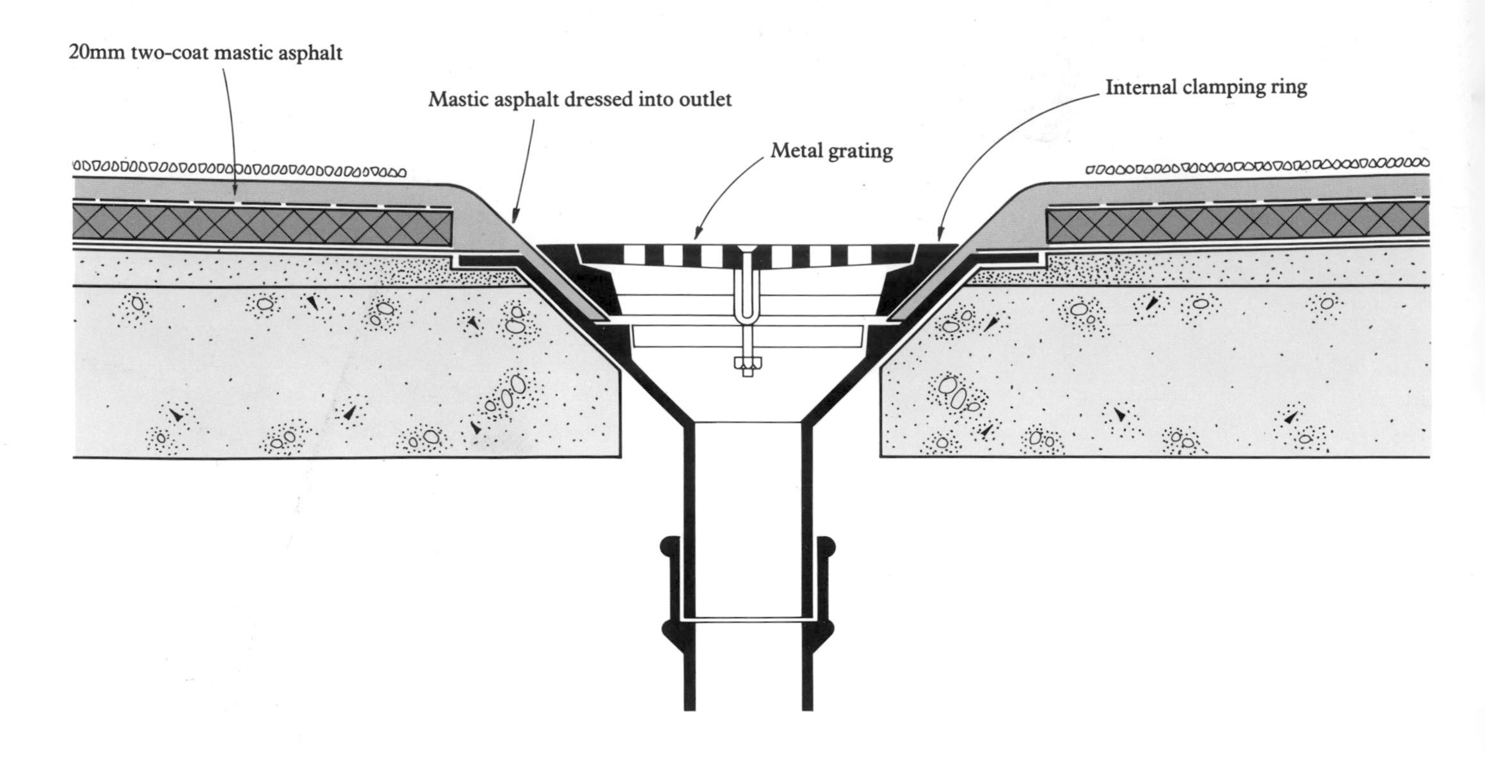

Skirting to cold pipe

Skirting to hot pipe

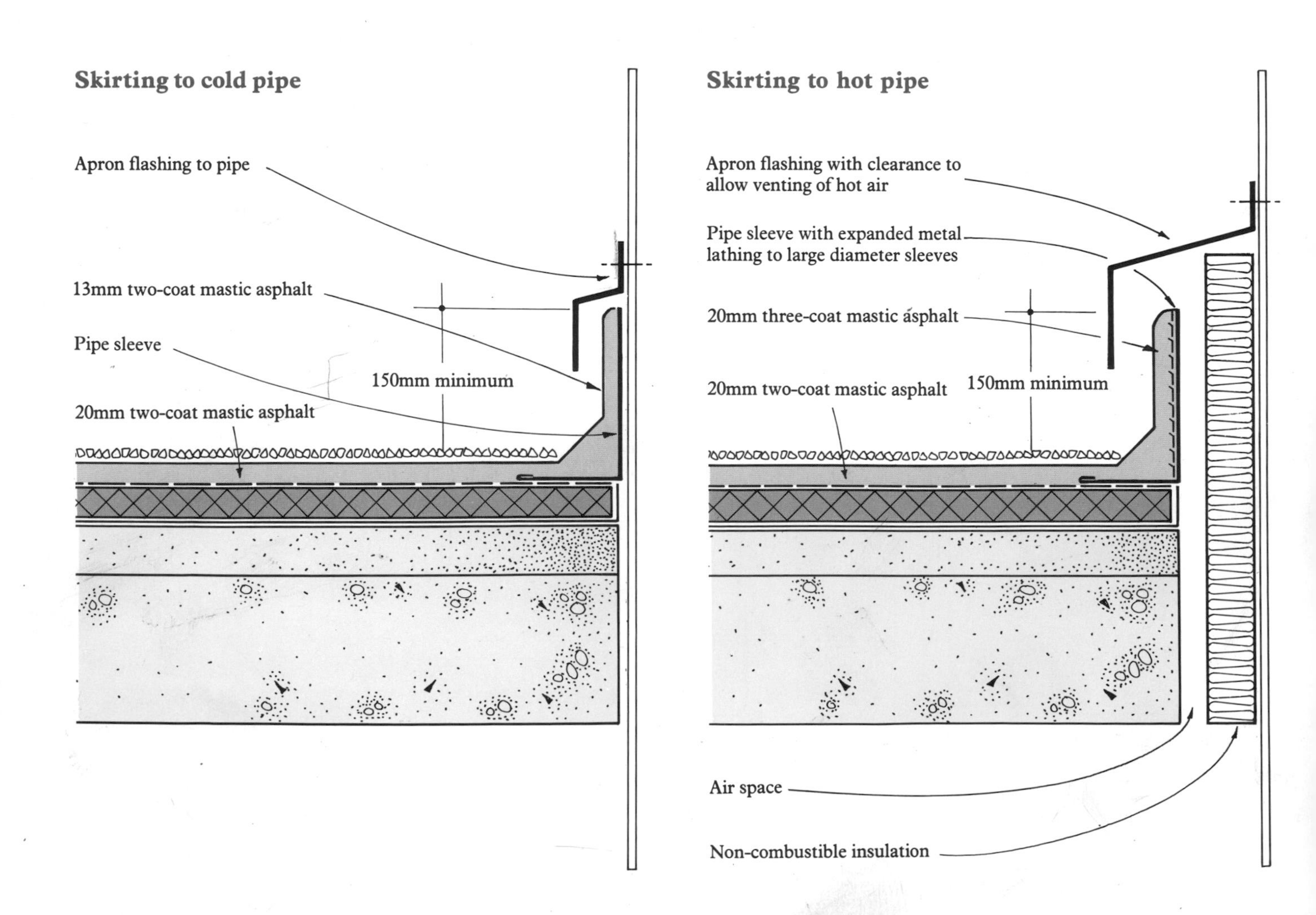

MASTIC ASPHALT TIMBER/WOODWOOL DECK

(Plywood illustrated)

Skirting to brick parapet

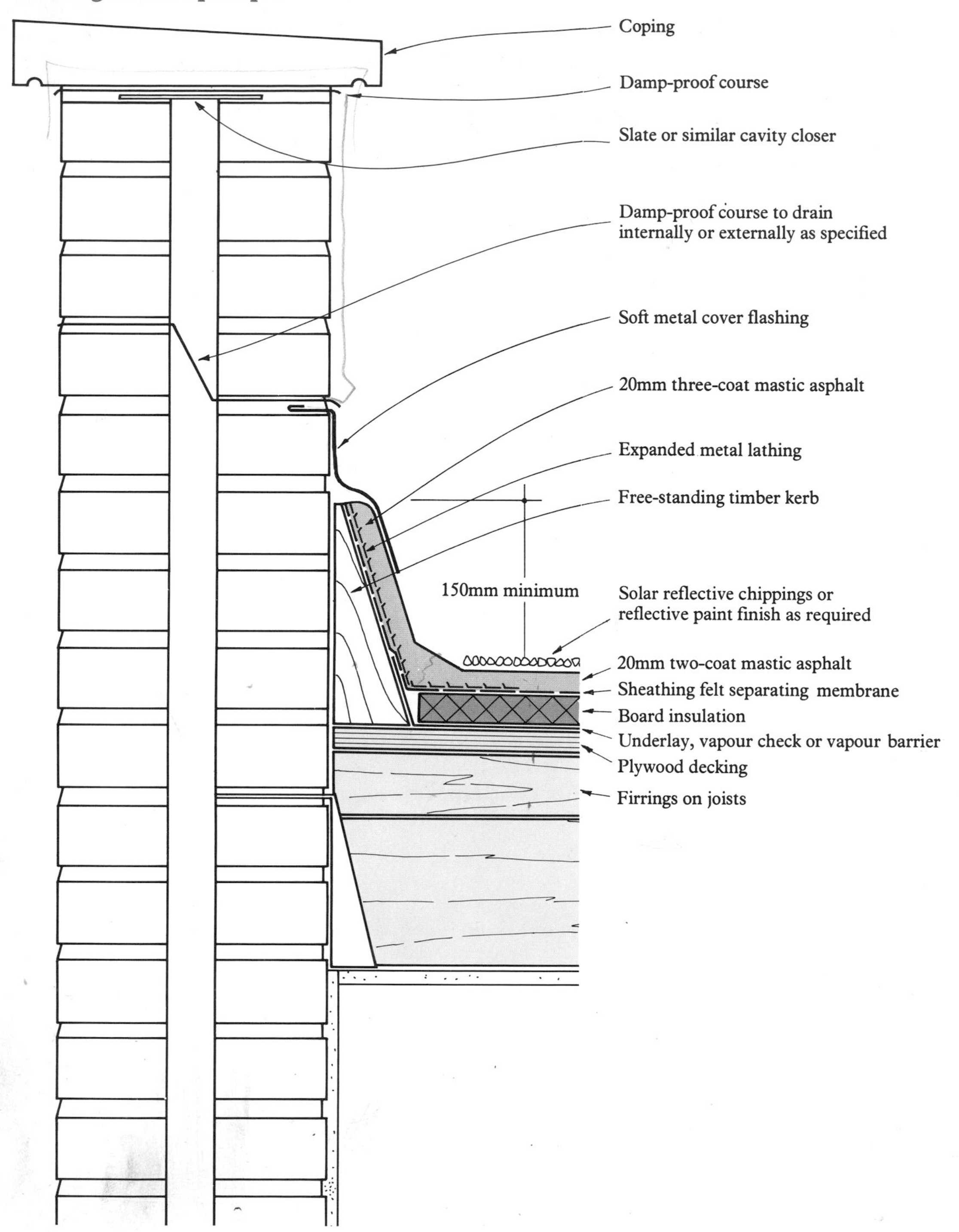

Check kerb and metal trim edge

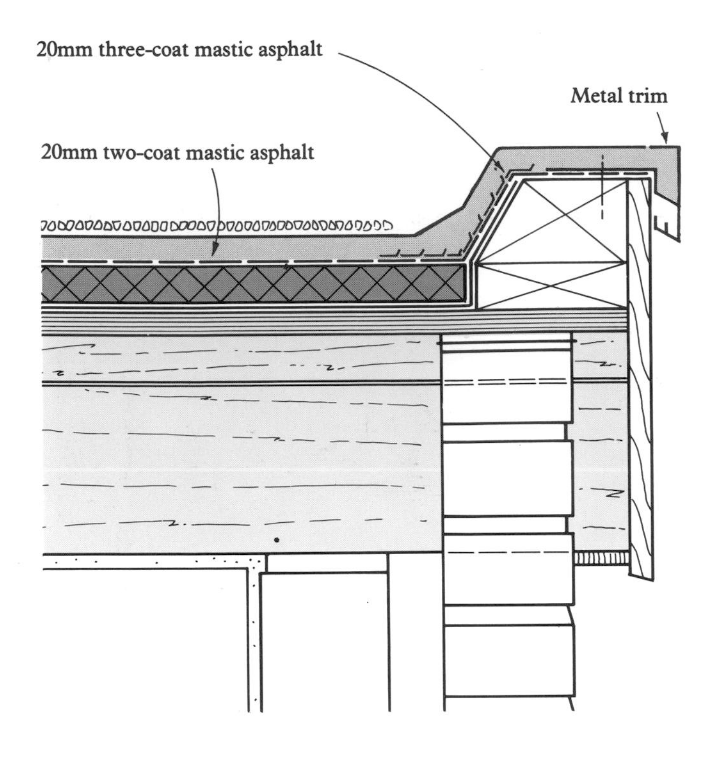

Asphalt apron and drip (optional asphalt water check)

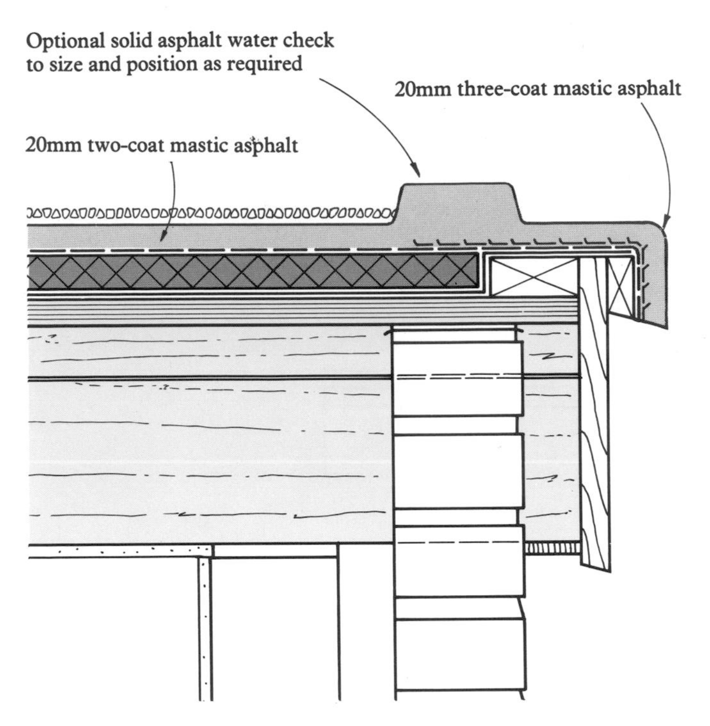

Asphalt apron and drip to external gutter

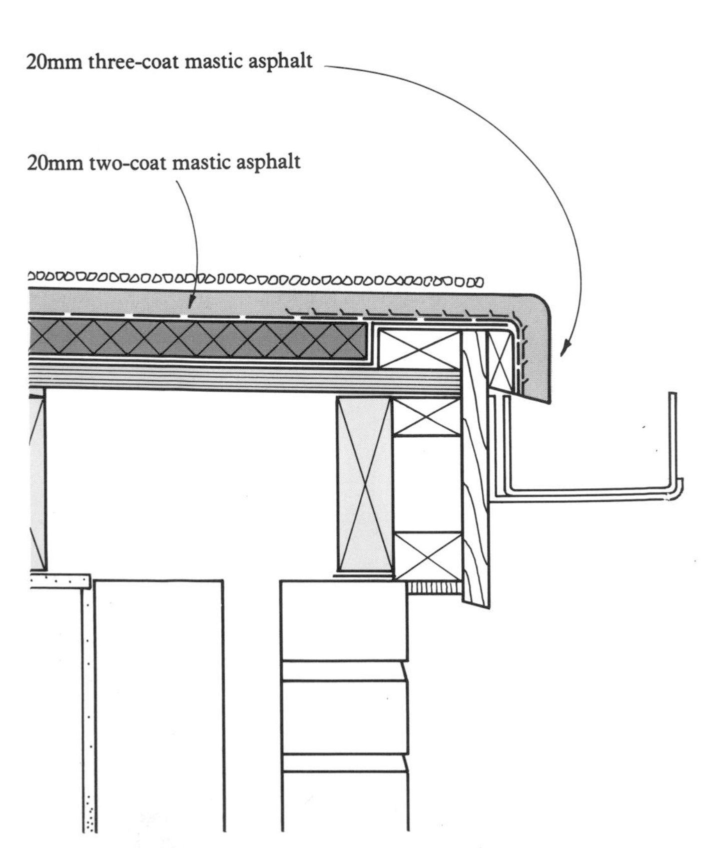

Roof outlet: proprietary flanged type

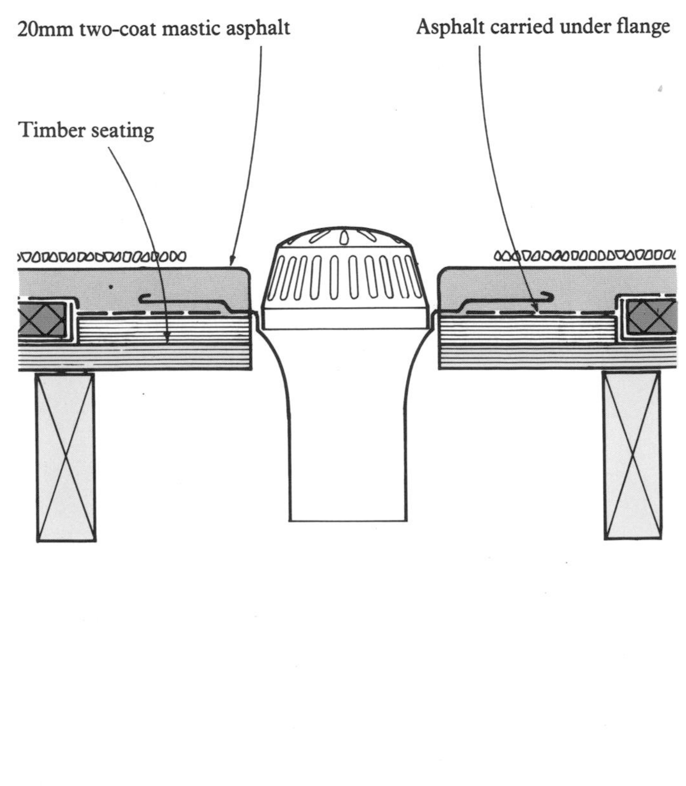

Skirting to metal rooflight or ventilator kerb

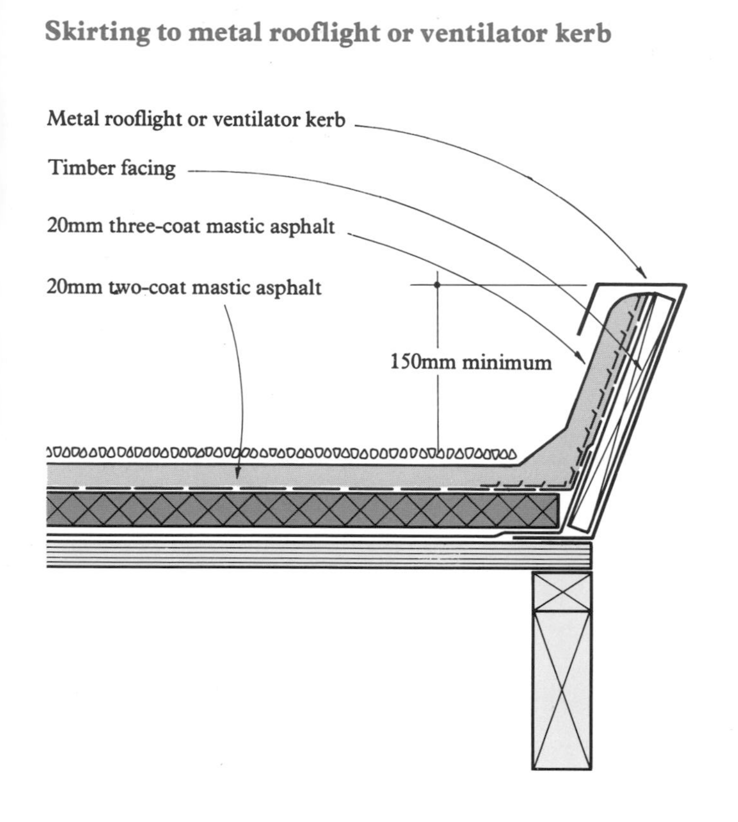

Skirting to cold pipe

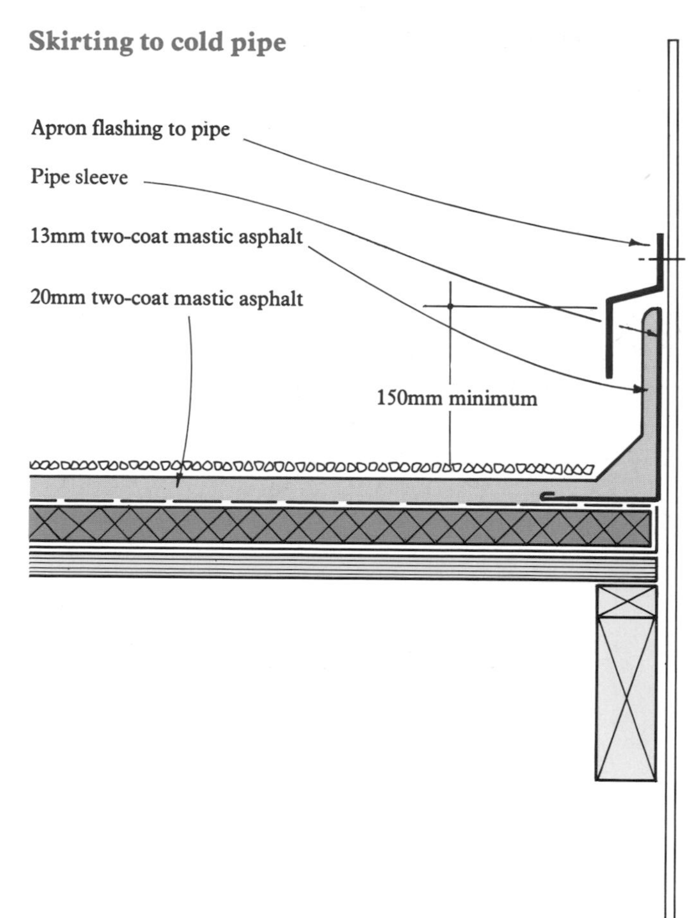

Skirting to hot pipe

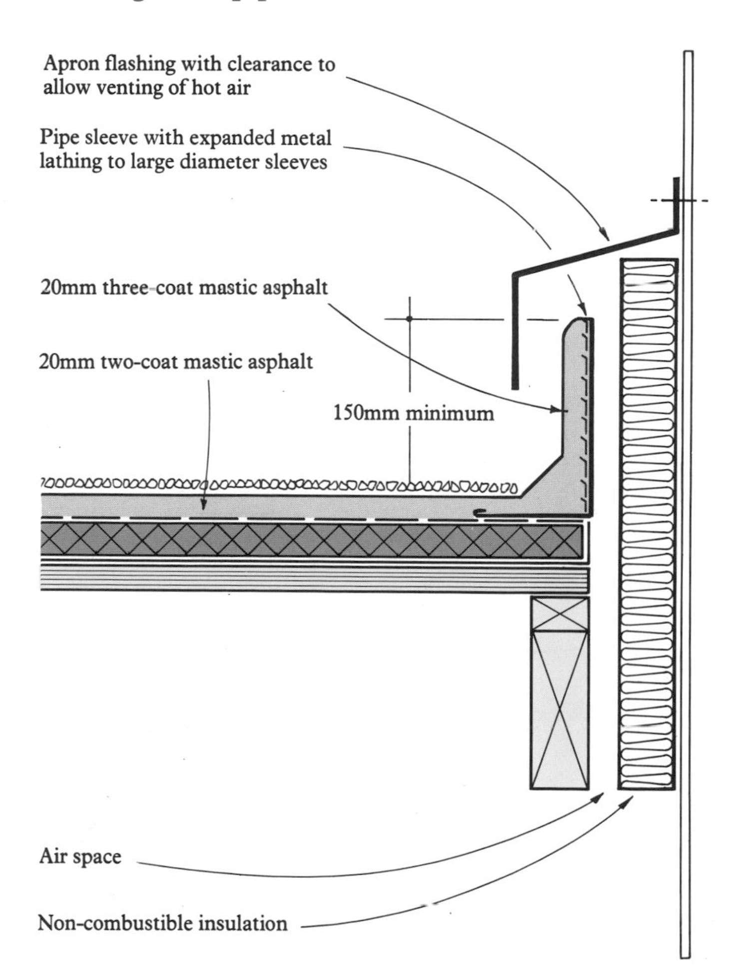

MASTIC ASPHALT METAL DECKING

Skirting to brick parapet

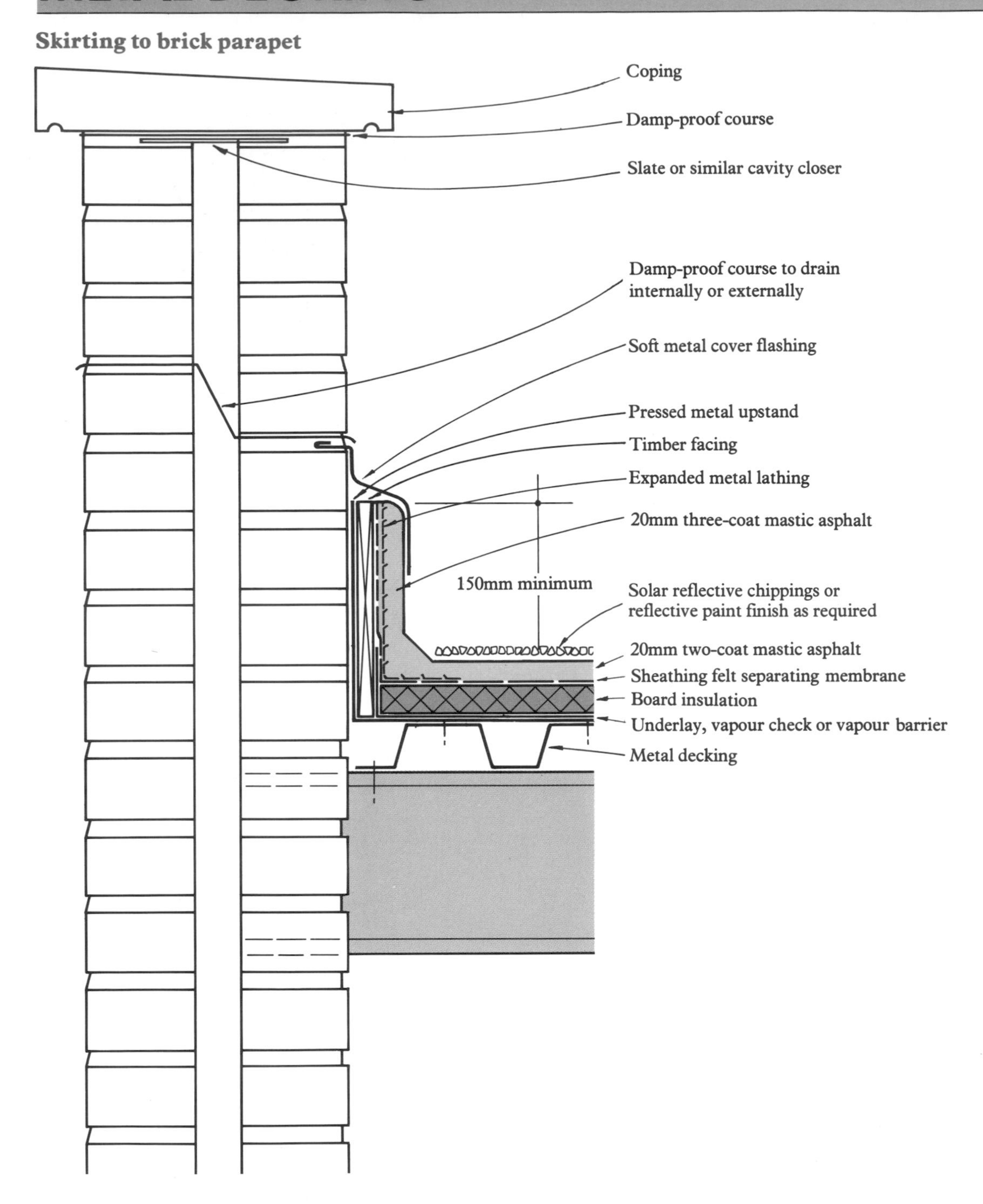

Skirting to cladding parapet

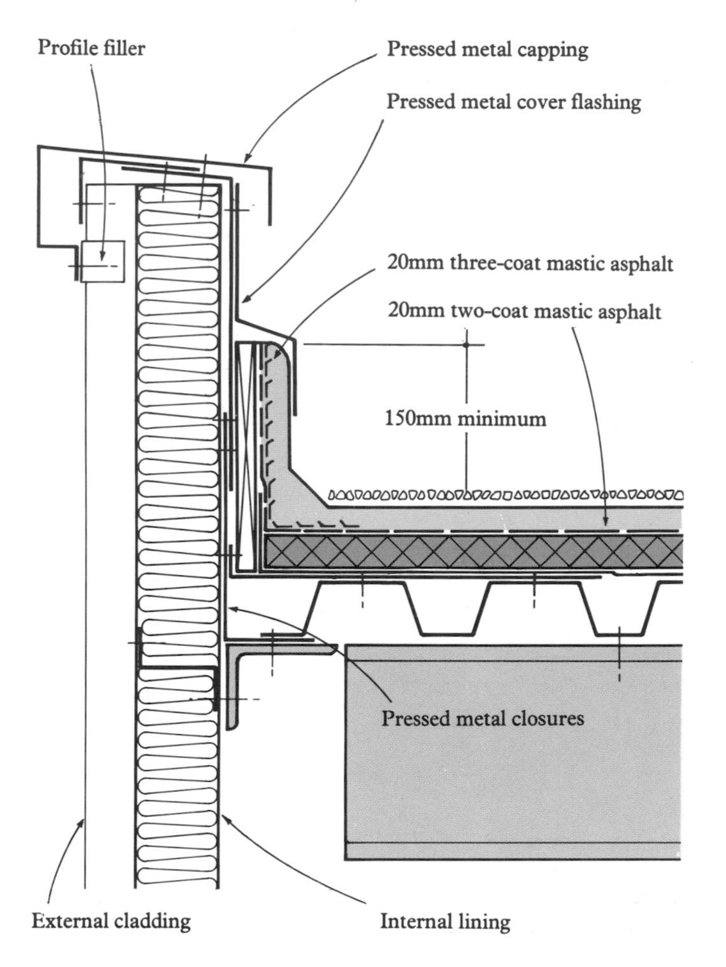

Skirting to cladding abutment

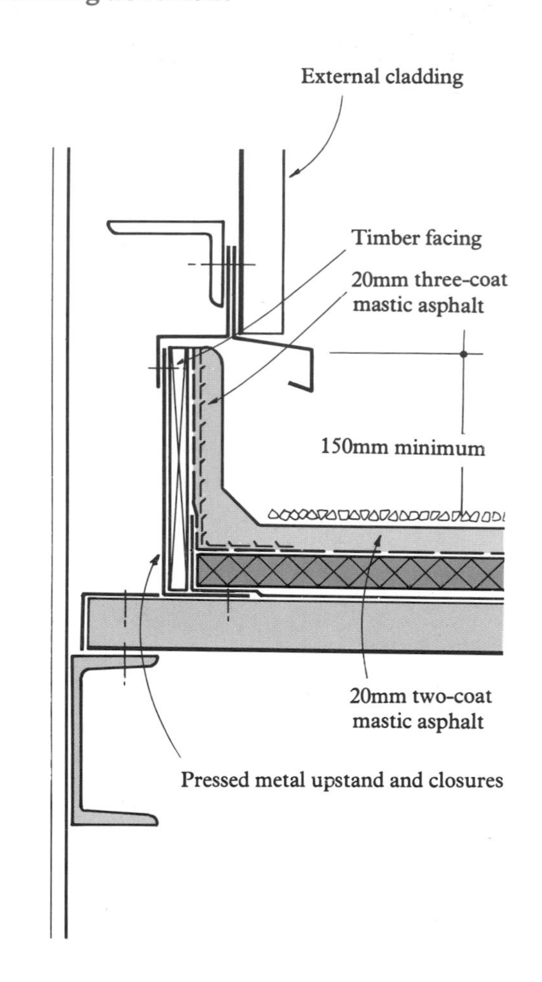

Asphalt check kerb and metal trim edge

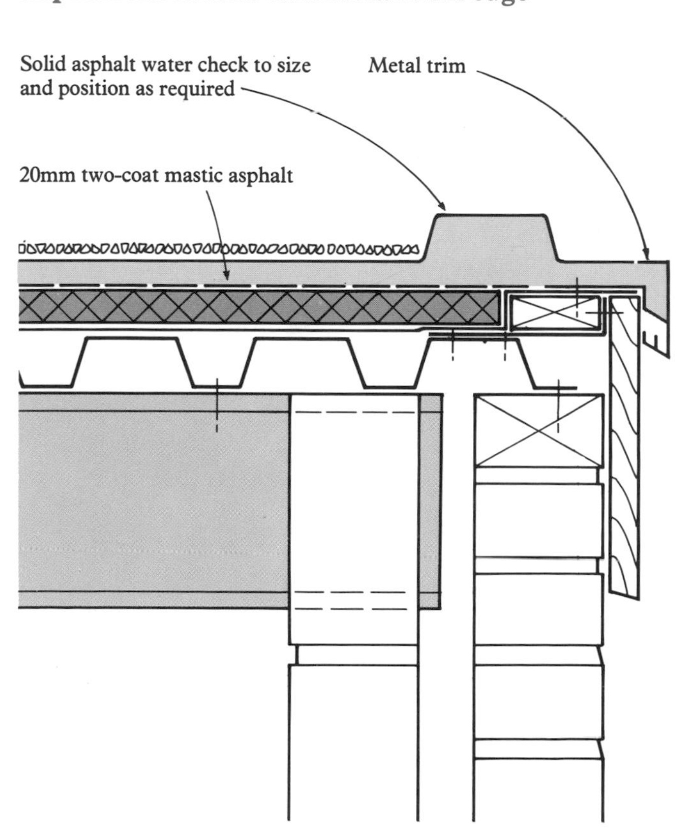

Asphalt apron and drip to external gutter

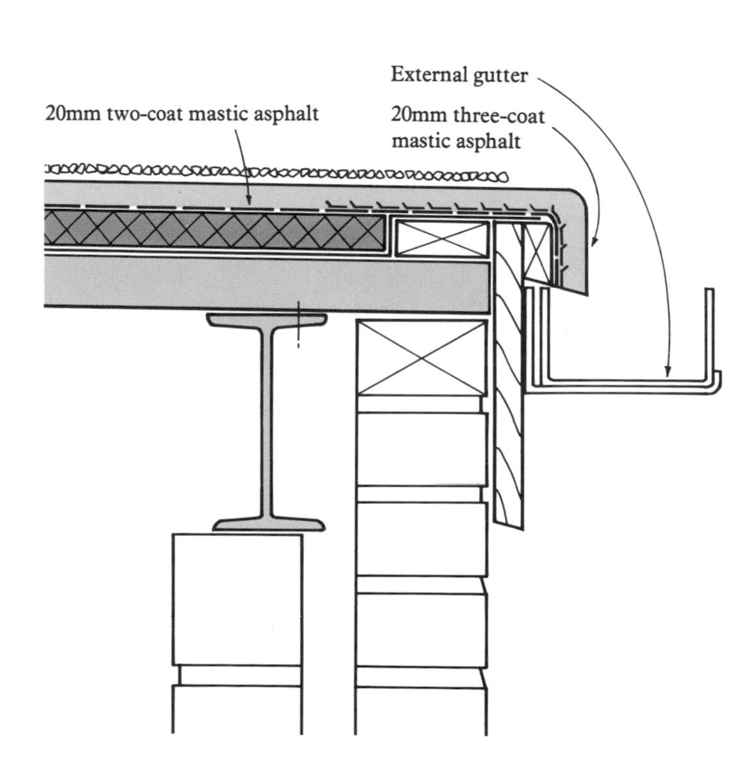

Roof outlet: proprietary flanged type

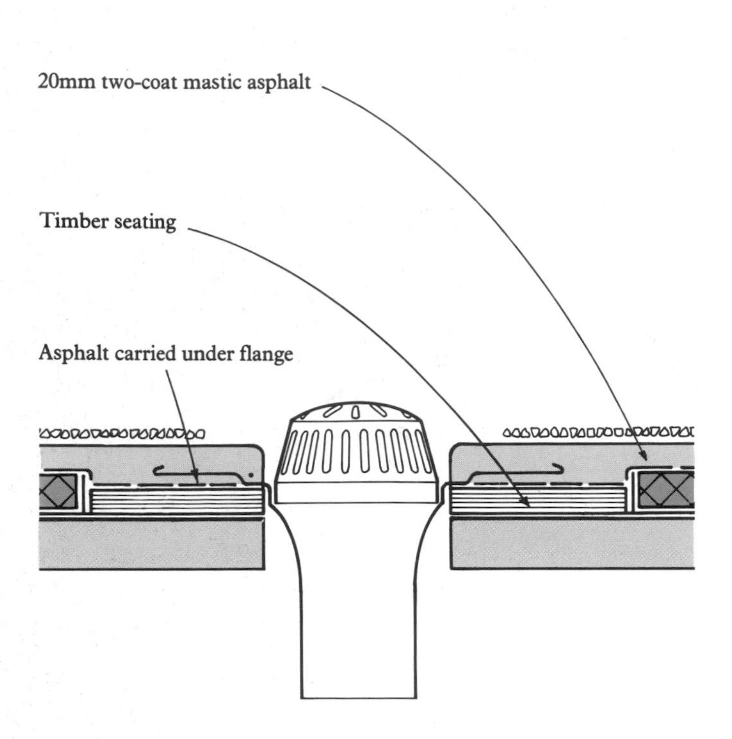

Skirting to metal rooflight or ventilator kerb

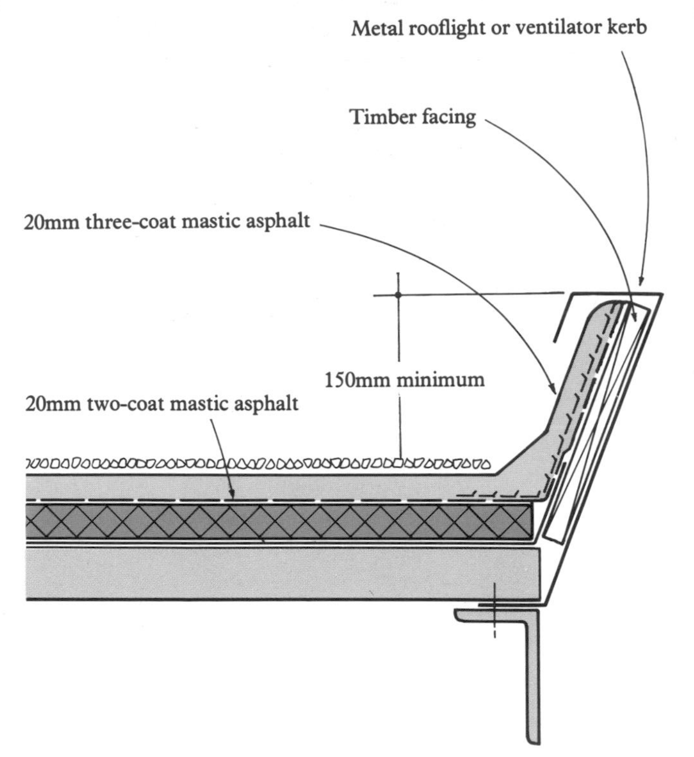

Skirting to cold pipe

Apron flashing to pipe

Pipe sleeve

20mm three-coat mastic asphalt

20mm two-coat mastic asphalt

150mm minimum

Skirting to hot pipe

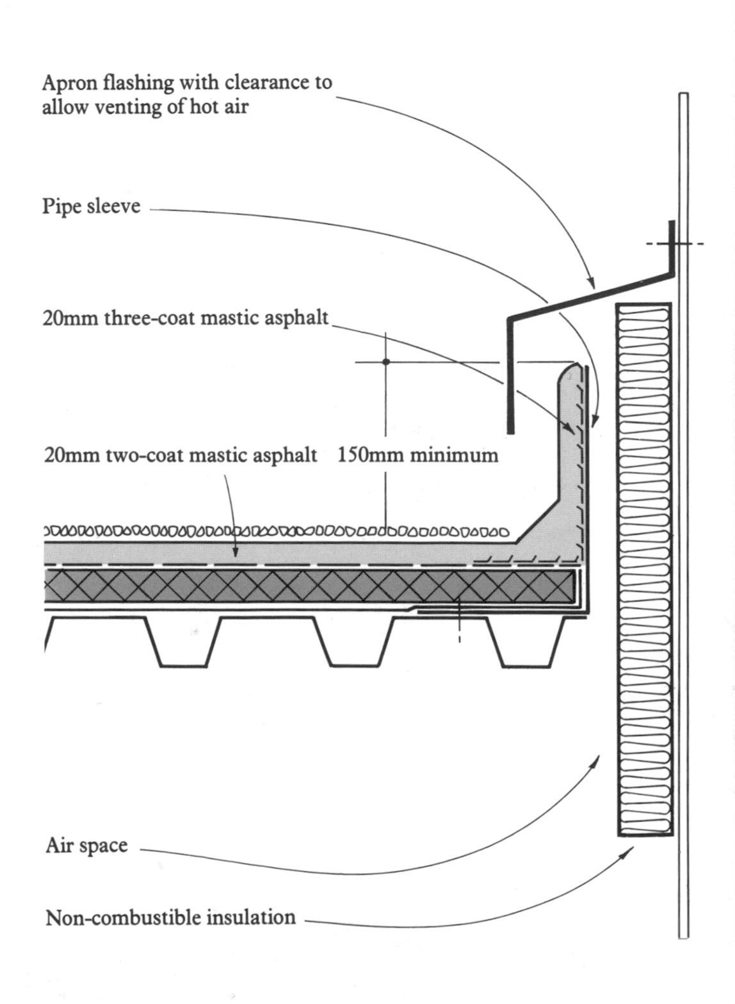

SECTION 6 MAINTENANCE AND RE-ROOFING

6.1 MAINTENANCE

ROUTINE MAINTENANCE

Flat roofs should be designed to avoid the need for maintenance as far as this is possible, but inevitably, some items of maintenance will occur. As a matter of good housekeeping the building owner should arrange for an inspection of all roofs and details at least once a year. A simple inspection by maintenance personnel will suffice, and it should follow a routine, to include the following:

General	Inspect for debris, leaves, nails, surplus building materials and stored goods. These should all be removed. Note the general condition of the roof and incidence of ridges and blisters. It is not normally wise to cut away or repair ridges or blisters unless they are in areas where regular traffic could cause damage, in which case repair should be carried out by a specialist contractor.
Drainage	Inspect the gutters and outlets individually, clean gratings or wire cages and renew where necessary.
Roof edge details	Inspect flashings, trims, cappings, and arrange for repair if they are loose, or if they have slipped out of position. Inspect the pointing which holds flashings and arrange for loose pointing to be repaired.
Chipping surface	If there are bare patches arising from displacement of loose chippings it will normally be sufficient to sweep them back into position. When the chippings have been displaced by wind scour, or traffic which is likely to be repeated, the chippings should be rebonded with a suitable bituminous mastic.
Reflective paint surfaces	If a paint coating has been included in the original specification, this should be renewed as necessary to restore the protective or reflective qualities. Regular and comprehensive re-coating programmes will be required, and this should have been agreed between the designer and the building owner during the design stages of the roof. In the case of mastic asphalt, a reflective paint coating may have been applied to reduce the surface temperatures and improve the weathering characteristics of the asphalt during the settling down period in the first few years. In this case re-coating may not be necessary.

REPORTING PROCEDURES

In case of leakage, an established reporting procedure is essential and the staff responsible should be clearly identified. This may seem obvious, but great frustration builds up when leakage is not dealt with and nobody seems responsible for doing anything about the problem. It is surprising how often a roof will be allowed to leak for years with no action taken. This early tolerance of small problems can culminate in the sudden demand for complete re-roofing.

A roofing contractor is placed in a very difficult situation if he is called in to look at a building which is reported to have been leaking for a long time. He should be called in as soon as the leakage has occurred, and be told exactly where the leak is entering the building, and if possible how long it takes from the onset of rain to the observation of leakage.

The building owner may prefer to call in a local roofing service known to be reliable, but in the first few years of the life of a roof it will be best to call in the original builder or roofer who will have records of the work done and experience of the original contract. It may also prove from inspection that failure is the result of condensation or a defect which should be repaired free of charge.

FAULT TRACING

The roofing contractor should first make an internal inspection and take measurements to enable the position of direct leaks to be pinpointed at roof level. If the reason for leakage is not immediately obvious at roof level, it will be necessary to observe the possible runs of water to the entry point. For example, if water is entering at a gutter run, the whole area of roof between the gutter and the ridge may need to be searched for the fault.

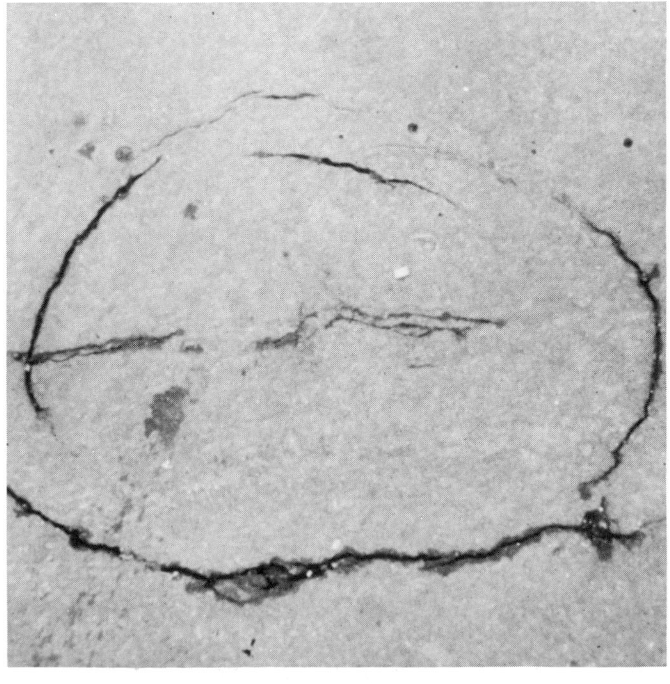

Often the fault will be visible or obvious, for example damage to asphalt

A systematic approach is essential, not only to find the fault, but also the cause of the problem.

The search for a fault will include consideration of possible defects in walls, damp proof courses, windows, openings, and flashings. If the fault is in the waterproofing, the search will be for a split or rupture. The most likely places to search are the bottom of skirtings, the waterproofing above panel joints and purlins, alongside gutters or at outlets and trims.

A record of the original roofing specification, including the method of bonding, type of insulation and type of deck will provide valuable clues in the search for the fault, particularly if it is clear that the design of the specification has not followed good practice. For example, large numbers of roofs have failed because the waterproofing has been fully bonded to chipboard. Here it is almost certain that the split will be above a panel joint, and it will be necessary to trace and inspect the lines of the joints.

A split in the roofing is sometimes easy to spot, but in cases of difficulty treading over the suspect area systematically may pump water or air bubbles back up through the split to give a visible indication of the position. If this fails, test cuts should be made at intervals up the slope of the roof from the leakage point to trace the flow of water under the membrane or under the insulation back to the source.

The surfacing can make investigation more difficult and surfacings which are easily removable are an advantage when inspection and repair are necessary. Loose laid paving slabs on proprietary corner supports are easily set aside, but concrete paving which is fully bonded in bitumen makes a major task of the inspection and repair. Similarly, hot applied gritting solutions can make stone chippings almost impossible to remove, and inspection and repair becomes more difficult.

Because of the difficulty of removing chippings, some building owners now insist that a chipping surface is not applied. Unfortunately this discards an aspect of the specification which improves durability, and it is better to design the specification in accordance with good practice in the first place, including the use of stone chippings where necessary, and thus reduce the incidence of leakage.

REPAIR PROCEDURES

Repair is probably the aspect of flat roofing which calls for the greatest roofing skills. Unfortunately, a large number of roofs are completely re-covered when skilled repair would have proved entirely satisfactory.

Once the faults in a roof are traced, a decision must be made on the extent of the repair: the specification of repair must take into account the form of failure which caused the leakage, and the extent of any movement.

BS 747 felts should not be used for repair work. They do not have the strength and elongation properties required. Hessian based roofing is a traditional repair material, but nowadays polyester base or woven glass base torch-on materials are most often used. These save the cost of setting up a boiler to heat bitumen. Hot bonding bitumen would only be used if a decision had been taken to re-cover a substantial area, and even then there are many advantages in adopting torch-on materials for the repair.

When a split is found along the joint line of an insulation board or deck unit, the repair should cover the entire line of the joint on which the split has occurred, to make sure that no further trouble arises on that obvious line of weakness. Small patches which cover only the visible split are seldom effective in the long term, as the split is likely to continue along the unrepaired sections. A small patch will, however, be appropriate on an area which has been damaged by a severe knock, and there is no question of splitting caused by movement of the insulation or deck.

As a general rule, local repairs should be carried out in two overlapping layers. Suitable dimensions are 150mm girth for the first layer, and 300mm girth for the second layer.

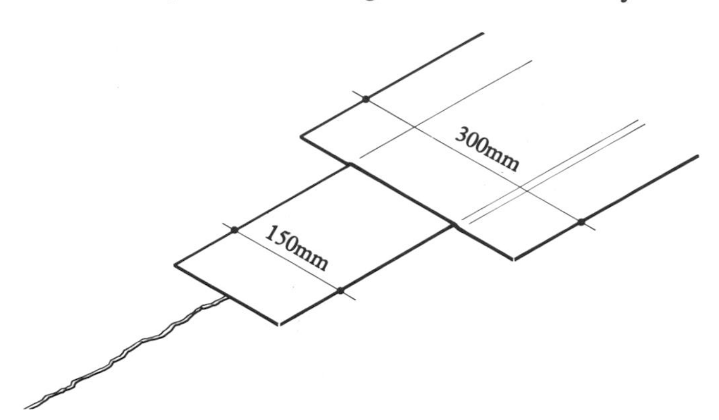

The experienced contractor will be able to judge from the condition of the existing membrane whether or not it is possible to make a permanent secure repair. This is not always possible, particularly if the original roofing is of BS 747 fibre or asbestos based materials. The cap sheets of these roofings become absorbent after long term exposure, and if not properly drained are likely to become saturated. It will be difficult to achieve a satisfactory bond to these materials, and the repair is not likely to be effective in the long term. For patch repairs it is sometimes possible to scrape away the fibres of a degraded cap sheet to expose the intermediate layer, which might be in a good enough condition to provide a good bond for the repair.

After repair, it is essential that the surfacing is replaced to provide protection to the repair and to the existing membrane in that area.

When repairing mineral surfaced roofing, great care must be taken to dry the surface. Primer should be applied if large areas of repair are to be carried out. Torch-on materials must be applied with plenty of heat to ensure a free flow of bitumen to fill the spaces between the mineral granules and form a good bond, even if this detracts from the appearance of the finished work.

Finally, a record should be kept of the date and location of the repair and the cause of failure. This will maintain a systematic approach to the maintenance of the roof and be a helpful guide in the event of future problems.

Occasions will arise when leakage cannot be traced and it is thought best to re-cover the entire suspect area. There are times when this is the only solution, but it is a haphazard approach and has proved a very unreliable method in practice. It is better to pursue the systematic investigation using test cuts as necessary.

Suggestions that a bituminous or plastic coating over the entire area will cure the problem should be treated with caution. This can be an expensive treatment, and may well prove ineffective, as coatings placed over a split in the roof will be subjected to the same movement that has already caused failure of the original membrane. The coating, being weaker and more vulnerable than the original waterproofing, is likely to fail in the same place.

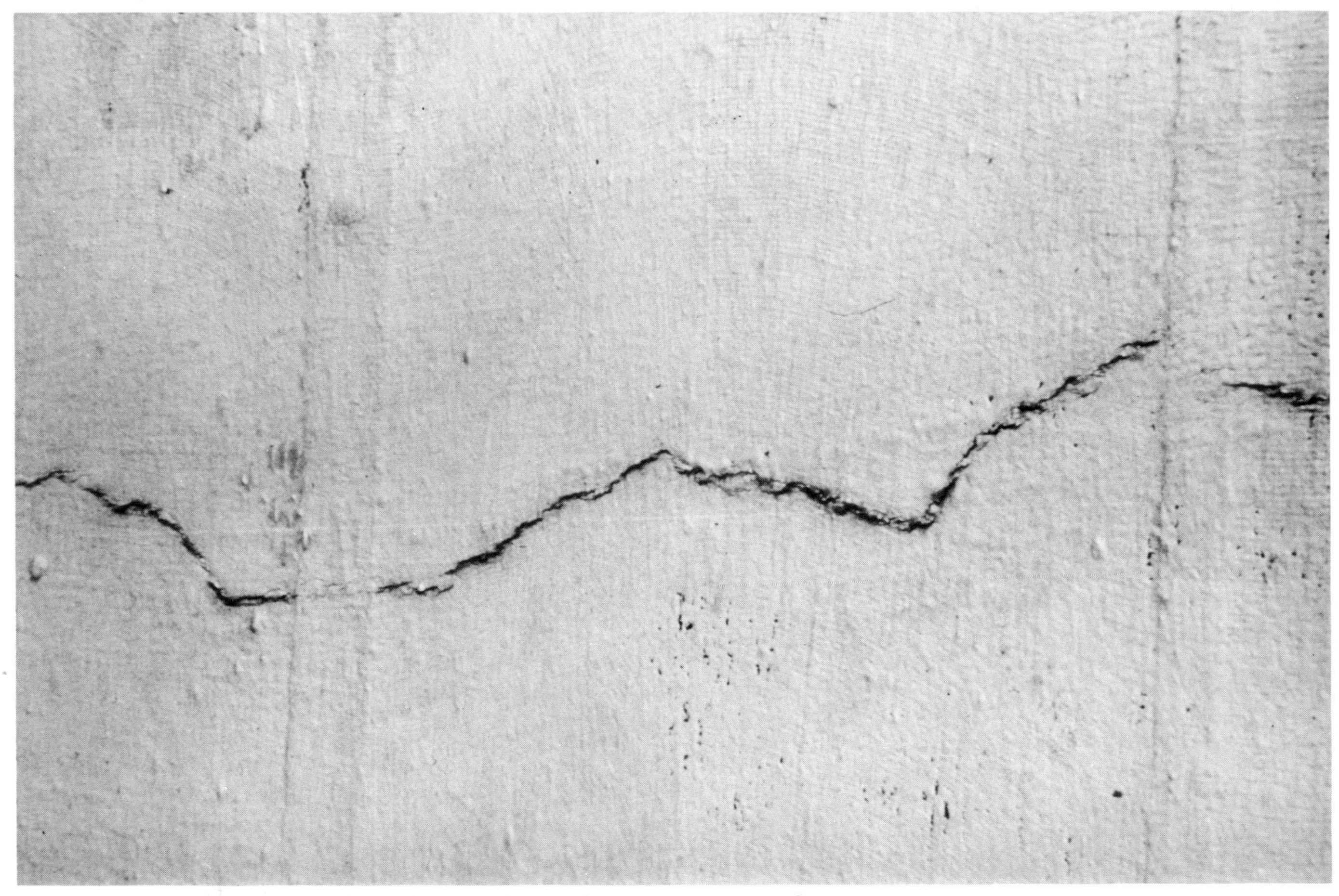

Light repair coatings are likely to fail in the same place as the original waterproofing.

For a successful repair with a lightweight or unreinforced coating it is necessary to trace the split and repair it with reinforcement before applying the coating. Only heavy duty or fully reinforced repair systems should be applied, and not many liquid applied systems would come into this category.

6.2 RE-ROOFING

There is as much re-roofing required in the UK as there is new roofing. The design of re-roofing specifications follows the same pattern as new work, but with added complications arising from the need to adapt, re-use or overlay existing materials, or to strip and replace the existing roof specification.

The designer no longer has complete freedom of design and may be restricted to the practice, good or bad, employed on the original building. Perhaps the only advantage is that the whole roof is time tested, and it may be reasonable to repeat the features which have proved successful, and modify those which have proved unsuccessful.

INSPECTION

An inspection is necessary to determine the limiting factors which will influence the design and selection of materials. The inspection must take into account the condition of the supports, the roof deck, the insulation, and the waterproofing.

Structure
The structure must be investigated to determine whether it is capable of taking additional load, and whether any special limitation on loads must be imposed during re-roofing, including the loads from the placing of boilers and the storage of new materials on the roof.

Similarly, the roof deck must be examined for condition and load bearing capacity. Sometimes it is impossible to gain access for the inspection from below and test cuts from above may prove necessary.

Timber decks and other degradable or non-durable decks should always be fully examined. If they cannot be inspected from below, it will be best to strip off the existing insulation and roofing entirely to ensure a full inspection of the roof deck during the progress of the works.

The load bearing performance of proprietary decks can be checked with manufacturers, but professional advice may be necessary in the case of concrete decks. Metal decks are easily checked by calculation methods, and a calculation service is normally provided by manufacturers or specialist roofing contractors.

The spans of metal decks are often limited by the criterion of deflection or stiffness, and there may well be a reserve of load carrying capacity. An overlay of new insulation and waterproofing will add extra load and add to permanent deflections, but it is also likely to improve the load distribution of the deck and increase the feeling of stiffness underfoot. Small increases in permanent deflection might therefore be accepted, provided the new covering is designed to improve the load distribution, and provided the ultimate load bearing capacity of the deck is adequate.

Insulation
If the existing insulation is moisture sensitive, it is likely that the parts near to the source of leakage are wet or degraded and will need replacing. Treading the area concerned will often reveal soft patches which require attention. Otherwise it will be necessary to take moisture readings or make test cuts near the source of leakage or at random over the roof.

Electronic, infra-red and other proprietary methods are also available to survey large roof areas. They give an indication of water content by temperature variations or by radiation methods. The process is expensive and the results not entirely certain, but no doubt occasions will occur when such a survey is justified. The work must be carried out by trained specialists who will submit a detailed estimate of the presence of water in the insulation.

A check may also be necessary on the efficiency of attachment of the existing insulation to the deck. This may be below standard either because it was applied to low standards in the first place, or because of breakdown or release of the bonding bitumen.

Waterproofing
If the original waterproofing is to be left in position, its condition will influence the design of the new work. The worse the condition of the existing waterproofing, the more it will be necessary to isolate the new work from the old.

The new specification must take account of the attachment of the original work, and this may influence designers towards complete removal of the existing work or to the addition of extra fixings or loading coats.

Detail work usually presents the greatest problem in re-roofing. Kerb heights and skirtings may be too low, and prevent the application of extra insulation, or even the application of extra waterproofing. Rooflights, patent glazing and ventilators must often be entirely removed to allow access for the formation of the new waterproofing detail. Kerb heights may need to be increased, and new flashings may be necessary.

If the original waterproofing is to be left in position, some of the detail work can also be retained, but as a general rule skirtings must be removed entirely and re-formed. Metal cappings and flashings can often be re-used, but bituminous flashings will not be suitable for re-use, and there is practically no chance of lifting and re-sealing them successfully.

Metal trims can be re-used, but the original roofing should be removed from the bonding face and the trims re-applied as for new work. Similarly, the original roofing should be removed from rainwater outlets and pipes, and the detail completed as for new work.

The conflict between good practice and the cost of achieving it can give the designer and roofer some awkward decisions. It is not always possible to be sure of success and a risk of further failure may have to be accepted.

When considering the design of the specification, the reason for the failure of the original must be understood. If old age is the reason, the re-roofing will be straightforward, as there will be little to change. If the roof has suffered premature failure, the re-roofing specification must take account of the factors involved.

Having established the condition of the existing roofing specification, a number of alternative approaches to re-roofing can be considered:

1 A single layer torch-on waterproofing applied over existing waterproofing.

2 A new waterproofing system applied over the existing waterproofing.

3 New insulation and waterproofing applied over the existing waterproofing.

4 Strip and replace the existing insulation and waterproofing.

5 Strip and replace the existing deck, insulation and waterproofing.

Single layer torch-on waterproofing

The single layer torch-on system fully bonds a new thick surface layer onto the existing waterproofing. The old waterproofing will act as a backing to the new and provide a measure of lap security.

This system is satisfactory for up-grading old built-up roofing and mastic asphalt roofing which is in basically sound condition and only giving occasional trouble. It is not normally the specification to use in an attempt to turn a widespread failure into a long-term success as the new layer will be subjected to all the movements of the original, which will be most severe when the roof to be recovered has failed from extensive fatigue. In practice, it will be asking rather too much of a single layer system to provide a long service life under these conditions. It will be better to separate the new waterproofing from the old by a layer of insulation.

If the single layer torch-on specification is chosen, there are a number of conditions which should apply:

The old roof covering must be basically sound, rot-free, not saturated with water, and the insulation or deck below must be in a generally dry condition.

The roof surface must be clean, dry, and firm. An application of primer may prove necessary to bind the surface and ensure good adhesion. The strength of bond should be tested during the progress of the work by pulling back on the new material after it has been applied and has cooled.

The existing surface must be generally free of blisters and ridges. Minor blistering would be acceptable providing the blisters can be cut away or repaired, to leave a clean, firm surface.

If the original waterproofing has a stone chipping finish, this must be entirely removed to leave a sound surface. Torch-on materials are applied with powerful torches, and it is likely that flames will come into contact with all the materials in the close vicinity of the work. Great care is therefore necessary during application to control the risk of fire and under some circumstances the use of torch-on materials would be inadvisable.

Whilst it is often satisfactory to leave the existing waterproofing in position at details, circumstances will arise when the existing materials must be cut away. If a chase is to be re-used, the old roofing will be cut away and the new skirting formed on what is probably a rough and uneven surface. Under these circumstances, a new single layer skirting may not prove satisfactory due to the difficulty of ensuring lap security, and a two-layer skirting formation will be preferred.

New waterproofing system over existing waterproofing

Old waterproof coverings may be re-roofed with the direct application of the full range of built-up roofing materials. The existing insulation must be sound and effective and the surface of the existing waterproofing must be firm, clean, and sound. It must be possible to remove all chippings and blisters to leave a satisfactory surface.

Particular precautions need to be taken with overnight seals, as large volumes of rainwater can be trapped on the existing waterproofing. This must be prevented from flowing along the interface between the new and the old roofing.

A part-bonded specification will be necessary, probably using BS 747 type 3G for the first layer. The surface of the existing membrane may need to be primed to ensure good adhesion through the perforations. The quality of adhesion can be tested during the progress of the work by pulling back on material which has been applied and cooled.

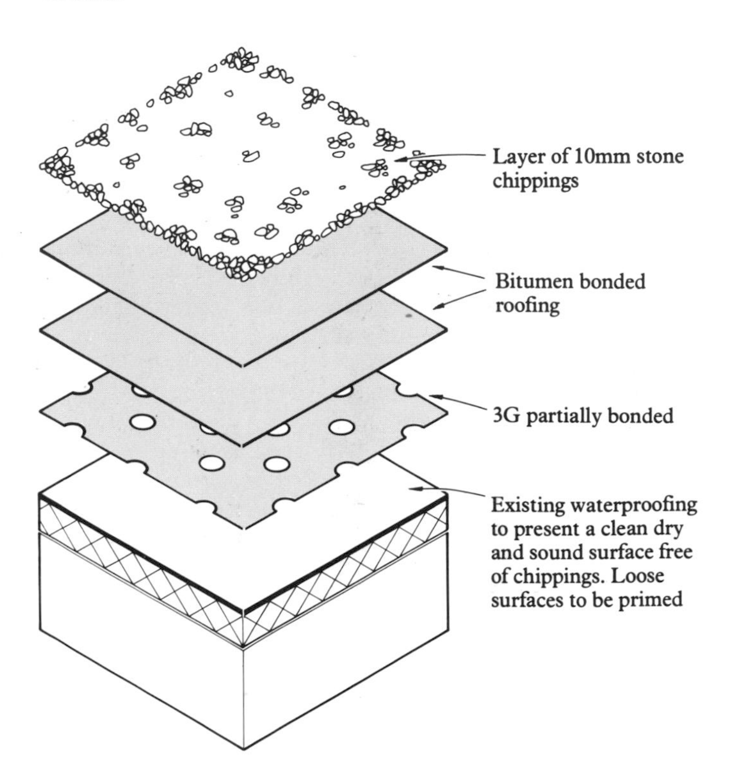

The design and selection of the specification will follow exactly the same procedures as described for new roofing, and an equivalent performance to that of a new roof should be achieved. The partial bond will isolate the new covering from movement to a useful extent, but if the old waterpoofing has failed through extensive movement or fatigue, or if the existing surface is badly broken down, it will be best to allow for the extra isolation and improved surface which is obtained by overlaying the existing waterproofing with a fresh layer of insulation.

Asphalt is not normally used over existing built-up roofing or asphalt roofs unless separated from the existing waterproofing by a new layer of insulation as described below.

New insulation and waterproofing over existing waterproofing

Old roofs can be re-roofed with confidence using an overlayer of new insulation, even if the old roof has suffered major movement or fatigue failure. It is only necessary for the existing insulation to be in sound condition.

The majority of roofers will choose this option. It provides an independent new waterproofing on a new substrate but avoids the awkward process of stripping away the original work, which could bring great problems of protection during the re-roofing, particularly if the building is to remain occupied.

The existing waterproofing is cleared of chippings if possible, but an advantage of this specification is that the new insulation covers major imperfections in the old surface, including small quantities of chippings which may be hard to remove. The surface should be dry, and a layer of roll roofing is applied in plenty of bitumen to cover the imperfections and fill up any voids. BS 747 roofing will be sufficient and it will not only present a satisfactory surface for the application of the insulation, but will turn the old roof into a vapour barrier if one is needed. The new insulation will be fully bonded to this new underlay and the application from then on is as for new roofing, using asphalt or built-up roofing fully bonded or partially bonded according to the insulation material.

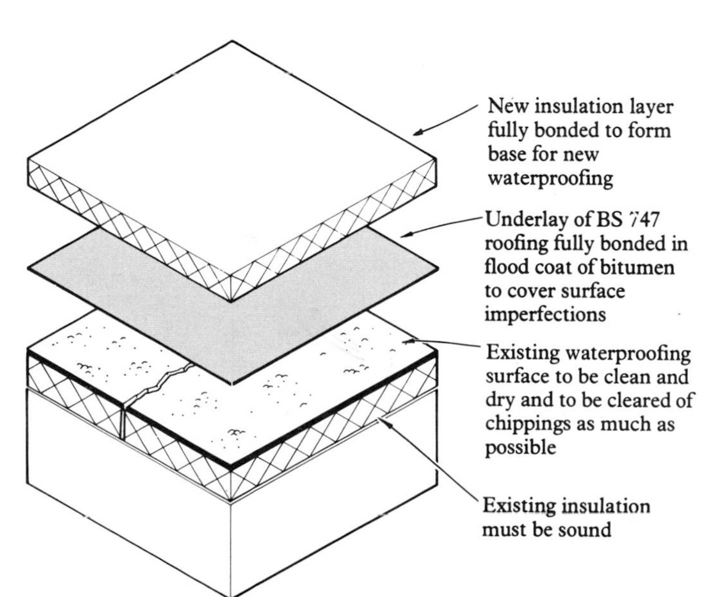

If the surface of the existing membrane is firm and clean and an underlay is not needed to act as a vapour barrier, the insulation may be bonded direct to the existing surface.

The original specification may already contain a vapour barrier, and if this has proved effective there should be no reason to add another. It is possible, however, that leakage may have caused the accumulation of significant amounts of trapped water between the joints of the existing insulation. If the insulation is water resistant, it may remain in good condition and still function efficiently, but it is necesary to ensure that the trapped moisture does not migrate as a vapour through the existing waterproofing to cause harm to the new insulation. A new insulation layer of wood fibreboard or similar moisture sensitive material is most at risk, and should always be applied to an underlay bonded to the original roofing, if there is a likelihood of significant trapped water.

Strip and replace existing insulation and waterproofing

A complete strip and re-lay will be required if the insulation is saturated or degraded, if it is necessary to inspect the roof deck, if the structure will not support the weight of an overlay in addition to the original materials, if the height of upstands limits the depth of extra covering, or if the attachment of existing materials is faulty.

The design and selection of the insulation and membrane should follow exactly the same procedures as for a new roof. It is only necessary to establish the condition of the deck surface after removal of the existing specification and allow for any necessary attention to make it suitable for the re-roofing. Lightweight screeds might be damaged by the stripping, and repair of the surface is likely to be necessary.

It should be remembered that the new insulation and roofing needs to be designed to current standards of moisture control, wind uplift allowances, fire and thermal performance.

Strip and replace existing deck, insulation and waterproofing

The need to replace the structure of a roof will arise only because of the severe deterioration or inadequacy of the roof deck. The design of the insulation and waterproofing specification is as for a new roof, but the form of failure of the original work must be investigated and completely understood to ensure the new design does not repeat the problemsof the old. For example, an unventilated cold roof constructed of timber which has failed because of condensation cannot be replaced without either a fundamental change to the ventilation, or a change to a warm roof construction. It is often difficult to tell with such a roof whether the rot arises from leakage or from condensation, and a full investigation of the problem must be carried out.

APPENDIX A

A.1 U-VALUE CALCULATION METHOD

The U-value is obtained from the total thermal resistance (R) of the roof structure which is calculated from the individual thermal resistance of each component of the roof.

$$U = \frac{1}{R_{si} + R_{so} + R_{cav} + R_1 + R_2 + R_3 \text{ etc}} \quad W/m^2{}^\circ C$$

Where Rsi = internal surface resistance
Rso = external surface resistance
Rcav = resistance of any cavity
from standard thermal resistance values
R_1, R_2, R_3 etc = thermal resistance of material, calculated from t/k where t is the thickness of material and k the thermal conductivity of the material.

Standard thermal properties of materials are given in Table 4.

Protected membrane roof
When selecting a thickness of extruded polystyrene board to achieve a specific U-value for a protected membrane roof, it is necessary to allow for loss of efficiency due to the effect of rainwater draining under the insulation. The calculated U-value may be adjusted by the addition of an increment which depends on the amount of thermal resistance below the level of the insulation, as shown in Table 1.

TABLE 1

Thermal resistance below insulation board relative to the total surface-to-surface thermal resistance	Thermal transmittance increment ($W/m^2{}^\circ C$)
0-5%	0.08
5.1-20%	0.06
20.1-40%	0.04
40.1-60%	0.02
more than 60%	0

* An increment of $0.08 W/m^2{}^\circ C$ should always be added when the thermal resistance of the structural roof is less than $0.1 m^2{}^\circ C/W$

EXAMPLES OF U-VALUE CALCULATIONS

Example 1
Calculation of the U-value of a roof construction of the following specification:

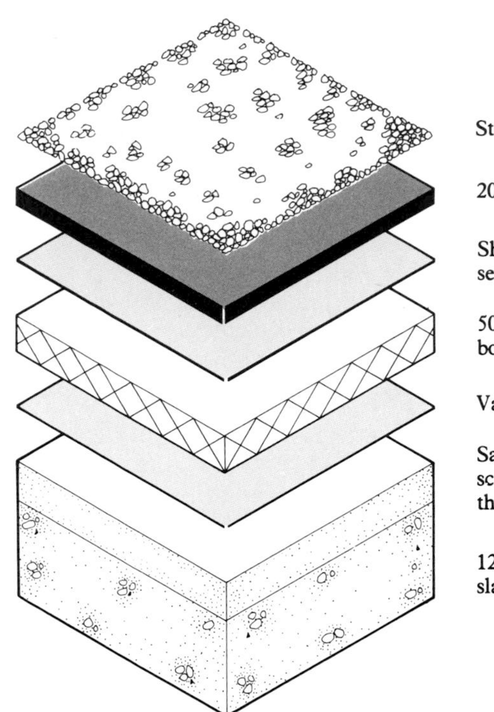

Calculate the thermal resistance of each component from t/k or take standard values from Table 4.

Roof element	Thickness t (m)	k-value (W/m°C)	Thermal resistance t/k (m²° C/W)
External surface resistance			0.045
Asphalt waterproofing			0.060
Cork insulation board	0.050	0.042	1.190
Vapour barrier			0.020
Sand and cement screed	0.050	1.40	0.036
Concrete slab	0.125	1.40	0.089
Internal surface resistance			0.105

Total resistance R = 1.545

Overall U-value 1/R = 0.64 W/m^{2}°C

Example 2 - How much insulation?
Details of construction are known, and a specific U-value is required. What thickness of insulation is necessary? The required R-value of the insulation component can be calculated and then multiplied by the k-value of insulation materials to find the thickness required.

Calculation of the thickness of insulation required to achieve an overall U-value of 0.6W/m^{2}°C in a roof constructed to the following specification:

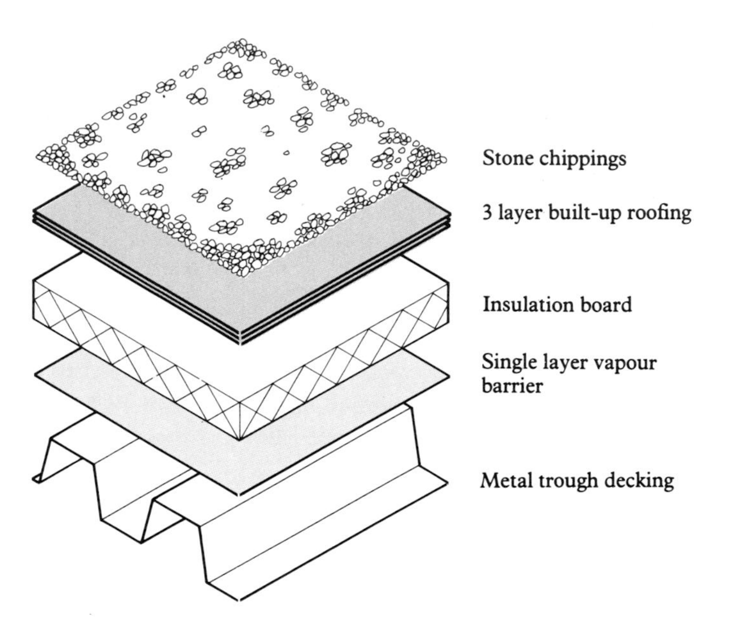

Roof element	Thermal resistance (m²°C/W)
External surface resistance	0.045
3-layer built-up roofing	0.06
Vapour barrier	0.020
Metal trough decking	0.000
Deck cavities	0.045
Internal surface resistance	0.105

Total thermal resistance without insulation: 0.275

U-value required is 0.6 W/m^{2}° C.

Therefore total thermal resistance (R) required is

$$\frac{1}{0.6} = 1.667$$

Subtracting the thermal resistance of the above elements 0.275 leaves a thermal resistance to be provided by the insulation of 1.392.

Thickness = k-value x resistance R

The minimum thickness required for cork insulation board would therefore be

k x R = 0.042 x 1.392 = 58mm
(nearest commercially available thickness 60mm).

Similarly, the minimum thickness for polyurethane insulation board would be

0.022 x 1.392 = 30.6mm
(nearest commercially available thickness 32mm).

A2 CONDENSATION RISK ANALYSIS

SURFACE CONDENSATION STEADY STATE CALCULATION

In the steady state calculation, the possibility of condensation is checked in the following stages to allow a comparison between the actual temperature at any point in a specification and the dew point temperature of the internal condition.

If the actual temperature at that point is lower than the dew point temperature, condensation will occur.

1. The external temperature normally used in the UK is -5°C.

   The internal temperature and relative humidity used are the maximum conditions that are likely to occur for an appreciable period of time.

2. The temperature gradient within the system must be calculated, either from the equation given in step 2 of moisture gain analysis below or by reading from a graph plotting temperature against the thermal resistance of the components of the roof structure.

   The relationship between temperature drop, and thermal resistance is a straight line and the internal and external temperatures give the two points from which the straight line graph is established.

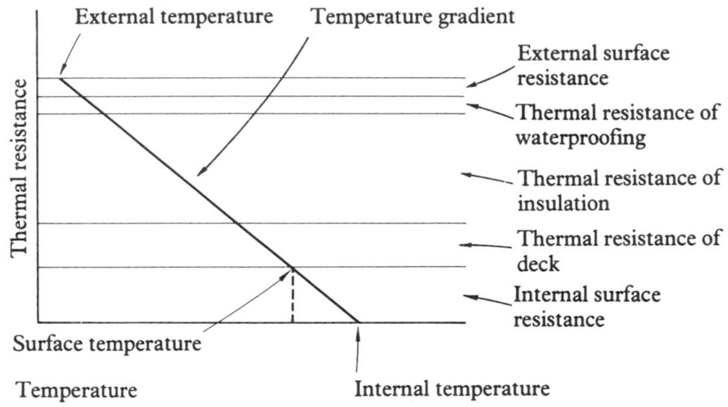

3. Calculate the dew point temperature for the internal conditions, using a psychrometric chart, or by reference to table 5. If the temperature at any point is lower than the dew point temperature, then condensation will occur at that position.

This calculation method is suitable for predicting surface condensation, including surface condensation in ceiling spaces.

The formation of interstitial condensation within the materials of constructions, in particular the insulation layer, is generally slow as the passage of water vapour is by diffusion only. In this case the steady state calculation is of little value and reference should be made to the moisture gain analysis calculation (page 201) which takes account of the rate of moisture movement and the quantity of interstitial condensation which would be formed.

Example 1
Check if surface condensation will occur for the specification illustrated with internal conditions of 20°C and 45% RH.

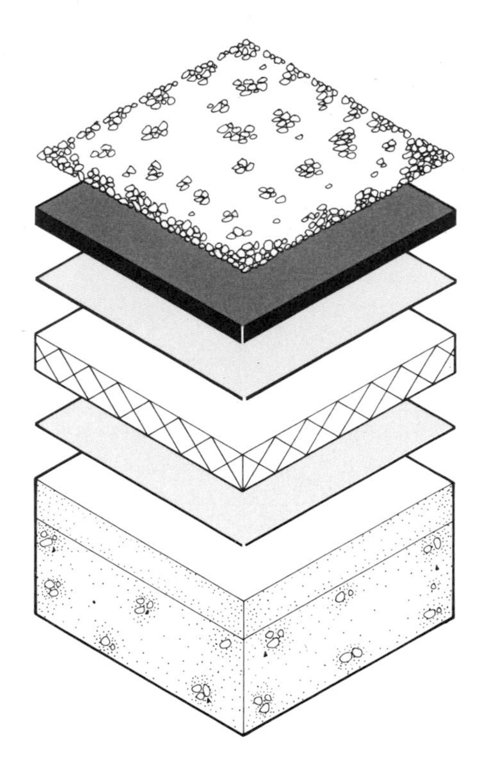

1. Calculate the thermal resistance of the specification

Roof element	Thickness t (m)	k-value (W/m°C)	Thermal resistance t/k (m^{2}°C/W)
External surface resistance			0.045
Asphalt waterproofing			0.060
Cork insulation board	0.050	0.042	1.190
Vapour barrier			0.020
Sand and cement screed	0.050	1.40	0.036
Concrete slab	0.125	1.40	0.089
Internal surface resistance			0.105

Total resistance R = 1.545

2. From the straight line Graph 1, plotted to show temperature against thermal resistance, a surface temperature of 18.3°C is indicated.

3. From Table 5 the dew point temperature at 20°C and 45% relative humidity is 7.7°C.

4. The surface temperature is therefore above the dew point temperature, and surface condensation will not occur.

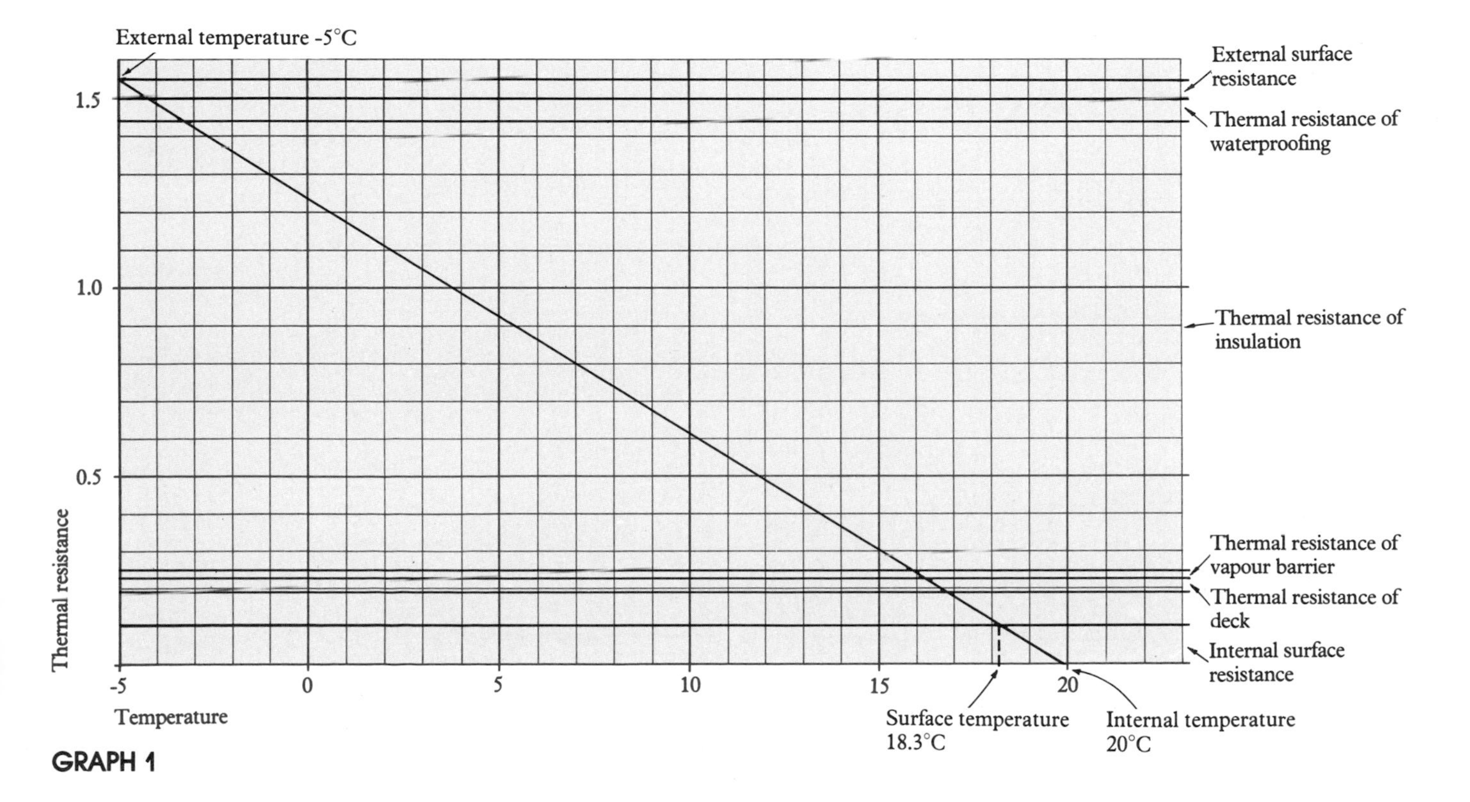

GRAPH 1

Example 2
Check if condensation will occur in the ceiling void for the specification illustrated below, with internal conditions of 20°C and 60% RH.

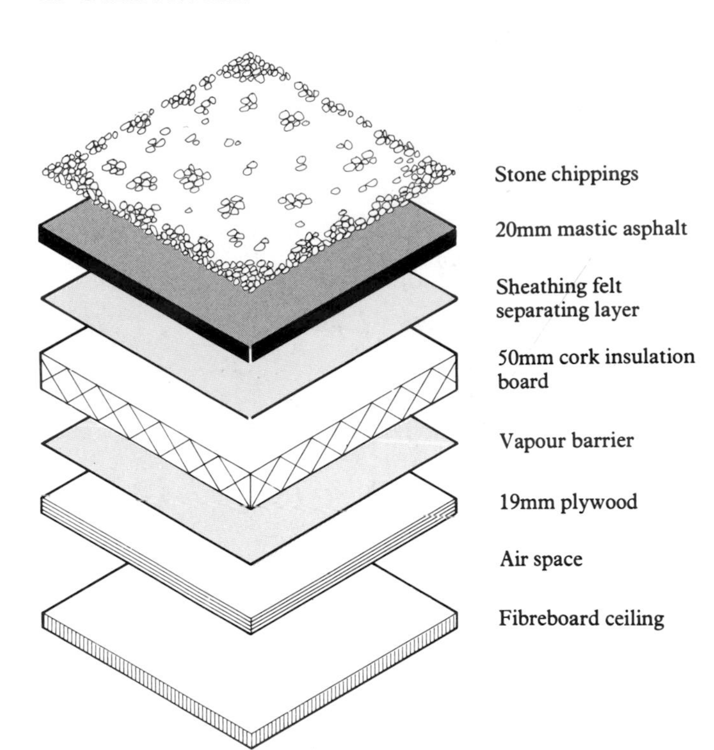

1. Calculate the thermal resistance of the specification

Roof element	Thickness t (m)	k-value (W/m°C)	Thermal resistance t/k (m^2°C/W)
External surface resistance			0.045
Asphalt waterproofing			0.060
Cork insulation board	0.050	0.042	1.190
Vapour barrier			0.020
Plywood	0.019	0.14	0.13
Air space			0.180
Glass fibre insulation	0.025	0.04	0.62
Fibreboard ceiling	0.012	0.05	0.24
Internal surface resistance			0.105

Total resistance R = 2.59

2. From the straight line Graph 2, plotted to show temperature against thermal resistance, a surface temperature of 9°C is indicated at the deck soffit.

3. From Table 5, the dew point temperature at 20°C and 60% relative humidity is 12°C.

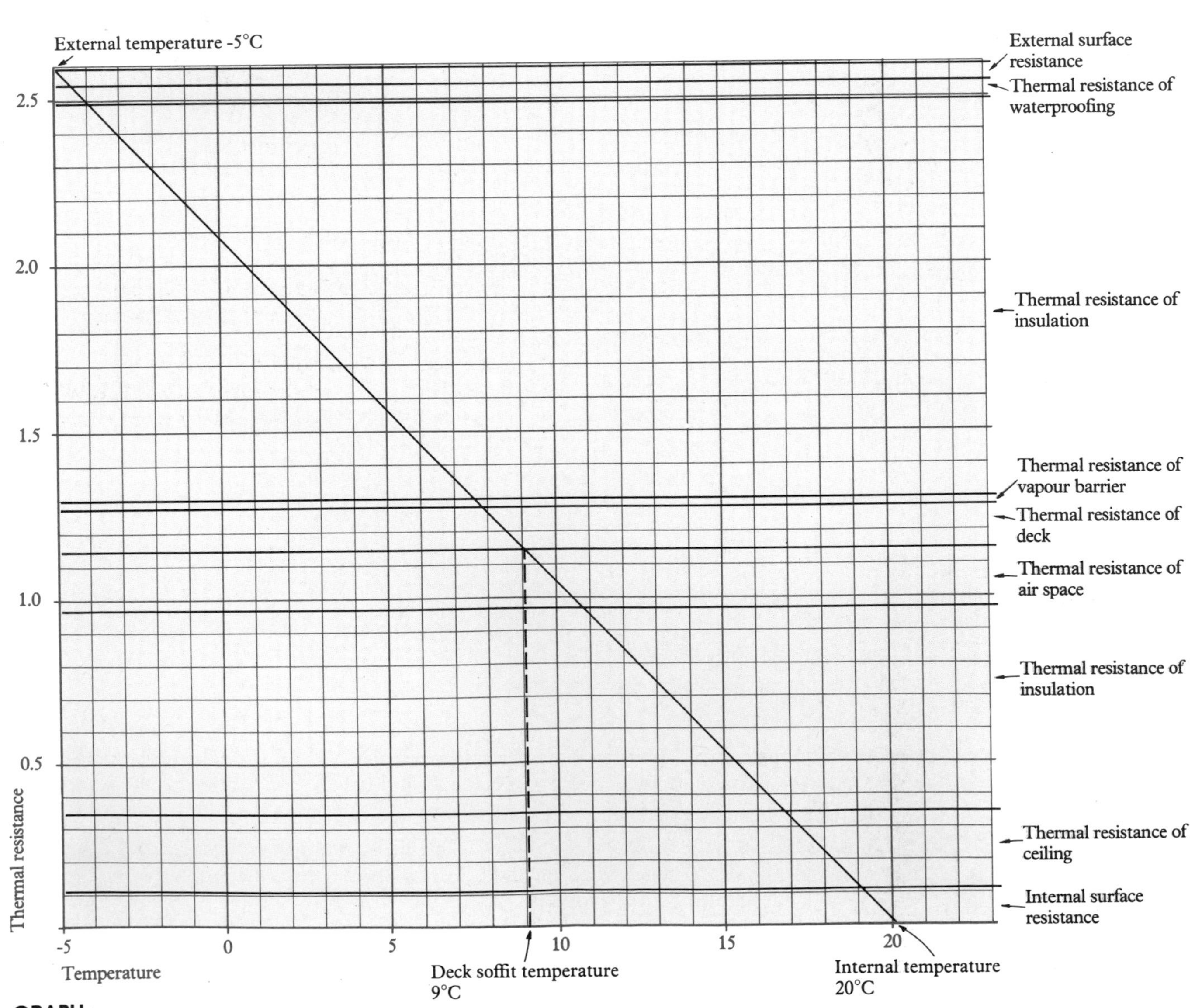

GRAPH

4. The deck soffit temperature is below the dew point temperature and condensation will therefore occur in the ceiling void. To stop condensation, the insulation above the deck should be increased.
The balance of insulation Table 6 shows the required minimum ratio between thermal resistance above the deck soffit and thermal resistance below the deck to avoid condensation.

In this example at 20°C and 60% relative humidity, the ratio is 2.1. The thermal resistance below the deck is 1.145m^{2}°C/W so the total thermal resistance required to prevent condensation in the ceiling void is 3.55 or a U-value of 0.28W/m^{2}°C. The thickness of insulation required can be calculated as described in example 2 of the U-value calculation method.

INTERSTITIAL CONDENSATION MOISTURE GAIN ANALYSIS

The calculation of moisture gain is valid for warm roof constructions when the moisture vapour movement is by diffusion. In the calculation method below, the waterproofing is assumed to be impermeable to vapour and the theoretical vapour pressure within the roof system may be taken as the vapour pressure within the building, with no significant loss of accuracy.

To determine the position of the condensation zone, the saturated vapour pressure at any point in the roof system is compared with the theoretical vapour pressure at the same point. Condensation will occur when the saturated vapour pressure is less than the theoretical vapour pressure.

The amount of condensation is calculated by assessing the rate at which moisture vapour arrives at this position.

The results of moisture gain analysis for typical warm roof specifications are given in Tables 1.6 to 1.11, pages 21 to 23. For specifications not covered by these tables the analysis may be carried out in the following stages:

1. Psychrometric conditions

External conditions
The outdoor notional conditions given below are assumed to apply anywhere in the United Kingdom.

	Temperature °C	Length of season (days)
Winter	-5	60
Summer	18	60

Internal conditions
In the absence of specific data for the building under consideration, the following indoor conditions may be assumed.

Type of building	Temperature °C	Relative humidity %
Houses and flats	20	55
Offices	20	40
Schools	20	50
Factories and heated warehouses	15	35
Textile factories	20	70
Swimming pool halls	25	70

These conditions are representative of temperature and relative humidity for a 60 day period.

2. Calculate the temperature at each interface between different materials in a roof system using the equation

$$t_x = t_i - (t_i - t_e) \times R_x U$$

where
t_x = temperature at interface x.
t_i = internal temperature
t_e = external temperature
R_x = thermal resistance from inside to interface x
U = U-value

3. Calculate the internal vapour pressure and the saturated vapour pressure at each interface using Table 7 or from a standard psychrometric chart.

4. If the internal vapour pressure is shown to be above the saturated vapour pressure at any interface (although of course, this is not possible in practice) then condensation will occur at that interface.

5. The amount of condensation or evaporation may be calculated from the equation

$$q = S \times \frac{(VP_i - SVP_x)}{VR_x}$$

where
VP_i = internal vapour pressure
VR_x = vapour resistance from inside to interface x
S = length of season (normally 60 days)
SVP_x = saturated vapour pressure at interface x

If the vapour resistance is in MNS/g, the vapour pressures in kilopascal (kPa), and the length of season in seconds, then the equation becomes

$$q = S \times 10^{-6} \times \frac{(VP_i - SVP_x)}{VR_x} \text{ kg/m}^2$$

6. *Acceptable levels of condensate*
For non-fibrous insulation materials, a maximum winter condensate of 0.5kg/m^2, and for fibrous insulation materials a maximum of 0.35kg/m^2 is accepted.
Although there should be full drying out of the condensate over the summer season, a calculated annual net retention of 5% of the maximum permissible winter condensate is allowable.

Example
Determine if a vapour barrier or vapour check is required for the specification below, over a heated warehouse.

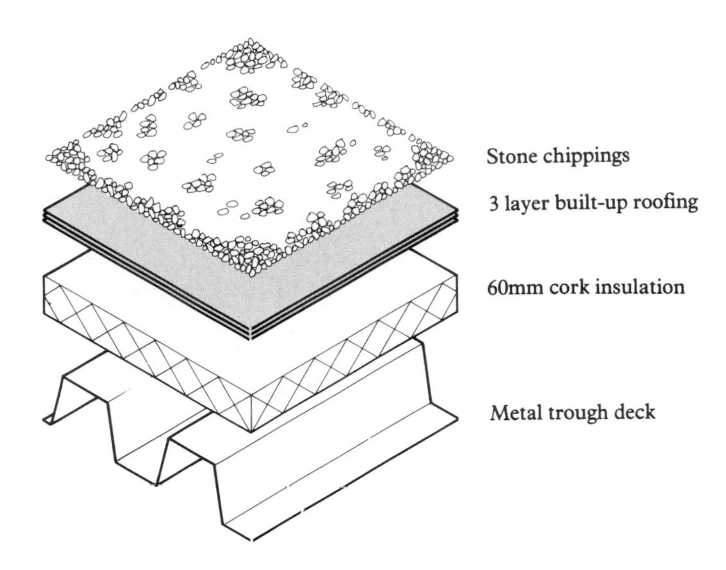

From the table of internal conditions, stage 1, the design conditions are taken as 15°C and 35%RH.

Calculation stages 2-5 must be completed. Detailed calculations are not shown, but the results of stages 2-3 for the given set of circumstances are shown in table 2.

From the table it can be seen that condensation will take place at the interface between the insulation and the waterproofing. From step 5 the amount of condensate over a 60 day winter period is shown to be 0.06kg/m^2 and the amount of evaporation over a 60 day summer period is 0.58kg/m^2. Therefore the condensate completely dries out over the summer period, and a vapour barrier is not required.

If the building use had been as a textile factory with conditions of 20°C and 70%RH internally, the results of the calculation would be as table 3. This shows that there would be 0.47kg/m^2 of condensate over the winter period. and only 0.18kg/m^2 drying out during the summer period. This would not be acceptable and further calculations will be needed, first including a vapour check in the specification and then, if that proves insufficient, a vapour barrier.

TABLE 2 INTERNAL TEMPERATURE 15°C 35% RH

	Thermal resistance m^2°C/W	Vapour resistance MNs/g	Winter Temperature °C	Saturated vapour pressure (k/Pa)	Vapour pressure (k/Pa)	Summer Temperature	Saturated vapour pressure (k/Pa)	Vapour pressure (k/Pa)
			-5	0.40		18.0	2.06	
External surface	0.045							
			-4.47	0.42		17.92	2.05	
Built-up roofing	0.06	Infinite						
			-3.75	0.45	0.60	17.81	2.04	0.60
60mm cork	1.43	3						
			13.22	1.52	0.60	15.27	1.73	0.60
Deck cavities	0.045	0						
			13.75	1.57	0.60	15.19	1.73	0.60
Metal deck	0	10						
			13.75	1.57	0.60	15.19	1.73	0.60
Internal surface	0.105	0						
			15.0	1.70	0.60	15.0	1.70	0.60

TABLE 3 INTERNAL TEMPERATURE 20°C 70% RH

	Thermal resistance m^2°C/W	Vapour resistance MNs/g	Winter Temperature °C	Saturated vapour pressure (k/Pa)	Vapour pressure (k/Pa)	Summer Temperature °C	Saturated vapour pressure (k/Pa)	Vapour pressure (k/Pa)
			-5.0	0.40		18.0	2.06	
External surface	0.045							
			-4.33	0.43		18.05	2.07	
Built-up roofing	0.06	Infinite						
			-3.44	0.46	1.64	18.13	2.08	1.64
60mm cork	1.43	3						
			17.77	2.03	1.64	19.82	2.31	1.64
Deck cavities	0.045	0						
			18.44	2.12	1.64	19.88	2.32	1.64
Metal deck	0	10						
			18.44	2.12	1.64	19.88	2.32	1.64
Internal surface	0.105	0						
			20.0	2.34	1.64	20.0	2.34	1.64

TABLE 4 THERMAL AND VAPOUR PROPERTIES

	Thermal conductivity (k-value) W/m°C	Thermal resistance (R-value) m²°C/W	Vapour resistivity MNs/gm	Vapour resistance MNs/g
Air surfaces				
Internal		0.105		0
External		0.045		0
Air cavities (unventilated)				
Low surface emissivity		0.18	5.2	
High surface emissivity		0.32	5.2	
Intermittent cavity (between metal deck and roofing specification)		0.045		0
Decks				
Dense concrete	1.4		210	
Lightweight aerated concrete	0.16		60	
Plywood and chipboard	0.14		520	
Timber boarded deck	0.14		210	
Woodwool (pre-screeded)*	0.093		15-40	
Metal decking	0			10
Vapour check/underlay				
Single layer bitumen felt on metal deck		0.02		100
Single layer bitumen felt fully supported		0.02		300
Vapour barrier				
Two layer built-up felt or metal lined felt		0.04		500
Insulations				
Cellular glass slab	0.045		Infinite	
Cork board	0.042		50	
Expanded polystyrene (bead board)	0.034		200	
Extruded polystyrene	0.031		1125	
Glass fibre roofboard	0.034		7	
Isocyanurate board	0.022			40
Mineral wool slab	0.034		7	
Perlite board	0.05		27	
Perlite bitumen screed	0.076		10	
Polyurethane board	0.022			55
Wood fibreboard	0.05		26	
Waterproofing				
Mastic asphalt		0.06		May be taken as impermeable
Built-up bitumen felt		0.06		

* For pre-felted woodwool, the combined properties of pre-screeded woodwool and single layer bitumen felt fully supported may be used, provided that the slab joints are taped.

A single layer of felt used as a vapour check on metal deck is likely to be damaged or perforated. The vapour resistance of this layer is taken as half that of a fully supported layer on concrete. The vapour resistance of a two layer vapour barrier also takes into account that normally the vapour barrier would be partially bonded. The vapour resistance of the waterproofing is so large compared with the resistance of the other components in the system, that its magnitude has very little effect on the calculation of the amount of condensate and the waterproofing may be taken as impermeable.

TABLE 5 DEW POINT TEMPERATURES

% RH	Temperature °C 15	16	17	18	19	20	21	22	23	24	25	26	27	28	29	30
30	-2.1	-1.4	-0.6	0.2	1.0	1.9	2.8	3.6	4.5	5.4	6.2	7.1	8.0	8.8	9.7	10.5
35	-0.3	0.6	1.4	2.3	3.2	4.1	5.0	5.8	6.7	7.6	8.5	9.4	10.2	11.1	12.0	12.9
40	1.5	2.4	3.3	4.2	5.1	6.0	6.9	7.8	8.7	9.6	10.5	11.4	12.3	13.2	14.0	14.9
45	3.2	4.1	5.0	5.9	6.8	7.7	8.6	9.5	10.4	11.3	12.2	13.2	14.1	15.0	15.9	16.8
50	4.7	5.6	6.5	7.4	8.4	9.3	10.2	11.1	12.0	12.9	13.9	14.8	15.7	16.6	17.5	18.4
55	6.0	7.0	7.9	8.8	9.8	10.7	11.6	12.6	13.5	14.4	15.3	16.3	17.2	18.1	19.0	20.0
60	7.3	8.2	9.2	10.1	11.1	12.0	12.9	13.9	14.8	15.8	16.7	17.6	18.6	19.5	20.5	21.4
65	8.5	9.4	10.4	11.3	12.3	13.2	14.2	15.1	16.1	17.0	18.0	18.9	19.9	20.8	21.7	22.7
70	9.6	10.5	11.5	12.5	13.4	14.4	15.3	16.3	17.2	18.2	19.2	20.1	21.1	22.0	23.0	23.9
75	10.6	11.6	12.5	13.5	14.5	15.4	16.4	17.4	18.3	19.3	20.3	21.2	22.2	23.2	24.1	25.1
80	11.6	12.6	13.5	14.5	15.5	16.4	17.4	18.4	19.4	20.3	21.3	22.3	23.3	24.2	25.2	26.2
85	12.5	13.5	14.5	15.4	16.4	17.4	18.4	19.4	20.3	21.3	22.3	23.3	24.3	25.2	26.2	27.2
90	13.4	14.4	15.3	16.3	17.3	18.3	19.3	20.3	21.3	22.3	23.2	24.2	25.2	26.2	27.2	28.2

TABLE 6 BALANCE OF INSULATION

% RH	Internal temperature °C 15	16	17	18	19	20	21	22	23	24	25	26	27	28	29	30
30	0.2	0.2	0.2	0.3	0.3	0.4	0.4	0.5	0.5	0.6	0.6	0.6	0.7	0.7	0.8	0.8
35	0.3	0.4	0.4	0.5	0.5	0.6	0.6	0.7	0.7	0.8	0.8	0.9	0.9	1.0	1.0	1.0
40	0.5	0.5	0.6	0.7	0.7	0.8	0.8	0.9	1.0	1.0	1.1	1.1	1.2	1.2	1.3	1.3
45	0.7	0.8	0.8	0.9	1.0	1.0	1.1	1.2	1.2	1.3	1.4	1.4	1.5	1.5	1.6	1.6
50	0.9	1.0	1.1	1.2	1.3	1.3	1.4	1.5	1.6	1.6	1.7	1.8	1.8	1.9	2.0	2.0
55	1.2	1.3	1.4	1.5	1.6	1.7	1.8	1.9	1.9	2.0	2.1	2.2	2.3	2.3	2.4	2.5
60	1.6	1.7	1.8	1.9	2.0	2.1	2.2	2.3	2.4	2.5	2.6	2.7	2.8	2.9	3.0	3.1
65	2.1	2.2	2.3	2.4	2.6	2.7	2.8	2.9	3.0	3.2	3.3	3.4	3.5	3.6	3.7	3.8
70	2.7	2.8	3.0	3.1	3.3	3.4	3.6	3.7	3.9	4.0	4.1	4.3	4.4	4.5	4.6	4.8
75	3.6	3.7	3.9	4.1	4.3	4.5	4.7	4.8	5.0	5.2	5.3	5.5	5.6	5.8	6.0	6.1
80	4.9	5.1	5.3	5.6	5.8	6.0	6.3	6.5	6.7	6.9	7.1	7.3	7.5	7.7	7.9	8.1
85	7.0	7.3	7.7	8.0	8.3	8.6	8.9	9.2	9.5	9.8	10.1	10.4	10.7	10.9	11.2	11.5
90	11.3	11.8	12.3	12.8	13.3	13.8	14.3	14.7	15.2	15.6	16.1	16.5	17.0	17.4	17.8	18.2

TABLE 7 VAPOUR PRESSURES (kPa)

% RH	Temperature °C -5	-4	-3	-2	-1	0	1	2	3	4	5	6	7	8	9	10	11	12
30	0.12	0.13	0.14	0.16	0.17	0.18	0.20	0.21	0.23	0.24	0.26	0.28	0.30	0.32	0.34	0.37	0.39	0.42
35	0.14	0.15	0.17	0.18	0.20	0.21	0.23	0.25	0.27	0.28	0.31	0.33	0.35	0.38	0.40	0.43	0.46	0.49
40	0.16	0.17	0.19	0.21	0.22	0.24	0.26	0.28	0.30	0.33	0.35	0.37	0.40	0.43	0.46	0.49	0.52	0.56
45	0.18	0.20	0.21	0.23	0.25	0.27	0.30	0.32	0.34	0.37	0.39	0.42	0.45	0.48	0.52	0.55	0.59	0.63
50	0.20	0.22	0.24	0.26	0.28	0.31	0.33	0.35	0.38	0.41	0.44	0.47	0.50	0.54	0.57	0.61	0.66	0.70
55	0.22	0.24	0.26	0.28	0.31	0.34	0.36	0.39	0.42	0.45	0.48	0.51	0.55	0.59	0.63	0.67	0.72	0.77
60	0.24	0.26	0.29	0.31	0.34	0.37	0.39	0.42	0.45	0.49	0.52	0.56	0.60	0.64	0.69	0.74	0.79	0.84
65	0.26	0.28	0.31	0.34	0.37	0.40	0.43	0.46	0.49	0.53	0.57	0.61	0.65	0.70	0.75	0.80	0.85	0.91
70	0.28	0.31	0.33	0.36	0.39	0.43	0.46	0.49	0.53	0.57	0.61	0.65	0.70	0.75	0.80	0.86	0.92	0.98
75	0.30	0.33	0.36	0.39	0.42	0.46	0.49	0.53	0.57	0.61	0.65	0.70	0.75	0.80	0.86	0.92	0.98	1.05
80	0.32	0.35	0.38	0.41	0.45	0.49	0.53	0.56	0.61	0.65	0.70	0.75	0.80	0.86	0.92	0.98	1.05	1.12
85	0.34	0.37	0.40	0.44	0.48	0.52	0.56	0.60	0.64	0.69	0.74	0.79	0.85	0.91	0.98	1.04	1.12	1.19
90	0.36	0.39	0.43	0.47	0.51	0.55	0.59	0.63	0.68	0.73	0.78	0.84	0.90	0.96	1.03	1.10	1.18	1.26
95	0.38	0.42	0.45	0.49	0.53	0.58	0.62	0.67	0.72	0.77	0.83	0.89	0.95	1.02	1.09	1.17	1.25	1.33
100*	0.40	0.44	0.48	0.52	0.56	0.61	0.66	0.71	0.76	0.81	0.87	0.93	1.00	1.07	1.15	1.23	1.31	1.40

TABLE 7 continued

% RH	Temperature °C 13	14	15	16	17	18	19	20	21	22	23	24	25	26	27	28	29	30
30	0.45	0.48	0.51	0.55	0.58	0.62	0.66	0.70	0.75	0.79	0.84	0.89	0.95	1.01	1.07	1.13	1.20	1.27
35	0.52	0.56	0.60	0.64	0.68	0.72	0.77	0.82	0.87	0.92	0.98	1.04	1.11	1.18	1.25	1.32	1.40	1.48
40	0.60	0.64	0.68	0.73	0.77	0.83	0.88	0.93	0.99	1.06	1.12	1.19	1.27	1.34	1.43	1.51	1.60	1.70
45	0.67	0.72	0.77	0.82	0.87	0.93	0.99	1.05	1.12	1.19	1.26	1.34	1.42	1.51	1.60	1.70	1.80	1.91
50	0.75	0.80	0.85	0.91	0.97	1.03	1.10	1.17	1.24	1.32	1.40	1.49	1.58	1.68	1.78	1.89	2.00	2.12
55	0.82	0.88	0.94	1.00	1.07	1.13	1.21	1.29	1.37	1.45	1.54	1.64	1.74	1.85	1.96	2.08	2.20	2.33
60	0.90	0.96	1.02	1.09	1.16	1.24	1.32	1.40	1.49	1.59	1.68	1.79	1.90	2.02	2.14	2.27	2.40	2.55
65	0.97	1.04	1.11	1.18	1.26	1.34	1.43	1.52	1.62	1.72	1.83	1.94	2.06	2.18	2.32	2.46	2.60	2.76
70	1.05	1.12	1.19	1.27	1.36	1.44	1.54	1.64	1.74	1.85	1.97	2.09	2.22	2.35	2.49	2.65	2.80	2.97
75	1.12	1.20	1.28	1.36	1.45	1.55	1.65	1.75	1.86	1.98	2.11	2.24	2.37	2.52	2.67	2.83	3.00	3.18
80	1.20	1.28	1.36	1.45	1.55	1.65	1.76	1.87	1.99	2.11	2.25	2.39	2.53	2.69	2.85	3.02	3.20	3.39
85	1.27	1.36	1.45	1.54	1.65	1.75	1.87	1.99	2.11	2.25	2.39	2.54	2.69	2.86	3.03	3.21	3.40	3.61
90	1.35	1.44	1.53	1.64	1.74	1.86	1.98	2.10	2.24	2.38	2.53	2.68	2.85	3.02	3.21	3.40	3.60	3.82
95	1.42	1.52	1.62	1.73	1.84	1.96	2.09	2.22	2.36	2.51	2.67	2.83	3.01	3.19	3.39	3.59	3.80	4.03
100*	1.50	1.60	1.70	1.82	1.94	2.06	2.20	2.34	2.49	2.64	2.81	2.98	3.17	3.36	3.56	3.78	4.00	4.24

*100% RH = saturated vapour pressure

REFERENCES

BRITISH STANDARDS AND CODES OF PRACTICE

BS 476 Fire tests on building materials and structures

Part 3:1958 External fire exposure roof test
Part 3:1975 External fire exposure roof test
Part 8:1972 Test methods and criteria for the fire resistance of elements of building construction

BS 747:1977 Specification for roofing felts

BS 988, 1076, 1097, 1451:1973 Mastic asphalt for building (limestone aggregate)

BS 1105:1981 Wood wool cement slabs up to 125mm thick

BS 1142 Fibre building boards
Part 1: 1971 Method of test
Part 3:1972 Insulating board (softboard)

BS 1162, 1418, 1410:1973 Mastic asphalt for building (natural rock asphalt aggregate)

BS 1446:1973 Mastic asphalt (natural rock asphalt fine aggregate) for roads and footways

BS 1447:1973 Mastic asphalt (limestone fine aggregate) for roads and footways

BS 1455:1972 Plywood manufactured from tropical hardwoods

BS 5669:1979 Specification for wood chipboard and methods of test for particle board

CP 3: Chapter V Loading

Part 1:1967 Dead and imposed loads
Part 2:1972 Wind loads

CP 144 Roof coverings

Part 3:1970 Built-up bitumen felt
Part 4:1970 Mastic asphalt

CP 308:1974 Drainage of roofs and paved areas

PHOTOGRAPHS

We would like to thank the following organisations for providing photographs:

Pages 29 and 170: MACEF
Page 121: Wiggins Teape

INDEX

C

D

E

F

G

O

P

R

S

T

U

V

W

TABLES

Falls and drainage Page

Thermal design

Wind

TARMAC BUILDING PRODUCTS LIMITED

Tarmac Building Products Limited
Ebury Gate
23 Lower Belgrave Street
London
SW1W 0NG

Telephone: 01-730 0055

Briggs Amasco Limited
Goodwyns Place
Tower Hill Road
Dorking
Surrey
RH4 2AW

Telephone: (0306) 885933

Permanite Limited
Mead Lane
Hertford
SG13 7AU

Telephone: (0992) 50511

Permanite Asphalt Limited
St Peters House
Gower Street
Derby
DE1 1FB

Telephone: (0332) 683124

Coolag Purlboard
Heysham Works
Middleton Road
Morecambe
Lancashire
LA3 3PP

Telephone: (0524) 55611

British Hydroflex Limited
Appley Lane
Appley Bridge
Wigan
Lancashire
WN6 9AB

Telephone: (025 75) 2333